Vegetation Dynamics Revealed by Remote Sensing and Its Feedback to Regional and Global Climate

Vegetation Dynamics Revealed by Remote Sensing and Its Feedback to Regional and Global Climate

Editors

Xuejia Wang
Tinghai Ou
Wenxin Zhang
Youhua Ran

MDPI • Basel • Beijing • Wuhan • Barcelona • Belgrade • Manchester • Tokyo • Cluj • Tianjin

Editors

Xuejia Wang
Lanzhou University
China

Tinghai Ou
University of Gothernburg
Sweden

Wenxin Zhang
Lund University
Sweden

Youhua Ran
Chinese Academy of Sciences
China

Editorial Office
MDPI
St. Alban-Anlage 66
4052 Basel, Switzerland

This is a reprint of articles from the Special Issue published online in the open access journal *Remote Sensing* (ISSN 2072-4292) (available at: https://www.mdpi.com/journal/remotesensing/special_issues/vegetation_dynamics_climate).

For citation purposes, cite each article independently as indicated on the article page online and as indicated below:

LastName, A.A.; LastName, B.B.; LastName, C.C. Article Title. *Journal Name* **Year**, *Volume Number*, Page Range.

ISBN 978-3-0365-5495-2 (Hbk)
ISBN 978-3-0365-5496-9 (PDF)

Cover image courtesy of Xuejia Wang

Contents

About the Editors

Xuejia Wang

Xuejia Wang is a climate scientist and Professor of Climate and Cryosphere Studies at the College of Earth and Environment Sciences, Lanzhou University, China. His research, which uses meteorological observations, multi-source re-analyses, remote sensing datasets, and climate models, focuses on climate change and its impacts on the cryosphere, as well as various land–surface interactions (e.g., vegetation, soil freezing–thawing processes, and land–surface albedo) and the atmosphere. Dr. Wang has a Ph.D in Atmospheric Physics and Atmospheric Environment from the University of the Chinese Academy of Sciences. Prior to joining Lanzhou University, he was an Associated Researcher at the Northwest Institute of Eco-Environment and Resources, Chinese Academy of Sciences, China. He has published more than 50 peer-reviewed research papers and book chapters, with over 920 total citations. Dr. Wang serves as a committee member of the Youth Working Group in China Cryosphere Science and is a member of several research societies, such as the Geographical Society of China, the Research Society of the Tibetan Plateau, and the Society of China Cryosphere Science.

Tinghai Ou

Tinghai Ou is a climate scientist and Principal Research Engineer (specializing in climate modeling) in the Department of Earth Sciences at the University of Gothenburg, Sweden. His research interests, primarily based on climate modeling, are in regional climate change, land–atmosphere interaction, climate dynamics, and downscaling. Dr. Ou has a Ph.D. in Natural Science, specializing in Physical Geography, from the University of Gothenburg. Before holding his current position at the University of Gothenburg, he was a Post-Doctoral Fellow in the Faculty of Earth Systems & Environmental Sciences at the Chonnam National University, South Korea. He was also a visiting researcher at Pennsylvania State University, USA.

Wenxin Zhang

Wenxin Zhang is a Researcher at the Department of Physical Geography and Ecosystem Science, Lund University, Sweden. His research interests, which use satellite observation, ground-based measurements and Earth system models, include vegetation–climate feedback, Earth system modeling, underground greenhouse gas transmission, ecological modeling, and plant root modeling. He received his Ph.D in 2015 from Lund University. After that, until July 2018, he undertook postdoc research at the University of Copenhagen. He has published more than 40 peer-reviewed papers, with more than 1500 total citations. He also serves as a reviewer for several journals, such as the first draft of *IPCC AR6 Group 2 Report*, *Earth Science System Data*, and *Agricultural and Forest Meteorology*, etc.

Youhua Ran

Youhua Ran is currently a professor at the Northwest Institute of Eco-Environment and Resources, Chinese Academy of Sciences. He received his Ph.D in 2017 and his Master's degree in 2009 in Cartography and Geographical Information System from the University of the Chinese Academy of Sciences. His research interests include the remote sensing of the cryosphere, validation of remote sensing products, and sustainable watershed development. He has published more than 80 peer-review papers, with a total of more than 3300 citations. Dr. Youhua is a scientific committee member, Digital Polar Committee of the International Society for Digital Earth (ISDE), Digital Mountain Committee of the ISDE.

Editorial

An Overview of Vegetation Dynamics Revealed by Remote Sensing and Its Feedback to Regional and Global Climate

Xuejia Wang [1,*], Tinghai Ou [2], Wenxin Zhang [3] and Youhua Ran [4]

[1] Key Laboratory of Western China's Environmental Systems (Ministry of Education), College of Earth and Environment Sciences, Lanzhou University, Lanzhou 730000, China
[2] Department of Earth Science, University of Gothenburg, 40530 Gothenburg, Sweden
[3] Department of Physical Geography and Ecosystem Science, Lund University, 22362 Lund, Sweden
[4] Heihe Remote Sensing Experimental Research Station, Northwest Institute of Eco-Environment and Resources, Chinese Academy of Sciences, Lanzhou 730000, China
* Correspondence: xjwang@lzb.ac.cn

Citation: Wang, X.; Ou, T.; Zhang, W.; Ran, Y. An Overview of Vegetation Dynamics Revealed by Remote Sensing and Its Feedback to Regional and Global Climate. *Remote Sens.* **2022**, *14*, 5275. https://doi.org/10.3390/rs14205275

Received: 13 October 2022
Accepted: 18 October 2022
Published: 21 October 2022

Publisher's Note: MDPI stays neutral with regard to jurisdictional claims in published maps and institutional affiliations.

1. Introduction

Vegetation, as one of the crucial underlying land surfaces, plays an important role in terrestrial ecosystems and the Earth's climate system through the alternation of its phenology, type, structure, and function. Vegetation responds to climate warming quite differently, such as greening and browning across different regions, which have been reported by many remote sensing studies [1–3]. Vegetation is an important and sensitive indicator of climate and environment evolutions, underscoring the need to better understand vegetation physiological and phenological responses, detect mechanisms of how changes in land surface properties (e.g., surface albedo and roughness length) are associated with vegetation dynamics, and reveal climate and ecological feedbacks of vegetation changes. The recent advances of satellite remote sensing techniques and their derived products provide unique opportunities to study vegetation dynamics and its feedback to regional and global climate systems. Moreover, some of the new generation of climate models, such as CMIP6 Earth system models, which include dynamic vegetation, are state-of-the-art tools for investigating the interactions between vegetation and climate change.

After an open call to the community, we received some interesting works based on remote sensing data from which we summarize 10 papers that are already published, which focus on vegetation changes and the associated drivers, the effect of extreme climate events on vegetation, land surface albedo related to vegetation change, plant fingerprint, and vegetation dynamics in climate models. These articles well represent the focus of the Special Issue, which aims to investigate vegetation dynamics and its response to climate change.

2. Overview of Contribution and Future Perspectives

Identifying the influencing factors, so-called detection and attribution, for vegetation changes is an important subject of current research. Dong et al. [4] analyzed the spatial-temporal changes of vegetation NDVI in the Loess Plateau between 2000–2015 and used the geographical detector model (GDM) to quantify its dominant factors from climate, environment, and anthropogenic factors. They revealed that NDVI increases more rapidly in the semi-humid area than in the semi-arid area. For the former, anthropogenic factors, such as the GDP density, land-use type, and population density, have a great effect on the NDVI increase, while for the latter, the climate and environment factors, such as precipitation, soil type, and vegetation type, have a great effect. Meanwhile, the interactions between factors enhance the effects on vegetation change. A similar piece of work was done by Li et al. [5] in temperate drylands, specifically the Inner Mongolia grasslands. With an upward trend in NDVI in the growing season over the period 2000–2018, GDM suggests that both nature and human activities exert significant effects on the NDVI changes, accounting for more than 15% of the variability. Interactions between precipitation and air temperature dominate the NDVI change, accounting for 39%. Taken together, these articles include an attribution model to resolve the dominant contributors to vegetation change.

One of the most noticeable vegetation responses to the rapid warming in northern high latitudes is changes in the timing of thermal growing seasons and phenological cycles of plants [2]. These changes may induce direct and legacy effects on ecosystem gross primary production. Based on three widely-used remote sensing products of GPP (gross primary productivity) at a spatial resolution of 0.05° over 2001–2018, Marsh and Zhang [6] found that legacy effects from spring temperature are most pronounced in summer, where the Arctic ecosystem productivity has been stimulated. Spring warming likely lessens the harsh climatic constraints that govern the Arctic tundra and extends the growing season length. Further south, legacy effects are mainly negative. This strengthens the hypothesis that enhanced vegetation growth in spring will increase plant water demand and stress in summer and autumn. Soil moisture is the dominant control of summer GPP in temperate regions. However, the dominant meteorological variables controlling vegetation growth are different among the three GPP products. Different biomes show disparate (positive or negative, even lagged) impacts for the three GPP products. Overall, this work quantitatively assesses the direct and legacy effects of spring warming on seasonal GPP, and it also highlights the need to address uncertainties among different methods that are used to estimate GPP.

Net primary productivity (NPP) is a variable that reflects the efficiency of vegetation fixation and conversion of light energy, and thus it is often used to monitor vegetation dynamics, such as plant growth, development, reproduction, and senescence. Based on the MODIS NPP product and environmental factors (air temperature, solar radiation, and soil moisture) derived from the atmospheric reanalysis data (ERA5, MERRA2, and NCEP2), Wang et al. [7] found that nearly 60% of the global areas showed a higher NPP that is associated with an increased elevation. Soil moisture has the largest uncertainty to explain either the spatial pattern or inter-annual variation of NPP, while air temperature has the smallest uncertainty among the three environmental factors. NPP shows an obvious elevation differentiation with an elevation of 3060 m as the demarcation point, which divides the elevation into low and high. Mean annual air temperature is the main driving that affects the elevation distribution of NPP. Their work implies that elevation is a crucial factor when quantifying the carbon sequestration capability of vegetation globally.

When it comes to the world's Third Pole (Tibetan Plateau, TP), changes in vegetation dynamics also play a critical role in terrestrial ecosystems and environments. Similar to the Arctic, the TP has also experienced rapid and amplified warming during recent decades [8]. Therefore, vegetation dynamics in the TP have been attached increasing attention as it profoundly influences the terrestrial carbon cycle and climate change. Land surface albedo directly affects the energy balance on the land surface. Li et al. [9] examined the spatial-temporal changes of land surface albedo dynamics and its influencing factors (snow cover and vegetation) in the Qilian Mountains, Northeastern TP, using multi-source remote sensing data. Annual average albedo showed a weak increasing trend from 2001 to 2020. Surface albedo is closely associated with land surface cover, and vegetation is significantly negatively correlated with albedo. The improvement of vegetation condition reduces the surface albedo in the edge areas. Therefore, surface albedo can also be used for monitoring land surface conditions. Deng et al. [10] used remote sensing products and a dynamic global vegetation model to study the TP vegetation dynamics and the associated climatic drivers. They customized the Community Land Surface Biogeochemical Dynamic Vegetation Model to simulate the TP vegetation distribution and carbon flux and improved the model's phenology representation using seasonal–deciduous phenology parameterization. The newly developed processes substantially improve the model to reproduce in situ observations on the TP. In addition to better simulations of spatial-temporal patterns of GPP in the TP, their work also showed different indications of dominant drivers between the remote sensing product and the terrestrial ecosystem model.

Remote sensing is also used to monitor land and forest change footprint, facilitating the protection of the fragile environment. In combination with 8203 scenes of multi-source remote sensing data, Yan and Wang [11] use the LandTrendr spectral-temporal segmentation algorithm to explore forest change footprint in the upper Indus Valley. This work suggests that the area of forest recovery is 1% more than that of disturbance between 1990–2020, in which 70% of disturbance appears between 1990 and 2001 and 60% of recovery appears between 1999 and 2012. Although little difference exists in the overall trend of forest disturbance and recovery, the significant differences remain in forest management status across different regions because of grazing, fire, commercial tree planting, and afforestation policies. Li et al. [12] investigated the effects of time interpolation on phenology trend estimation in the mid-high latitudes of the northern hemisphere between 2001–2019, using a daily NDVI generated based on the moderate resolution imaging spectroradiometer (MODIS) MCD43A4 daily surface reflectance data over 120 selected sites. They found that there are nonignorable effects of the time interpolation on trend estimation, even though the effects are not significant. The effects of the time interpolation on trend estimation have shown significant differences among different vegetation types, with significant effects on vegetation types with apparent seasonal changes, such as deciduous broadleaf forests, and no significant effects among vegetation types with weak seasonal changes, such as evergreen needleleaf forests. In addition, the selection of extraction methods also affected trend estimation.

In recent years, climate extremes have been frequently reported by literature and media across the globe. Vegetation in response to climate extremes has been arousing general concern. For the response of vegetation to the heatwave, Dong et al. [13] examined the impact of heatwaves on vegetation growth rate on the TP from 2000 to 2020 using MODIS Nadir Bidirectional Reflectance Distribution Function Adjusted Reflectance (NBAR) based NDVI and EVI, microwave-based surface soil moisture, and long-term meteorological data. They found that the significant increase in the frequency of heatwaves only occurs in August during the last two decades. During heatwave periods, the soil moisture and precipitation are significantly lower than the corresponding multi-year average value. The temperature stress and water limitation caused by heatwave slow the vegetation growth on the TP but the sensitivity of alpine vegetation on heatwave is higher in June than in July and August. Wang et al. [14] investigated the effects of climate extremes on vegetation at multi-time scales using NDVI during 1982–2015 in Guangxi, China. They found that there are clear seasonal differences in the trend of NDVI in Guangxi, with the strongest greening in spring and February. On an annual scale, the NDVI is generally significantly correlated with extreme temperature indices, while there is no significant correlation between NDVI and most of the extreme precipitation indices used. On seasonal and monthly scales, the correlations between NDVI and extreme temperature and precipitation indices vary in months. Overall, these works are of great significance for the understanding of vegetation response to increasing extreme weather events under the background of rapid climate change.

The above studies advance the understanding of vegetation changes and their driving factors to a large extent. However, there is still some gap between our expectations and the currently collected papers. In addition to the response of vegetation to climate change, we also expect to look at some progress in the feedback of vegetation dynamics to ecosystems (e.g., carbon stocks, water, and soil conservation) and the climate systems. Therefore, we plan to reopen the Special Issue (named Specific Issue II: https://www.mdpi.com/journal/remotesensing/special_issues/6201BU8J59) to the community to collect recent progress involving this.

Author Contributions: Conceptualization, X.W. and W.Z.; writing—original draft preparation, X.W., T.O., W.Z. and Y.R.; writing—review and editing, X.W., T.O., W.Z. and Y.R. All authors have read and agreed to the published version of the manuscript.

Funding: This work was funded by the Start-up Funds for Introduced Talent at Lanzhou University (561120217). W.Z. acknowledged the Swedish Research Council (Veteskapsrådet) start grant (2020-05338), T. O would like to acknowledge the Swedish Foundation for International Cooperation in Research and Higher Education (CH2019-8377).

Acknowledgments: We thank all authors, reviewers, and assistant editors for their contribution to the Special Issue entitled "Vegetation Dynamics Revealed by Remote Sensing and Its Feedback to Regional and Global Climate".

Conflicts of Interest: The authors declare no conflict of interest.

References

1. Pang, G.; Wang, X.; Yang, M. Using the NDVI to identify variations in, and responses of, vegetation to climate change on the Tibetan Plateau from 1982 to 2012. *Quat. Int.* **2016**, *444*, 87–96. [CrossRef]
2. Piao, S.; Liu, Q.; Chen, A.; Janssens, I.; Fu, Y.; Dai, J.; Liu, L.; Lian, X.; Shen, M.; Zhu, X. Plant phenology and global climate change: Current progresses and challenges. *Glob. Chang. Biol.* **2019**, *25*, 1922–1940. [CrossRef] [PubMed]
3. Cortés, J.; Mahecha, M.D.; Reichstein, M.; Myneni, R.B.; Chen, C.; Brenning, A. Where Are Global Vegetation Greening and Browning Trends Significant? *Geophys. Res. Lett.* **2021**, *48*, e2020GL091496. [CrossRef]
4. Dong, Y.; Yin, D.; Li, X.; Huang, J.; Su, W.; Li, X.; Wang, H. Spatial–Temporal Evolution of Vegetation NDVI in Association with Climatic, Environmental and Anthropogenic Factors in the Loess Plateau, China during 2000–2015: Quantitative Analysis Based on Geographical Detector Model. *Remote Sens.* **2021**, *13*, 4380. [CrossRef]
5. Li, S.; Li, X.; Gong, J.; Dang, D.; Dou, H.; Lyu, X. Quantitative Analysis of Natural and Anthropogenic Factors Influencing Vegetation NDVI Changes in Temperate Drylands from a Spatial Stratified Heterogeneity Perspective: A Case Study of Inner Mongolia Grasslands, China. *Remote Sens.* **2022**, *14*, 3320. [CrossRef]
6. Marsh, H.; Zhang, W. Direct and Legacy Effects of Spring Temperature Anomalies on Seasonal Productivity in Northern Ecosystems. *Remote Sens.* **2022**, *14*, 2007. [CrossRef]
7. Wang, Z.; Wang, H.; Wang, T.; Wang, L.; Huang, X.; Zheng, K.; Liu, X. Effects of Environmental Factors on the Changes in MODIS NPP along DEM in Global Terrestrial Ecosystems over the Last Two Decades. *Remote Sens.* **2022**, *14*, 713. [CrossRef]
8. Wang, X.; Ran, Y.; Pang, G.; Chen, D.; Su, B.; Chen, R.; Li, X.; Chen, H.W.; Yang, M.; Gou, X.; et al. Contrasting characteristics, changes, and linkages of permafrost between the Arctic and the Third Pole. *Earth-Sci. Rev.* **2022**, *230*, 104042. [CrossRef]
9. Li, J.; Pang, G.; Wang, X.; Liu, F.; Zhang, Y. Spatiotemporal Dynamics of Land Surface Albedo and Its Influencing Factors in the Qilian Mountains, Northeastern Tibetan Plateau. *Remote Sens.* **2022**, *14*, 1922. [CrossRef]
10. Deng, M.; Meng, X.; Lu, Y.; Li, Z.; Zhao, L.; Niu, H.; Chen, H.; Shang, L.; Wang, S.; Sheng, D. The Response of Vegetation to Regional Climate Change on the Tibetan Plateau Based on Remote Sensing Products and the Dynamic Global Vegetation Model. *Remote Sens.* **2022**, *14*, 3337. [CrossRef]
11. Yan, X.; Wang, J. The Forest Change Footprint of the Upper Indus Valley, from 1990 to 2020. *Remote Sens.* **2022**, *14*, 744. [CrossRef]
12. Li, X.; Zhu, W.; Xie, Z.; Zhan, P.; Huang, X.; Sun, L.; Duan, Z. Assessing the Effects of Time Interpolation of NDVI Composites on Phenology Trend Estimation. *Remote Sens.* **2021**, *13*, 5018. [CrossRef]
13. Dong, C.; Wang, X.; Ran, Y.; Nawaz, Z. Heatwaves Significantly Slow the Vegetation Growth Rate on the Tibetan Plateau. *Remote Sens.* **2022**, *14*, 2402. [CrossRef]
14. Wang, L.; Hu, F.; Miao, Y.; Zhang, C.; Zhang, L.; Luo, M. Changes in Vegetation Dynamics and Relations with Extreme Climate on Multiple Time Scales in Guangxi, China. *Remote Sens.* **2022**, *14*, 2013. [CrossRef]

Article

Spatial–Temporal Evolution of Vegetation NDVI in Association with Climatic, Environmental and Anthropogenic Factors in the Loess Plateau, China during 2000–2015: Quantitative Analysis Based on Geographical Detector Model

Yi Dong [1], Dongqin Yin [1,2,3,*], Xiang Li [4,5,6], Jianxi Huang [1,2], Wei Su [1,2], Xuecao Li [1,2] and Hongshuo Wang [1,2]

1   College of Land Science and Technology, China Agricultural University, Beijing 100083, China; dongyi@cau.edu.cn (Y.D.); jxhuang@cau.edu.cn (J.H.); suwei@cau.edu.cn (W.S.); xuecaoli@cau.edu.cn (X.L.); hswang@cau.edu.cn (H.W.)
2   Key Laboratory of Remote Sensing for Agri-Hazards, Ministry of Agriculture and Rural Affairs, Beijing 100083, China
3   State Key Laboratory of Hydro-Science and Engineering, Department of Hydraulic Engineering, Tsinghua University, Beijing 100083, China
4   State Key Laboratory of Simulation and Regulation of Water Cycle in River Basin, China Institute of Water Resources and Hydropower Research, Beijing 100038, China; lixiang@iwhr.com
5   Key Laboratory of Sediment Science and Northern River Regulation, Ministry of Water Resources, Beijing 100038, China
6   State Key Laboratory of Plateau Ecology and Agriculture, Qinghai University, Xining 810016, China
*   Correspondence: dongqin.yin@cau.edu.cn

Citation: Dong, Y.; Yin, D.; Li, X.; Huang, J.; Su, W.; Li, X.; Wang, H. Spatial–Temporal Evolution of Vegetation NDVI in Association with Climatic, Environmental and Anthropogenic Factors in the Loess Plateau, China during 2000–2015: Quantitative Analysis Based on Geographical Detector Model. *Remote Sens.* **2021**, *13*, 4380. https://doi.org/10.3390/rs13214380

Academic Editors: Xuejia Wang, Tinghai Ou, Wenxin Zhang and Youhua Ran

Received: 7 October 2021
Accepted: 28 October 2021
Published: 30 October 2021

Publisher's Note: MDPI stays neutral with regard to jurisdictional claims in published maps and institutional affiliations.

Abstract: In the Loess Plateau (LP) of China, the vegetation degradation and soil erosion problems have been shown to be curbed after the implementation of the Grain for Green program. In this study, the LP is divided into the northwestern semi-arid area and the southeastern semi-humid area using the 400 mm isohyet. The spatial–temporal evolution of the vegetation NDVI during 2000–2015 are analyzed, and the driving forces (including factors of climate, environment, and human activities) of the evolution are quantitatively identified using the geographical detector model (GDM). The results showed that the annual mean NDVI in the entire LP was 0.529, and it decreased from the semi-humid area (0.619) to the semi-arid area (0.346). The mean value of the coefficient of variation of the NDVI was 0.1406, and it increased from the semi-humid area (0.1165) to the semi-arid area (0.1926). The annual NDVI growth rate in the entire LP was 0.0079, with the NDVI growing faster in the semi-humid area (0.0093) than in the semi-arid area (0.0049). The largest increments of the NDVI were from grassland, farmland, and woodland. The GDM results revealed that changes in the spatial distribution of the NDVI could be primarily explained by the climatic and environmental factors in the semi-arid area, such as precipitation, soil type, and vegetation type, while the changes were mainly explained by the anthropogenic factors in the semi-humid area, such as the GDP density, land-use type, and population density. The interactive analysis showed that interactions between factors strengthened the impacts on the vegetation change compared with an individual factor. Furthermore, the ranges/types of factors suitable for vegetation growth were determined. The conclusions of this study have important implications for the formulation and implementation of ecological conservation and restoration strategies in different regions of the LP.

Keywords: Loess Plateau; China; normalized difference vegetation index (NDVI); spatial–temporal evolution; geographical detector model; driving forces

1. Introduction

Vegetation plays an indispensable role in regional terrestrial ecosystems, and constitutes an essential link between soil, water, and atmosphere. Moreover, vegetation is the material basis for the survival of surface organisms [1,2]. Vegetation coverage effectively reduces the surface soil erosion caused by exogenous forces such as wind, diminishes the splash erosion caused by raindrops, alleviates the hydrodynamic erosion of rivers, and improves the soil environment. Therefore, it is of crucial importance to explore the vegetation coverage changes and dynamics for the soil erosion prevention and control, ecological environmental protection, and sustainable social and economic development [3,4]

At present, the main monitoring method used in large-scale vegetation coverage change research is based on satellite remote sensing because of its wide spatial range and gradually improving resolution, which effectively makes up for the shortcomings of traditional monitoring methods [5]. The normalized difference vegetation index (NDVI) is strongly associated with the vegetation coverage, leaf area index, biomass, and land use, which can comprehensively reflect the vegetation's greenness, photosynthesis intensity, and vegetation metabolism intensity [6,7]. The NDVI can be used to quantitatively evaluate the regional vegetation coverage and growth, which is considered to be an effective indicator for monitoring terrestrial vegetation changes, and thus, has been widely used in research and management in various fields, such as agriculture and ecology [8].

The growth process of vegetation is affected by multiple factors. Temperature and precipitation are directly related to global climate change and are generally regarded as the most important natural factors affecting vegetation growth and changes in the long-term [9,10]. Since the 20th century, intense human activities, which are characterized by industrialization and urbanization, have brought about rapid population and economic growth, rapid changes in land use, and ecological and environmental problems such as vegetation degradation and soil erosion, resulting in significant impacts on vegetation growth and changes in the short-term [11–13].

The Loess Plateau (LP) of China is located in the semi-arid/humid zones, the local ecological environment of which is extremely fragile. Vegetation degradation and soil erosion have been particularly prominent in this region throughout history, making the LP main sediment source (nearly 90%) of the Yellow River (middle and lower reaches), once the most sand-laden river in the world [14]. Exploring the evolution of vegetation and its driving forces on the LP cannot only help formulate strategies on ecological restoration for this area, but also help predict the sediment and tackle the sediment related problems (such as reservoir sedimentation and flood control) in the Yellow River. Since the 1980s, particularly after 2000 when the Grain for Green program (GGP) was implemented in this area, the restoration of vegetation has been witnessed, and accordingly, the expansion trend of the soil erosion has been curbed [15]. In addition, a new stage of the GGP was launched in 2014 to consolidate the achievements [16].

Many scholars have studied the causes of the vegetation coverage changes on the LP. There are limitations in the existing research, which mainly include the following. (1) Many studies used NDVI data from original Global Inventory Modeling and Mapping Studies (GIMMS) or GIMMS NDVI data expanded from Moderate Resolution Imaging Spectroradiometer (MODIS) and Satellite Pour l'Observation de la Terre (SPOT) Vegetation (VGT) NDVI [6,17–19]. The spatial resolution of 8 km leads to serious mixed pixels and insufficient detail capture, making it difficult for the results to show the actual spatial–temporal changes in the vegetation coverage. (2) Several studies used statistical methods, such as correlation or regression, by assuming that the vegetation growth is linearly related to the potential factors with time, but they ignored the spatial heterogeneity [20,21]. Some evidence has suggested that there is no strict linear correlation between the factors and the vegetation growth under the climate change. A statistical linear correlation model may not be able to accurately describe the internal relationship [22,23]. (3) Several studies only considered the influences of climatic factors on vegetation coverage changes, and they neglected other environmental factors. Actually, certain environmental factors (e.g., altitude, slope, and slope aspect) also have key impacts on vegetation growth and changes. For instance, the altitude affects the temperature, precipitation, and soil, thus affecting the vertical distribution of the vegetation. The slope aspect affects the degree to which the slope receives various hydrothermal conditions, which has a certain impact on the type and distribution of the vegetation. Moreover, the steep slopes mean that there may be serious soil erosion, which inhibits the vegetation growth. However, many environmental factors cannot be quantified using statistical methods, which limits the exploration of the effects of environmental factors on vegetation changes [24,25]. (4) Most previous studies merely evaluated the individual effects of the factors without quantitatively evaluating the interactive effects between multiple factors on vegetation changes. (5) The majority of previous studies focused on the entire area and local area of the LP (e.g., relevant research was carried out on Shannxi Province in the LP [26]), little comparative research has been conducted for different climatic regions, and the research conclusions are not conducive to guiding policymaking for different regions.

In this study, attempts were made to address the above research gaps. SPOT NDVI products were used to evaluate the vegetation coverage and growth. The spatial–temporal evolution of the NDVI on the LP during 2000–2015 was explained by the Geographical Detector Model (GDM) [27]. Based on the spatial analysis of variance, the GDM was shown to avoid linear assumption between variables and has a clear physical meaning, which reflects the explanatory power of an individual independent variable acting alone or of multiple independent variables interacting on a dependent variable. Moreover, the GDM can utilize a variety of type variables such as geomorphic type, soil type, vegetation type, and land-use type, and thus, is superior to the traditional statistical methods. The GDM is widely adopted in the exploration of the spatial differentiation characteristics of issues in natural and social sciences. For instance, the GDM was applied to study the explanatory powers of different factors on vegetation changes in Shannxi, Mongolia, Sichuan, and the Heihe River Basin [26,28–30].

The study begins by introducing the study area, method and data in Section 2. Then, the linear regression, coefficient of variation and transfer matrix models are used to identify the spatial–temporal evolution of the NDVI on the semi-arid/humid areas of the LP during 2000–2015 in Section 3. The influences of climatic (precipitation and temperature), environmental (altitude, slope, slope aspect, geomorphic type, soil type, and vegetation type), and anthropogenic factors (GDP density, population density, land-use type) on the spatial–temporal evolution of the vegetation NDVI on the semi-arid/humid areas of the LP are identified with the factor, interaction, risk and ecological detectors of the GDM. Section 4 further compares the results in the semi-arid and sub-humid areas, proposes suggestions for formulating and implementing ecological protection and restoration strategies in different areas, and discusses the connections and distinctions with other studies as well as the possible future work. Section 5 summarizes the main findings of the paper. The achievements gained from this study will benefit policy makers and administrative managers in their strategy development and implementation.

2. Data and Methods

2.1. Study Area

The LP is located north of central China, with a total area of 6.49×10^5 km^2 ($107°54'$–$114°33'$ E, $33°41'$–$41°16'$ N), as shown in Figure 1. It is the transitional zone from the coastal humid region to the inland arid region, and from forest to grassland. The terrain is generally low in the southeast and high in the northwest, including the mountain area, the loess hilly area, the loess tableland area, and the valley plain area. The LP has a temperate monsoon climate, with an annual mean temperature of 3.6–14.3 °C and precipitation of 200–800 mm. Both the temperature and precipitation increase from the northwest to the southeast. The annual and daily temperature ranges are large. Furthermore, the seasonal variation of precipitation is large. Heavy rains occur frequently during summer, resulting in the soil erosion, floods, landslides, debris flows and other disasters. The total annual radiation is 50.2×10^4 J/cm^2 to 67.0×10^4 J/cm^2, with a long illumination time and high radiation. About 200 rivers have their headwaters on the LP, including the Tao River, Zuli River, Qingshui River, Huangfuchuan River, Kuye River, Wuding River, Beiluo River Wei River, Qin River, Fen River, etc. The study area includes the provinces (or autonomous regions) of Qinghai, Gansu, Ningxia, Inner Mongolia, Shaanxi, Shanxi, and Henan. Throughout history, the overall vegetation coverage has been low and soil erosion has been very serious on the LP due to the special natural and geographical environment, complex vegetation types, and severe impacts of human activities. Since the 1980s, a series of measures for soil and water conservation have been taken to control the soil erosion. Particularly, after 2000, when the GGP was implemented, the regional ecological environment has significantly improved.

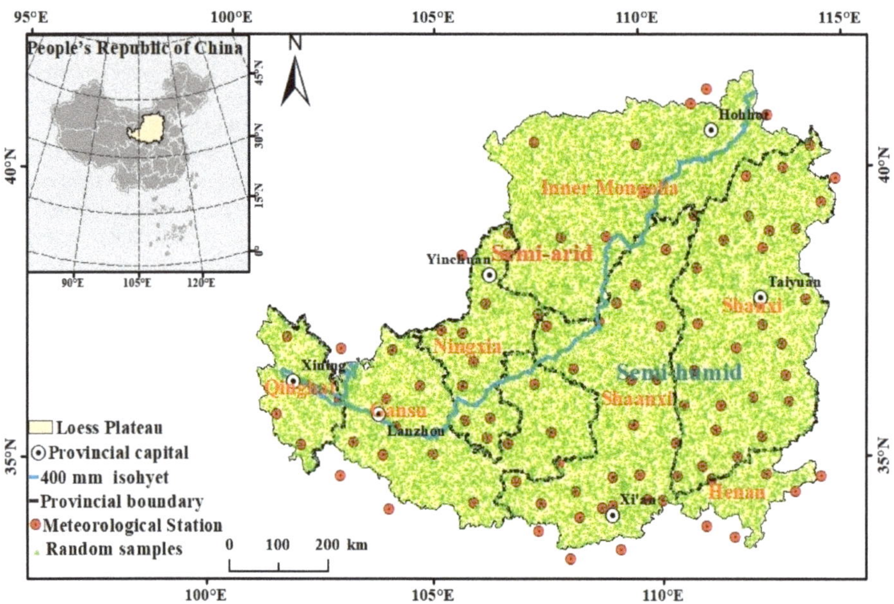

Figure 1. Study area.

The study area was divided into semi-arid and semi-humid areas using the 400 mm isohyet (calculated using the annual mean precipitation data for the LP during 2000–2015) to carry out the comparative analysis. It should be noted that the 400 mm isohyet is of great significance to the physical geography and socio-economic regionalization in China, and it coincides with the Huhuanyong Line. In addition to dividing the climatic zones, this line divides the forest vegetation from the grassland vegetation and the densely populated area from the sparsely populated area [31,32].

2.2. Geographical Detector Model (GDM)

The GDM is proposed based on spatial differentiation theory and geographic information system (GIS) spatial analysis technique [27]. It is usually employed to study the factors affecting spatial hierarchical heterogeneities and the underlying mechanisms. The model assumes similar spatial pattern between the independent and dependent variables if these variables are spatially correlated [33]. The GDM consists of four detectors, including the factor, interaction, risk and ecological detectors.

(1) The factor detector can be utilized to analyze the spatial heterogeneity and to quantify the explanatory power (measured by the q value) of different impact factors X to dependent variable Y.

$$q = 1 - \frac{\sum_{h=1}^{L} N_h \sigma_h^2}{N \sigma^2} \tag{1}$$

where $q \in [0,1]$, and with the increase in the q value, the explanatory power is expected to be stronger; h represents the stratum of variable Y (or factor X), $h \in [1, L]$; N_h is the unit number of stratum h; N is the unit number in all the strata; σ_h^2 represents the variance of variable Y of stratum h; σ^2 is the variance of variable Y of all the strata.

(2) The interaction detector can be used to determine the explanatory power of interaction between two factors (say X_1 and X_2) on the spatial heterogeneity of variable Y, and to judge whether the interactive effect on variable Y would be enhanced or weakened. The steps are as follows. (1) Compute the respective q value of factors X_1 and X_2 to variable Y (q_{X1} and q_{X2}). (2) Compute the q value of the interaction between X_1 and X_2 to Y ($q_{X1\&X2}$). (3) Compare q_{X1}, q_{X2}, and $q_{X1\&X2}$. If $max(q_{X1},q_{X2}) < q_{X1\&X2} < q_{X1} + q_{X2}$ is true, the factors X_1 and X_2 enhance each other (bi-enhance). If $q_{X1\&X2} > q_{X1} + q_{X2}$ is true, the nonlinearity of two factors is enhanced (nonlinear enhancement). The interaction criteria are presented in Table 1.

Table 1. Definition of interaction detector.

Description	Interaction
$min(q_{X1},q_{X2}) < q_{X1\&X2} < max(q_{X1},q_{X2})$	Weaken, uni-
$q_{X1\&X2} < min(q_{X1},q_{X2})$	Weaken, nonlinear
$q_{X1\&X2} > max(q_{X1},q_{X2})$	Enhance, bi-
$q_{X1\&X2} > q_{X1} + q_{X2}$	Enhance, nonlinear
$q_{X1\&X2} = q_{X1} + q_{X2}$	Independent

(3) The risk detector is utilized to judge the difference of significance between the attribute mean values of two strata with the t statistic:

$$t_{\overline{y}_{h=1}-\overline{y}_{h=2}} = \frac{\overline{Y}_{h=1} - \overline{Y}_{h=2}}{\left[\dfrac{Var(\overline{Y}_{h=1})}{n_{h=1}} + \dfrac{Var(\overline{Y}_{h=2})}{n_{h=2}} \right]^{0.5}} \tag{2}$$

where $\overline{Y}_h$ is the attribute mean value within stratum h; n_h is the number of samples within stratum h; Var is the variance.

(4) The ecological detector is developed to compare the explanatory powers of two factors, X_1 and X_2, on variable Y with the F statistic:

$$F = \frac{N_{X1}(N_{x2} - 1)SIV_{X1}}{N_{X2}(N_{x1} - 1)SIV_{X2}} \tag{3}$$

$$SIV_{X1} = \sum_{h=1}^{L1} N_h\sigma_h^2, SIV_{X2} = \sum_{h=1}^{L2} N_h\sigma_h^2 \tag{4}$$

where N_{X1} and N_{X2} are the sample numbers of factors X_1 and X_2, respectively; SIV_{X1} and SIV_{X2} are the sum of the internal variances of the strata from factors X_1 and X_2, respectively; and L_1 and L_2 are the strata numbers of factors X_1 and X_2, respectively.

2.3. Data Description

In this study, multi-source data (Table 2) were collected, mainly including the following.

(1) Vector data: The shape file data for the LP were downloaded from the Resource and Environment Science and Data Center (RESDC), Chinese Academy of Sciences (available from http://www.resdc.cn, accessed on 27 October 2021).

(2) Vegetation NDVI data: The SPOT NDVI data were used in this study (Figure 2), which were obtained from the RESDC. The time span is 2000–2015, the time step is 1 year, and the spatial resolution is 1 km $\times$ 1 km. It should be noted that the annual maximum synthesis method was used to obtain the annual value of NDVI, and therefore, there was no area with NDVI < 0.

Table 2. Climatic, environmental and anthropogenic factors.

Category	Factor	Unit	Code	Range/Type
Climatic	Precipitation	mm	X_1	<250, 250 to 350, 350 to 450, 450 to 550, 550 to 650, >650
	Temperature	°C	X_2	<0, 0 to 3, 3 to 6, 6 to 9, 9 to 12, >12
Environmental	Altitude	m	X_3	90 to 790, 790 to 1228, 1228 to 1611, 1611 to 2136, 2136 to 2963, 2963 to 4914
	Slop	degree	X_4	0 to 5, 5 to 10, 10 to 15, 15 to 20, 20 to 25, >25
	Slope aspect	type	X_5	no slope aspect (−1), east slope (67.5° to 112.5°), west slope (247.5° to 292.5°), south slope (157.5° to 202.5°), north slope (0° to 22.5° and 337.5° to 360°), southeast slope (112.5° to 157.5°), northeast slope (22.5° to 67.5°), southwest slope (202.5° to 247.5°), and northwest slope (292.5° to 337.5°)
	Geomorphic type	type	X_6	plain, platform, hill, small undulating mountain, medium undulating mountain, large undulating mountain
	Soil type	type	X_7	alpine soil, anthropogenic soil, saline alkali soil, hydrogenetic soil, semi-hydrogenetic soil, primary soil, desert soil, arid soil, calcareous soil, semi-eluvial soil, and eluvial soil
	Vegetation type	type	X_8	cultivated vegetation, meadow, grass, grassland, desert, shrub, broad-leaved forest, and coniferous forest
Anthropogenic	GDP density	CNY/km^2	X_9	<200, 200 to 500, 500 to 1000, 1000 to 3000, 3000 to 5000, >5000
	Population density	population/km^2	X_{10}	<100, 100 to 200, 200 to 500, 500 to 1000, 1000 to 2000, >2000
	Land-use type	type	X_{11}	farmland, woodland, grassland, water bodies, construction land, unused land

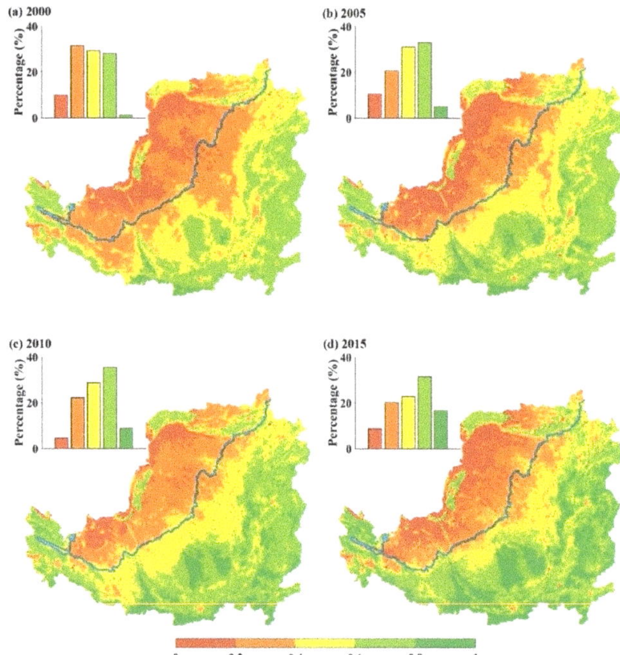

Figure 2. Spatial distribution of NDVI on Loess Plateau during 2000–2015.

According to [34], the NDVI of the LP was divided into 5 intervals using the equal interval method, i.e., 0 to 0.2, 0.2 to 0.4, 0.4 to 0.6, 0.6 to 0.8, and 0.8 to 1.0, which represent 5 types of vegetation, i.e., low, relatively low, medium, relatively high, and high vegetation coverages, respectively.

(3) Meteorological data: The monthly mean precipitation and temperature from 101 meteorological stations on the LP and in its surrounding areas during 2000–2015 were collected for use, which were downloaded from the China Meteorological Data Service

Center (available from http://data.cma.cn/site/index.html, accessed on 27 October 2021). The annual value of the data from each meteorological station was calculated. Moreover, ANUSPLIN v4.3 was used to perform spatial interpolation of the meteorological station data, which were divided into 6 intervals using the equal interval method. Therefore, annual precipitation and temperature grid data (Figures 3 and 4) with the same projection mode, spatial resolution, and time series as the NDVI data were obtained. It should be noted that ANUSPLIN is a widely used software application for spatial interpolation, which was developed on the basis of the thin plate smoothing splines theory. The main feature of ANUSPLIN is that the topographic information can be used in the spatial interpolation of the meteorological data [35].

(4) DEM data: The DEM data were downloaded from the Shuttle Radar Topography Mission (available from http://srtm.csi.cgiar.org/, accessed on 27 October 2021). According to [28], the natural breakpoint method was used to divide the elevation data into 6 categories, including 90 to 790 m, 790 to 1228 m, 1228 to 1611 m, 1611 to 2136 m, 2136 to 2963 m, and 2963 to 4914 m (Figure 5a).

Based on the DEM data, the slope and slope aspect data for the study area were extracted using GIS spatial analysis tools. According to the policy of the Grain for Green [36], the slope was divided into 6 categories, including 0° to 5°, 5° to 10°, 10° to 15°, 15° to 20°, 20° to 25°, >25° (Figure 5b).

The slope aspect was divided into 9 categories, which are denoted as no slope aspect (−1), east slope (67.5° to 112.5°), west slope (247.5° to 292.5°), south slope (157.5° to 202.5°), north slope (0° to 22.5° and 337.5° to 360°), southeast slope (112.5° to 157.5°), northeast slope (22.5° to 67.5°), southwest slope (202.5° to 247.5°), and northwest slope (292.5° to 337.5°) (Figure 5c).

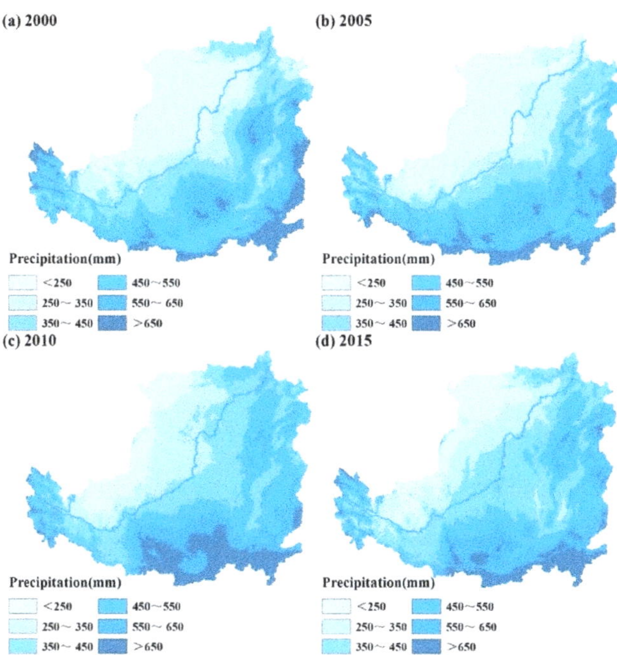

Figure 3. Spatial distribution of precipitation on Loess Plateau during 2000–2015.

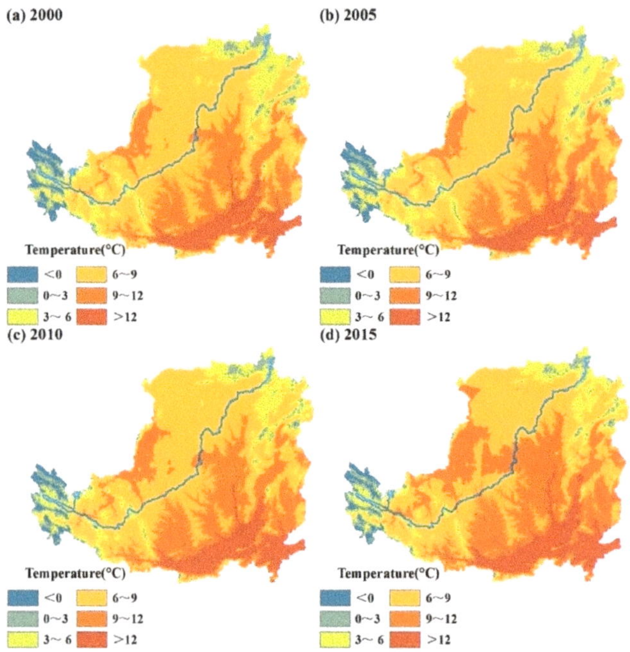

Figure 4. Spatial distribution of temperature on Loess Plateau during 2000–2015.

(5) Geomorphic type data: These data were obtained from the RESDC. Referring to the Geomorphological Atlas of the People's Republic of China (1:1 million) (2009), the geomorphology was divided into 6 types, including plain, platform, hill, small undulating mountain, medium undulating mountain, and large undulating mountain (Figure 5d).

(6) Soil type data: These data were downloaded from the RESDC. Referring to the 1:1 Million Soil Map of the People's Republic of China (1995), the soil was divided into 11 types, including alpine soil, anthropogenic soil, saline alkali soil, hydrogenetic soil, semi-hydrogenetic soil, primary soil, desert soil, arid soil, calcareous soil, semi-eluvial soil, and eluvial soil (Figure 5e).

(7) Vegetation type data: These data were obtained from the RESDC. Referring to the 1:1 Million Vegetation Atlas of China (2001), the vegetation was divided into 8 types, including cultivated vegetation, meadow, grass, grassland, desert, shrub, broad-leaved forest, and coniferous forest (Figure 5f).

(8) Socio-economic data: These data were obtained from the RESDC. The GDP density and population density data for China in 2000, 2005, 2010, and 2015 were obtained. The GDP density and population density were divided into 6 categories, respectively (Figures 6 and 7).

(9) Land-use type data: These data were downloaded from the RESDC. The spatial resolution is 1 km × 1 km. There are 6 land-use types, including unused land, construction land, water bodies, grassland, woodland, and farmland (Figure 8), with a total of 36 transfer combinations.

The influences of the various factors on the spatial–temporal evolution of the NDVI on the LP were calculated and analyzed with the GDM. The various factors were classified into climatic, environmental, and anthropogenic aspects. For the climatic factors, precipitation, and temperature were selected as proxy variables. For the environmental factors, altitude, slope, slope aspect, geomorphic type, soil type, and vegetation type were selected as proxy variables. For the anthropogenic factors, the proxy variables were the GDP density, population density, and land-use type.

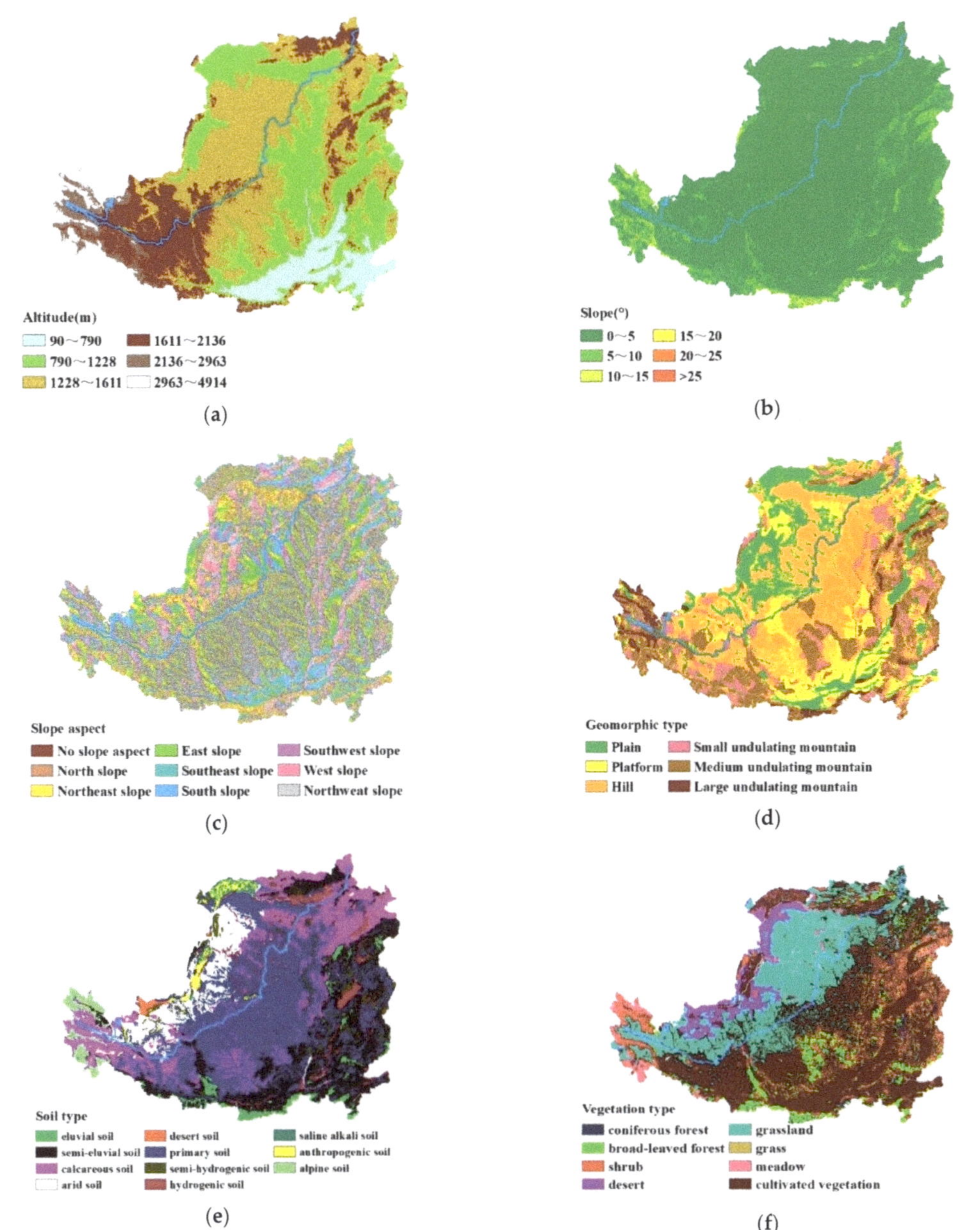

Figure 5. Ranges/types of environmental factors on Loess Plateau. (**a**) Altitude. (**b**) Slope. (**c**) Slope aspect. (**d**) Geomorphic type. (**e**) Soil type. (**f**) Vegetation type.

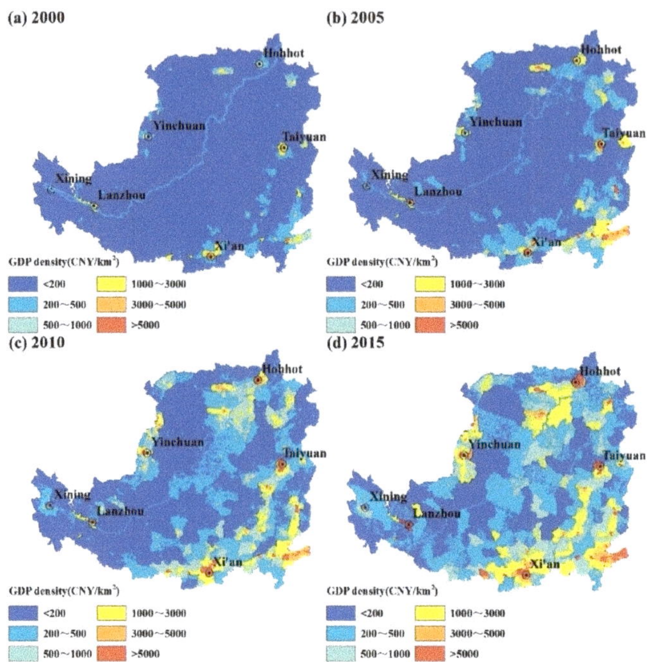

Figure 6. Spatial distribution of GDP density on Loess Plateau during 2000–2015.

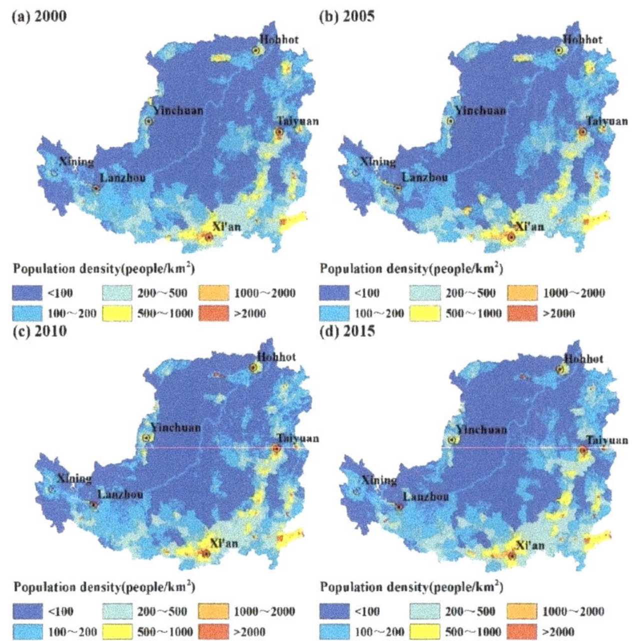

Figure 7. Spatial distribution of population density on Loess Plateau during 2000–2015.

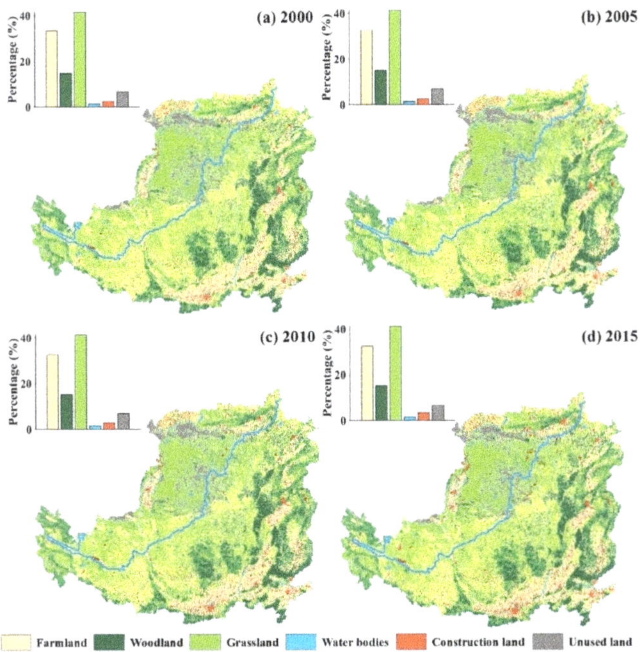

Figure 8. Spatial distribution of land-use type on Loess Plateau during 2000–2015.

The following points should be noted: (1) Although environmental factors do not change in the short-term, these factors play a key role in vegetation growth and changes. Thus, they were also considered in this study according to [37]. (2) The data for each factor were discretized since the input of the GDM needs to be discrete data. (3) The data for the factors were converted into grid data with the 1 km × 1 km spatial resolution (same with the NDVI data) in order to facilitate calculation and analysis. (4) Due to the large area of the LP, the GDM's calculation capacity would be exceeded if all of the grid data were used. Therefore, the sampling method was used and 32,000 sampling points were evenly selected in space to carry out the calculations and analysis (Figure 1), and the consistency of the results of each calculation could be ensured. (5) Two types of inputs were considered in the analysis of the explanatory power of different factors for the spatial–temporal evolution of the NDVI on the LP using the GDM. First, the typical annual values of 2000, 2005, 2010, and 2015 were used to drive the GDM to analyze the influence of various factors on the spatial distribution state of the NDVI. Second, the differences between the annual mean values from 2000 to 2005 and those from 2010 to 2015 were calculated, which were used to drive the GDM to analyze the influence of various factors on the spatial distribution change of the NDVI.

3. Results

3.1. Spatial–Temporal Evolution in NDVI

(1) Grid scale area and area transfer of the NDVI and the driving forces:

The vegetation types on the LP are shown in Figure 5f. In the semi-arid area, the vegetation types were mainly grassland, cultivated vegetation, and desert, accounting for 15.9, 7.2 and 5.0% of the total area, respectively. While in the semi-humid area, cultivated vegetation, grassland, shrub, and broad-leaved forest were widely distributed, accounting for 37.8, 9.2, 6.7 and 5.4% of the total area, respectively.

The grid scale area transfer matrix of the NDVI on the LP during 2000–2015 is shown in Table 3. In 2000, over the entire LP, the coverage areas of the low, relatively low, medium, relatively high, and high vegetation accounted for, respectively, 10.0, 31.5, 29.3, 28.1 and 1.3% of the total area; in the semi-arid area, these types of vegetation accounted for 9.4, 16.0, 5.1, 1.2 and 0.01%, respectively; and in the semi-humid area, they accounted for 0.5, 15.5, 24.2, 26.9 and 1.3%, respectively. In 2015, over the entire LP, the coverage areas of the low, relatively low, medium, relatively high, and high vegetation accounted for 8.7, 20.3, 22.9, 31.5 and 16.7%, respectively; in the semi-arid area, they accounted for 8.5, 14.5, 5.0, 3.3 and 0.4%, respectively; and in the semi-humid area, they accounted for 0.3, 5.8, 17.9, 28.1 and 16.3%, respectively. The proportion of the coverage areas with an NDVI of greater than 0.6 increased significantly, and the proportions of the relatively low vegetation converted to medium vegetation, medium vegetation converted to relatively high vegetation, and relatively high vegetation converted to high vegetation accounted for the largest proportions. The areas of these transitions accounted for, respectively, 10.8, 15.8 and 14.1% of the total area.

Table 3. Grid scale area transfer matrix of NDVI on Loess Plateau during 2000–2015 (km^2).

2000	2015					
	[0, 0.2]	[0.2, 0.4]	[0.4, 0.6]	[0.6, 0.8]	[0.8, 1]	Total
[0, 0.2]	40,824 (6.31%)	22,376 (3.46%)	988 (0.15%)	135 (0.02%)	10 (~0.00%)	64,333 (9.95%)
[0.2, 0.4]	15,375 (2.38%)	99,000 (15.31%)	69,499 (10.75%)	19,363 (3.00%)	144 (0.02%)	203,381 (31.46%)
[0.4, 0.6]	235 (0.04%)	8792 (1.36%)	69,526 (10.75%)	102,374 (15.84%)	8322 (1.29%)	189,249 (29.27%)
[0.6, 0.8]	12 (~0.00%)	792 (0.12%)	7717 (1.19%)	81,535 (12.61%)	91,338 (14.13%)	181,394 (28.06%)
[0.8, 1]	0 (0.00%)	6 (~0.00%)	9 (~0.00%)	110 (0.02%)	8015 (1.24%)	8140 (1.26%)
Total	56,446 (8.73%)	130,966 (20.26%)	147,739 (22.85%)	203,517 (31.48%)	107,829 (16.68%)	646,497 (100%)

The land-use type on the LP during 2000–2015 is shown in Figure 8, and the grid scale area transfer matrix is shown in Table 4. In 2000, over the entire LP, the proportions of the farmland, woodland, grassland, water bodies, construction land, and unused land were 33.3, 14.7, 41.5, 1.4, 2.5 and 6.7%, respectively; in the semi-arid area, they were 7.1, 1.2, 16.6, 0.7, 0.8 and 5.3%, respectively; and in the semi-humid area, they were 26.2, 13.5, 24.9, 0.8%, 1.7 and 1.4%, respectively. In 2015, over the entire LP, the proportions of farmland, woodland, grassland, water bodies, construction land, and unused land were 32.4, 15.1, 41.2, 1.5, 3.3 and 6.6%, respectively; in the semi-arid area, they were 6.9, 1.3, 16.4, 0.7, 1.1 and 5.2%, respectively; and in the semi-humid area, they were 25.4, 13.8, 24.8, 0.8, 2.1 and 1.4%, respectively. Overall, the area of the change in land-use types on the LP was not significant, and it was mainly in the 0.94% decrease in farmland, the 0.44% increase in woodland, and the 0.82% increase in construction land. To a certain extent, these changes reflect the impacts of human activities in the study area, such as the implementation of the GGP and urbanization-related development.

The difference in the NDVI of the different land-use type transfers on the LP during 2000–2015 is shown in Figure 9. It should be noted that Figure 9 was obtained by subtracting the NDVI data for 2000 from that for 2015 on the grid scale, and the differences in the NDVI of the different land-use type transfers were then counted. Figure 9a shows the mean difference in the NDVI of all grids of different land-use type transfers. Most of the transfers had a positive impact on the NDVI, except for the transfers of farmland and water bodies to construction land, which decreased the NDVI slightly (reduction rates of −0.0173 and −0.0016, respectively). Moreover, the largest increment of the NDVI was due to farmland transferred into woodland (increment of 0.1411), woodland transferred into farmland (increment of 0.1359), and unused land transferred into farmland (increment of 0.1338). Figure 9b shows the total difference in the NDVI of all grids of different land-use type transfers. It can be seen that the largest increments of the NDVI were due to grassland, farmland, and woodland, and these land-use types did not change. The increments were 28,539.55, 25,545.42, and 12,450.86, respectively. The probable reasons are as follows. On the LP, the precipitation is limited and there exists a gap between agricultural water demand and supply. In the past, there used to be large areas of farmland that could not be irrigated adequately and were greatly affected by the precipitation. With the implementation of the GGP, some infertile areas of farmland that could not be irrigated adequately were returned to the woodland or grassland, and what remained was adequately irrigated or fertile areas of farmland that were less affected by the precipitation. Moreover, with the development of agricultural technologies and the optimization and adjustment of planting structures, the crop yields on the LP were continuously increased.

Table 4. Grid scale area transfer matrix of land-use type on Loess Plateau during 2000–2015 (km^2).

2000	2015						
	Farmland	**Woodland**	**Grassland**	**Water Bodies**	**Construction Land**	**Unused Land**	**Total**
Farmland	207,681 (31.99%)	1667 (0.26%)	2739 (0.42%)	647 (0.10%)	3042 (0.47%)	414 (0.06%)	216,190 (33.30%)
Woodland	139 (0.02%)	94,495 (14.56%)	328 (0.05%)	66 (0.01%)	263 (0.04%)	82 (0.01%)	95,373 (14.69%)
Grassland	1455 (0.22%)	1696 (0.26%)	262,300 (40.40%)	330 (0.05%)	1676 (0.26%)	1607 (0.25%)	269,064 (41.45%)
Water bodies	317 (0.05%)	36 (0.01%)	222 (0.03%)	8312 (1.28%)	84 (0.01%)	258 (0.04%)	9229 (1.42%)
Construction land	17 (~0.00%)	15 (~0.00%)	41 (0.01%)	19 (~0.00%)	15,777 (2.43%)	10 (~0.00%)	15,879 (2.45%)
Unused land	448 (0.07%)	337 (0.05%)	1631 (0.25%)	211 (0.03%)	4189 (0.06%)	40,413 (6.23%)	43,458 (6.69%)
Total	210,057 (32.36%)	98,246 (15.13%)	267,261 (41.17%)	9585 (1.48%)	21,260 (3.27%)	42,784 (6.59%)	649,193 (100%)

(2) Spatial changes in the NDVI on the grid scale:

The spatial changes in the NDVI on the grid scale on the LP during 2000–2015 are shown in Figure 10. Figure 10a indicates that the annual mean value of the NDVI on the LP was 0.529, with an uneven spatial distribution, i.e., decreasing from the southeastern to the northwestern areas. The high and low values were mainly distributed in the semi-humid and semi-arid areas, with annual mean values of 0.619 and 0.346, respectively.

Figure 10b shows that the mean value of the coefficient of variation of the NDVI was 0.1406 on the LP, with an uneven spatial distribution, i.e., increasing from the southeastern semi-humid area (mean value of 0.1165) to the northwestern semi-arid area (mean value of 0.1926). There was 7.0% of the total area with $0 < CV < 0.05$, mainly distributed in the eastern and southern regions. The area with $0.05 \leq CV < 0.1$ accounted for 25.9% of the total area. The area with $0.1 \leq CV < 0.15$ accounted for 25.7% of the total area. Moreover, there was 41.4% of the total area with $0.15 \leq CV < 0.2$ and $CV \geq 0.2$, mainly distributed in the semi-arid area as well as the junction of the semi-arid/humid areas.

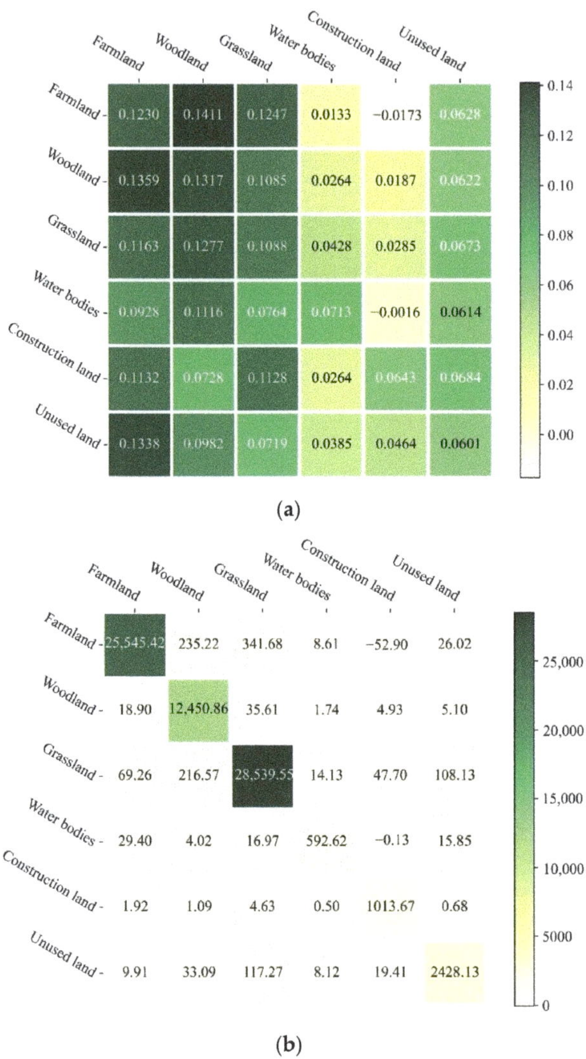

Figure 9. Difference in NDVI of different land-use type transfers during 2000–2015. (**a**) Mean difference in NDVI of all grids of different land-use type transfers. (**b**) Total difference in NDVI of all grids of different land-use type transfers.

Figure 10c,d show that of the total area, the positive area accounted for 91.5% while the negative area accounted for 8.5%, which was obtained by subtracting the NDVI in 2000 from that in 2015. The NDVI in most areas of the LP (92.8% of the total area) increased from 2000 to 2015, indicating that the ecological environment in the region significantly improved. In addition, 70.0% of the area passed the significance test ($p < 0.05$); and 68.6% of the area increased significantly, while 1.4% of the area decreased significantly. The decrease in the NDVI was mainly concentrated in the semi-arid area as well as the urban area with rapid economic development and a large population concentration.

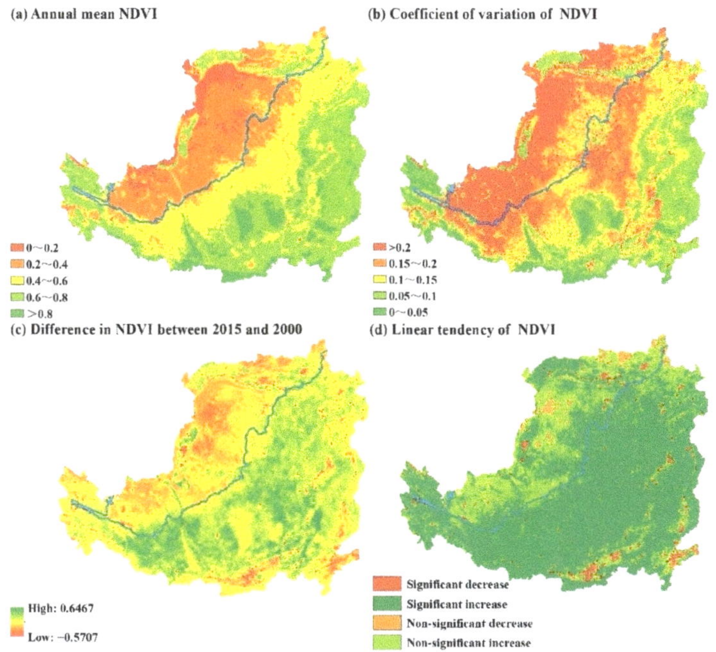

Figure 10. Spatial change in NDVI on grid scale on Loess Plateau during 2000–2015.

(3) Temporal changes in the NDVI on the regional scale:

The temporal changes in the NDVI on the regional scale on the LP during 2000–2015 are shown in Figure 11. The NDVI on the LP exhibited a fluctuating upward trend. The annual NDVI growth rates of the entire LP, the semi-arid area, and the semi-humid area were 0.0079, 0.0049, and 0.0093, respectively, indicating that the growth rate of the NDVI in the semi-humid area was higher than in the semi-arid area after the GGP was implemented.

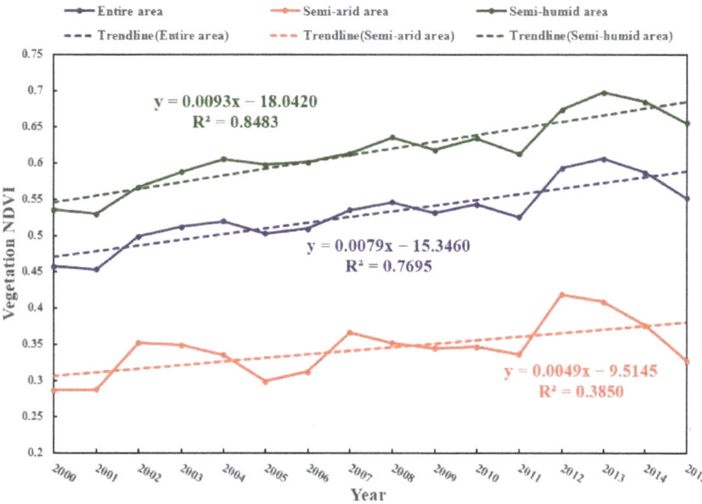

Figure 11. Temporal change in NDVI on regional scale on Loess Plateau during 2000–2015.

3.2. Individual Effects of Factors on NDVI

(1) Results of typical years:

Using the typical annual values of 2000, 2005, 2010, and 2015 to drive the factor detector, the individual effects of the factors (represented by the q value) could be obtained, as shown in Figure 12.

For the entire LP, the order of the annual mean q values was precipitation ($q = 0.5985$) > vegetation type ($q = 0.4790$) > soil type ($q = 0.3346$) > land-use type ($q = 0.2697$) > geomorphic type ($q = 0.2341$) > temperature ($q = 0.1469$) > altitude ($q = 0.1203$) > population density ($q = 0.1132$) > slope ($q = 0.0649$) > GDP density ($q = 0.0411$) > slope aspect ($q = 0.0012$).

For the semi-arid area, the order of the annual mean q values was precipitation ($q = 0.4796$) > vegetation type ($q = 0.3906$) > soil type ($q = 0.3409$) > land-use type ($q = 0.2368$) > geomorphic type ($q = 0.2013$) > temperature ($q = 0.1851$) > altitude ($q = 0.1695$) > population density ($q = 0.0954$) > slope ($q = 0.0774$) > GDP density ($q = 0.0110$) > slope aspect ($q = 0.0019$).

For the semi-humid area, the order of the annual mean q values was geomorphic type ($q = 0.2597$) > vegetation type ($q = 0.2466$) > precipitation ($q = 0.2250$) > land-use type ($q = 0.2053$) > soil type ($q = 0.1637$) > temperature ($q = 0.0965$) > population density ($q = 0.0783$) > slope ($q = 0.0541$) > altitude ($q = 0.0512$) > GDP density ($q = 0.0397$) > slope aspect ($q = 0.0086$).

(2) Results of annual mean differences:

Using the differences between the annual mean values from 2000 to 2005 and those from 2010 to 2015 to drive the factor detector, the individual effects of the factors (represented by the q value) could be obtained, as shown in Figure 13.

For the entire LP, the order of the q values was soil type ($q = 0.1287$) > vegetation type ($q = 0.1142$) > land-use type ($q = 0.0729$) > temperature ($q = 0.0680$) > geomorphic type ($q = 0.0532$) > precipitation ($q = 0.0508$) > GDP density ($q = 0.0463$) > altitude ($q = 0.0357$) > population density ($q = 0.0224$) > slope ($q = 0.0020$) > slope aspect ($q = 0.0010$).

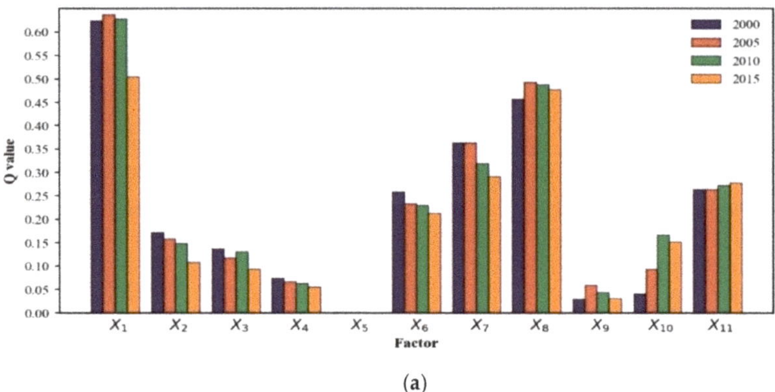

(a)

Figure 12. *Cont.*

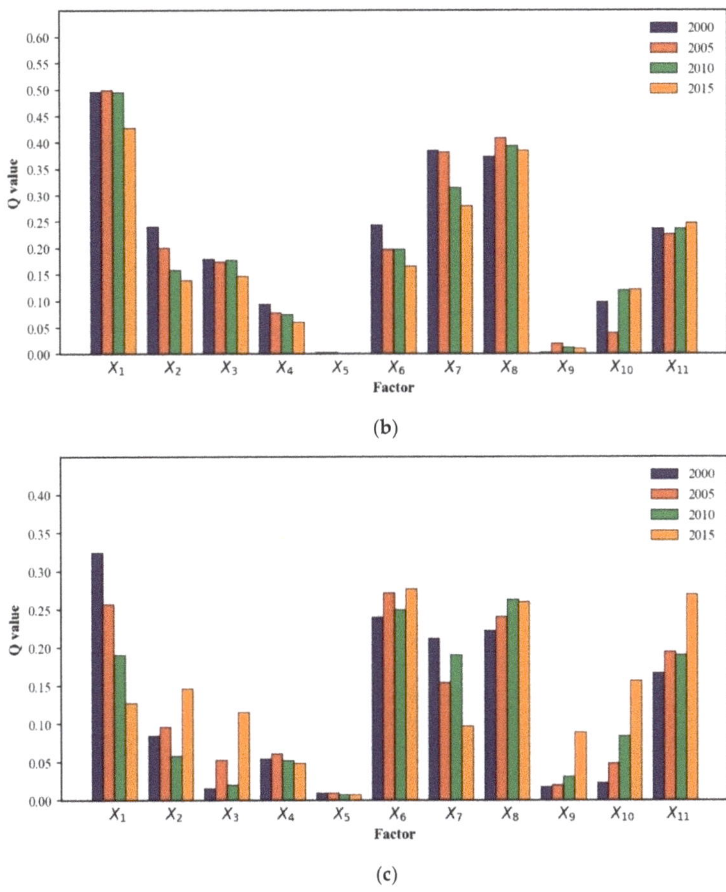

(b)

(c)

Figure 12. Effects of individual factors derived from the factor detector with inputs of typical annual values. (**a**) Entire area. (**b**) Semi-arid area. (**c**) Semi-humid area.

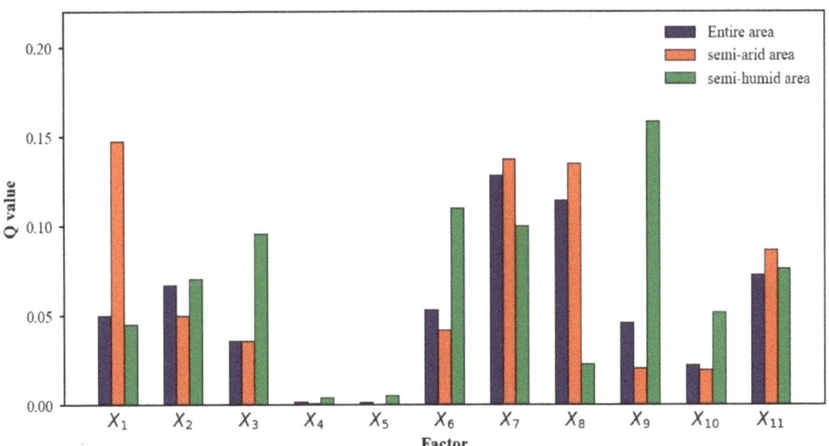

Figure 13. Effects of individual factors derived from the factor detector with inputs of differences between annual mean values.

For the semi-arid area, the order of the q values was precipitation ($q = 0.1476$) > soil type ($q = 0.1373$) > vegetation type ($q = 0.1353$) > land-use type ($q = 0.0864$) > temperature ($q = 0.0500$) > geomorphic type ($q = 0.0417$) > altitude ($q = 0.0358$) > GDP density ($q = 0.0202$) > population density ($q = 0.0195$) > slope aspect ($q = 0.0030$) > slope ($q = 0.0010$).

For the semi-humid area, the order of the q values was GDP density ($q = 0.1583$) > geomorphic type ($q = 0.1101$) > soil type ($q = 0.0999$) > altitude ($q = 0.0955$) > land-use type ($q = 0.0761$) > temperature ($q = 0.0705$) > population density ($q = 0.0520$) > precipitation ($q = 0.0452$) > vegetation type ($q = 0.0227$) > slope aspect ($q = 0.0049$) > slope ($q = 0.0040$).

In summary, the explanatory powers of the factors on the spatial distribution state of the NDVI in typical years and the spatial distribution change of the NDVI during 2000–2015 were different. In addition, the explanatory powers were different for the entire LP, and for the semi-arid/humid areas. With respect to the spatial distribution state of the NDVI in typical years, the decisive climatic factor was precipitation, the decisive environmental factors were geomorphic type, soil type, and vegetation type, and the decisive anthropogenic factor was land-use type. With respect to the spatial distribution change of the NDVI during 2000–2015, in the semi-arid area, the climatic and environmental factors were the decisive factors, including precipitation, soil type, and vegetation type; the impacts of anthropogenic factors, such as the GDP density, land-use type, and population density, were more significant in the semi-humid area.

3.3. Interactive Effects between Factors on NDVI

(1) Results of typical years:

Using the typical annual values of 2000, 2005, 2010, and 2015 to drive the interaction detector, the interactive effects of the factors (represented by the q value) could be obtained, as shown in Figure 14. For the entire LP and the semi-arid/humid areas, the interactive effects between factors were greater than their individual effects, indicating that none of the factors acted independently, but they had a certain enhancement effect, including nonlinear enhancement and bi-enhancement.

For the entire LP, 26.4% of the interactive factor combinations exhibited nonlinear enhancement and 73.6% exhibited bi-enhancement. The interactive effect between precipitation and vegetation type was the strongest (mean annual?$q = 0.7034$), followed by precipitation and soil type (mean annual?$q = 0.6987$).

For the semi-arid area, 32.3% of the interactive factor combinations exhibited nonlinear enhancement and 67.7% exhibited bi-enhancement. The interactive effect between precipitation and soil type was the strongest (mean annual?$q = 0.6375$), followed by precipitation and vegetation type (mean annual?$q = 0.6235$).

For the semi-humid area, 44.1% of the interactive factor combinations exhibited nonlinear enhancement and 55.9% exhibited bi-enhancement. The interactive effect between precipitation and geomorphic type was the strongest (mean annual?$q = 0.3961$), followed by precipitation and vegetation type (mean annual?$q = 0.3701$).

(2) Results of annual mean differences:

Using the differences between the annual mean values from 2000 to 2005 and those from 2010 to 2015 to drive the interaction detector, the interactive effects of the factors (represented by the q value) could be obtained, as shown in Figure 15.

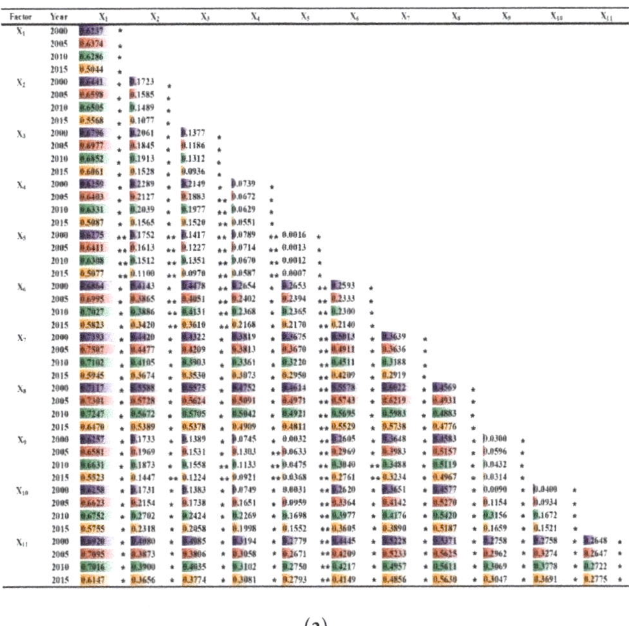

(a)

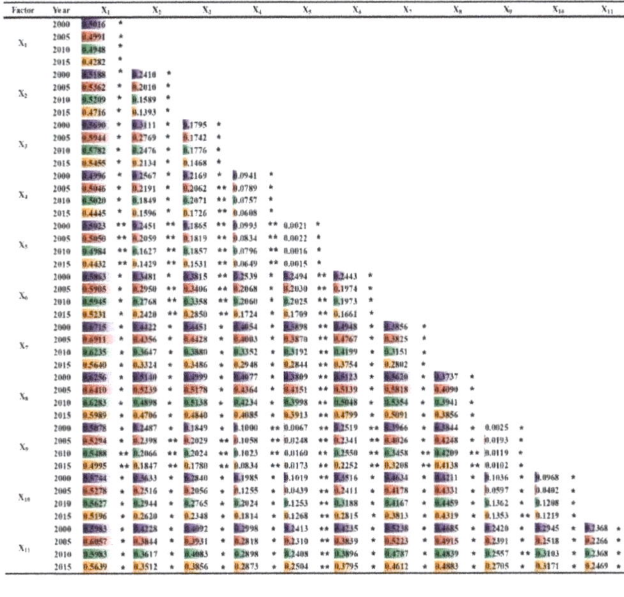

(b)

Figure 14. *Cont.*

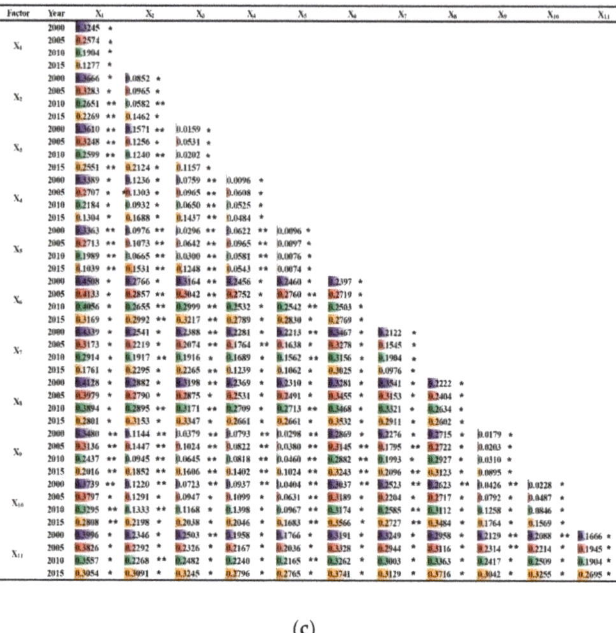

Factor	Year	X_1	X_2	X_3	X_4	X_5	X_6	X_7	X_8	X_9	X_{10}	X_{11}
X_1	2000	0.3245 *										
	2005	0.2574 *										
	2010	0.1904 *										
	2015	0.1277 *										
X_2	2000	0.3666 *	0.0852 *									
	2005	0.3283 *	0.0965 *									
	2010	0.2651 **	0.0582 **									
	2015	0.2269 **	0.1462 *									
X_3	2000	0.3610 **	0.1571 **	0.0159 *								
	2005	0.3248 **	0.1256 *	0.0531 *								
	2010	0.2599 **	0.1240 **	0.0202 *								
	2015	0.2551 **	0.2124 *	0.1157 *								
X_4	2000	0.3389 *	0.1236 *	0.0759 **	0.0096 *							
	2005	0.2707 *	0.1303 *	0.0965 **	0.0608 *							
	2010	0.2184 *	0.0932 *	0.0650 **	0.0525 *							
	2015	0.1304 *	0.1688 *	0.1437 **	0.0484 *							
X_5	2000	0.3363 **	0.0976 **	0.0296 **	0.0622 **	0.0096 *						
	2005	0.2713 **	0.1073 **	0.0642 **	0.0965 **	0.0097 *						
	2010	0.1989 **	0.0665 **	0.0300 **	0.0581 **	0.0076 *						
	2015	0.1039 **	0.1531 **	0.1248 **	0.0543 **	0.0074 *						
X_6	2000	0.4508 *	0.2766 *	0.3164 **	0.2456 *	0.2460 *	0.2397 *					
	2005	0.4133 *	0.2857 **	0.3042 **	0.2752 *	0.2760 **	0.2719 *					
	2010	0.4056 *	0.2655 **	0.2999 **	0.2532 *	0.2542 **	0.2503 *					
	2015	0.3169 *	0.2992 **	0.3217 **	0.2789 *	0.2830 *	0.2769 *					
X_7	2000	0.4339 *	0.2541 *	0.2388 **	0.2281 *	0.2213 **	0.3467 *	0.2122 *				
	2005	0.3173 *	0.2219 *	0.2074 **	0.1764 **	0.1638 *	0.3278 *	0.1545 *				
	2010	0.2914 *	0.1917 **	0.1916 *	0.1689 *	0.1562 **	0.3156 *	0.1904 *				
	2015	0.1761 *	0.2295 *	0.2265 **	0.1239 *	0.1062 *	0.3025 *	0.0976 *				
X_8	2000	0.4128 *	0.2882 *	0.3198 **	0.2369 *	0.2310 *	0.3281 *	0.3541 *	0.2222 *			
	2005	0.3979 *	0.2790 *	0.2875 *	0.2531 *	0.2491 *	0.3455 *	0.3153 *	0.2404 *			
	2010	0.3894 *	0.2895 **	0.3171 **	0.2709 *	0.2713 **	0.3468 *	0.3321 *	0.2634 *			
	2015	0.2801 *	0.3153 *	0.3347 *	0.2661 *	0.2661 *	0.3532 *	0.2911 *	0.2602 *			
X_9	2000	0.3480 **	0.1144 **	0.0379 **	0.0793 **	0.0298 **	0.2869 *	0.2276 *	0.2715 *	0.0179 *		
	2005	0.3136 **	0.1447 **	0.1024 **	0.0822 **	0.0380 **	0.3145 **	0.1795 **	0.2722 *	0.0203 *		
	2010	0.2437 **	0.0945 **	0.0645 **	0.0818 **	0.0460 **	0.2882 **	0.1993 *	0.2927 *	0.0310 *		
	2015	0.2016 **	0.1852 **	0.1606 **	0.1402 **	0.1024 **	0.3243 **	0.2096 **	0.3123 *	0.0895 *		
X_{10}	2000	0.3739 **	0.1220 **	0.0723 **	0.0937 **	0.0404 **	0.3037 **	0.2523 **	0.2623 **	0.0426 **	0.0228 *	
	2005	0.3797 *	0.1291 *	0.0947 *	0.1099 *	0.0631 **	0.3189 *	0.2204 *	0.2717 *	0.0792 *	0.0487 *	
	2010	0.3295 **	0.1333 **	0.1168 *	0.1398 *	0.0967 **	0.3174 *	0.2585 **	0.3112 *	0.1258 *	0.0846 *	
	2015	0.2808 **	0.2198 *	0.2038 *	0.2046 *	0.1683 **	0.3566 *	0.2727 **	0.3484 *	0.1764 *	0.1569 *	
X_{11}	2000	0.3996 *	0.2346 *	0.2503 **	0.1958 *	0.1766 *	0.3191 *	0.3249 *	0.2958 *	0.2129 **	0.2088 **	0.1666 *
	2005	0.3826 *	0.2292 *	0.2326 *	0.2167 *	0.2036 *	0.3328 *	0.2944 *	0.3116 *	0.2314 **	0.2214 *	0.1945 *
	2010	0.3587 *	0.2268 **	0.2482 *	0.2240 *	0.2165 **	0.3262 *	0.3003 *	0.3363 *	0.2417 *	0.2509 *	0.1904 *
	2015	0.3054 *	0.3091 *	0.3245 *	0.2796 *	0.2765 *	0.3741 *	0.3129 *	0.3716 *	0.3042 *	0.3255 *	0.2695 *

(c)

Figure 14. Effects of interactive factors derived from the interaction detector with inputs of typical annual values. (Note: * indicates bi-enhance; ** indicates nonlinear enhance). (**a**) Entire area. (**b**) Semi-arid area. (**c**) Semi-humid area.

For the entire LP, 90.9% of the interactive factor combinations exhibited nonlinear enhancement and 9.1% exhibited bi-enhancement. The interactive effect between soil type and temperature was the strongest ($q = 0.2194$), followed by soil type and vegetation type ($q = 0.2164$).

For the semi-arid area, 83.6% of the interactive factor combinations exhibited non-linear enhancement and 16.4% exhibited bi-enhancement. The interactive effect between precipitation and soil type was the strongest ($q = 0.2438$), followed by soil type and land-use type ($q = 0.2379$).

For the semi-humid area, 61.8% of the interactive factor combinations exhibited non-linear enhancement and 38.2% exhibited bi-enhancement. The interactive effect between GDP density and geomorphic type was the strongest ($q = 0.2470$), followed by GDP density and soil type ($q = 0.2300$).

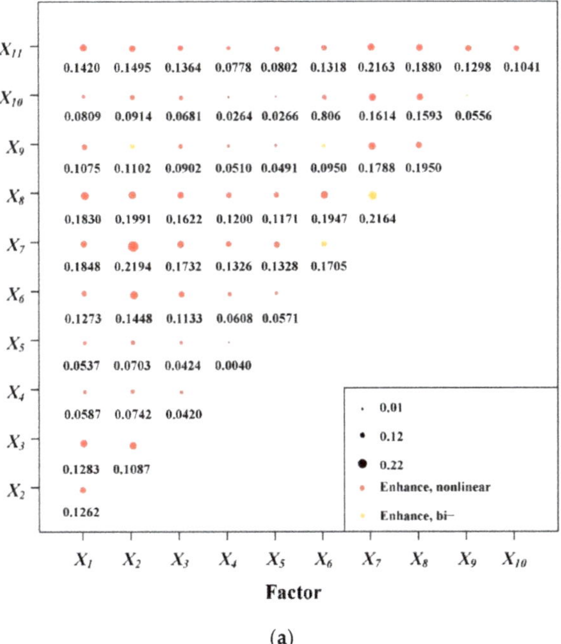

(a)

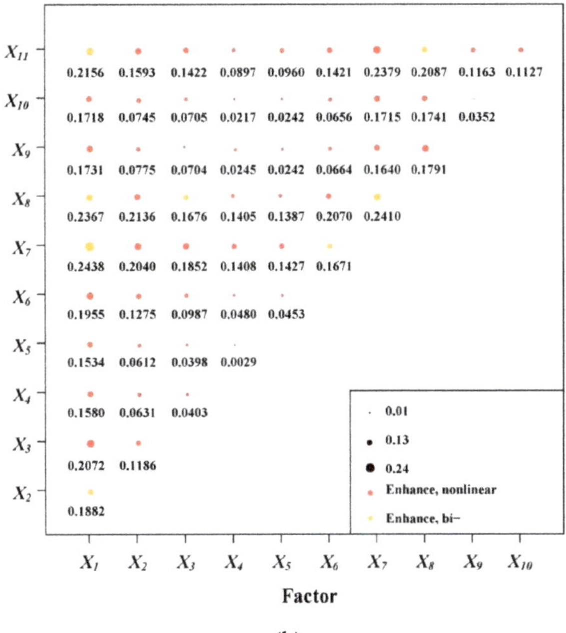

(b)

Figure 15. *Cont.*

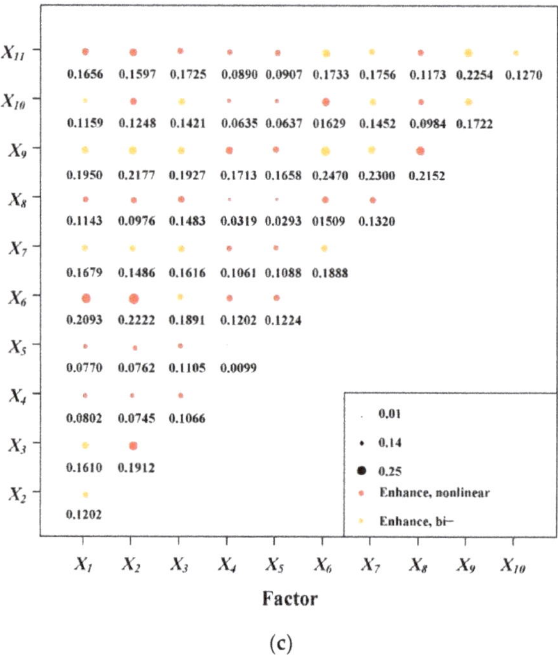

Figure 15. Effects of interactive factors derived from the interaction detector with inputs of differences between annual mean values. (**a**) Entire area. (**b**) Semi-arid area. (**c**) Semi-humid area.

3.4. Ranges or Types of Factors for NDVI

Using the typical annual values for 2000, 2005, 2010, and 2015 to drive the risk detector, the ranges or the types of factors for the NDVI could be obtained, as shown in Figure 16. The ranges or the types of factors for the NDVI were different in different years in the entire LP, and in the semi-arid/humid areas.

For the entire LP, the suitable precipitation range for the NDVI was greater than 650 mm (as the precipitation increased, the NDVI increased), the temperature range was smaller than 0 °C (as the temperature increased, the NDVI decreased first and then increased), the altitude ranged from 90 to 790 m or from 2963 to 4914 m (as the altitude increased, the NDVI decreased first and then increased), the slope ranged from 15° to 20°, the slope aspect was no slope aspect, the geomorphic type was large undulating mountain, the soil type was eluvial soil, the vegetation type was broad-leaved forest, the GDP density ranged from 500 to 1000 CNY/km², the population density ranged from 200 to 500 people/km², and the land-use type was woodland.

For the semi-arid area, the suitable precipitation for the NDVI ranged from 450 to 550 mm, the temperature range was smaller than 0 °C, the altitude ranged from 2963 to 4914 m, the slope ranged from 20° to 25°, the slope aspect was no slope aspect, the geomorphic type was large undulating mountain, the soil type was anthropogenic soil, the vegetation type was cultivated vegetation, the GDP density ranged from 500 to 1000 CNY/km², the population density ranged from 200 to 500 people/km², and the land-use type was farmland.

For the semi-humid area, the suitable precipitation range for the NDVI was greater than 650 mm, the temperature range was smaller than 0 °C, the altitude ranged from 2963 to 4914 m, the slope ranged from 15° to 20°, the slope aspect was no slope aspect, the geomorphic type was large undulating mountain, the soil type was eluvial soil, the vegetation type was broad-leaved forest, the GDP density ranged from 3000 to 5000 CNY/km², the population density ranged from 200 to 500 people/km², and the land-use type was woodland.

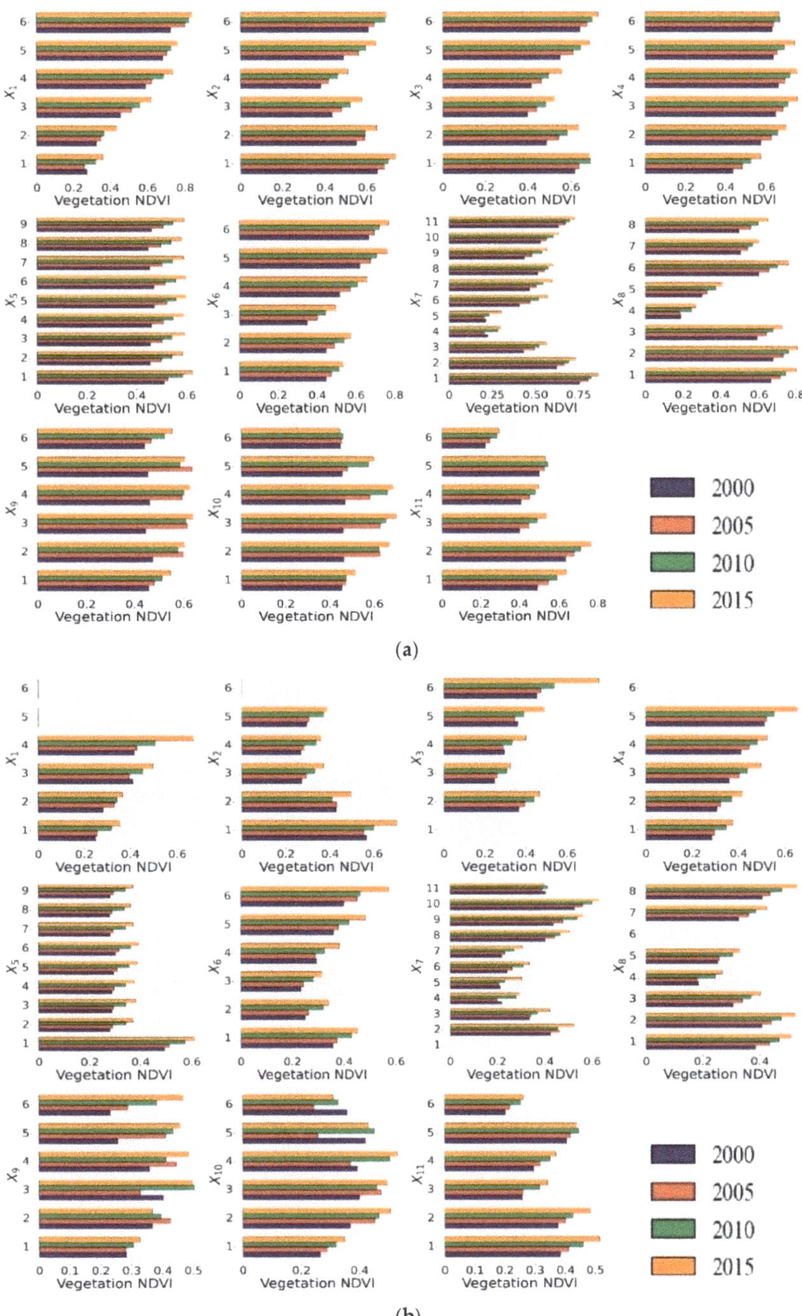

Figure 16. *Cont.*

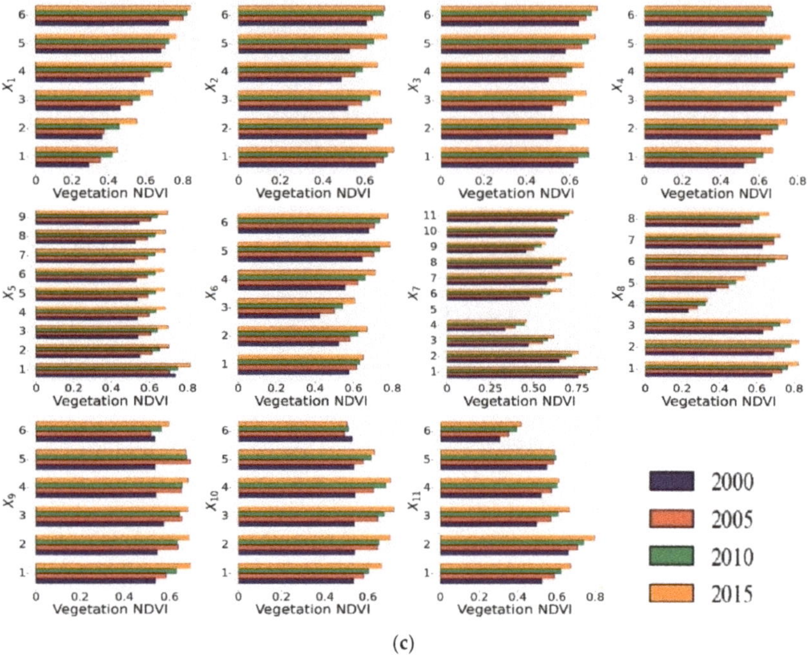

(c)

Figure 16. Ranges or types of factors for NDVI derived from the risk detector with inputs of typical annual values. (**a**) Entire area. (**b**) Semi-arid area. (**c**) Semi-humid area.

3.5. Differences of Significance between Factors on NDVI

Using the typical annual values for 2000, 2005, 2010, and 2015 to drive the ecological detector, the differences of significance between factors for the NDVI could be obtained, as shown in Table 5. For the entire LP, the differences were not significant between the precipitation and other factors in the different years (i.e., relatively certain). The significant differences between the soil type and the other factors did not vary in the different years (i.e., relatively certain). The soil type showed no significant differences with precipitation, GDP density, population density, and land-use type, but it exhibited significant differences with other factors. The differences of significance varied between the other pairs of factors in the different years (i.e., relatively uncertain).

Table 5. Statistical significance of factors derived from the ecological detector with inputs of typical annual values.

	X_1	X_2	X_3	X_4	X_5	X_6	X_7	X_8	X_9	X_{10}	X_{11}
X_1											
X_2	N										
X_3	N	N									
X_4	N	N	N								
X_5	N	N	N/Y	N							
X_6	N	Y	Y	Y	Y						
X_7	N	Y	Y	Y	Y	Y					
X_8	N	Y	N/Y	Y	Y	Y	Y				
X_9	N	N	N/Y	N/Y	N/Y	N/Y	N	N			
X_{10}	N	N/Y	N/Y	N/Y	N/Y	N/Y	N	N	N/Y		
X_{11}	N	Y	Y	Y	Y	Y	N	N	Y	Y	

Note: Y indicates that there is a significant difference in the effect of two factors on NDVI (confidence is 95%); N indicates no significant difference; N/Y indicates that there is (or is no) significant difference across different years.

4. Discussion

4.1. Comparison between Semi-Arid and Semi-Humid Areas

Climatic factors influence the environmental conditions of the LP; while environmental factors determine the range and intensity of the human activities. The three types of factors are not independent of each other, and they interact with each other to influence the spatial distribution of the NDVI and its evolution over time on the LP. Based on the above results, there were significant spatial differences in the NDVI and its driving forces on both sides of the 400 mm isohyet on the LP.

For the semi-arid area to the northwest of the 400 mm isohyet, a number of observations were made. (1) The analysis of the spatial–temporal evolution revealed that the precipitation is relatively small and the temperature is relatively low, with annual mean values of 294 mm and 7.6 °C, respectively. The geomorphic types are mainly hills and plains (accounting for 12.3 and 11.0%, respectively). The soil types are mainly primary soil and arid soil (accounting for 11.1 and 8.0%, respectively). The main vegetation types are grassland, cultivated vegetation, and desert (accounting for 15.9, 7.2 and 5.0%, respectively). The grassland and farmland are the main land-use types (accounting for 16.4% and 6.9% in 2015, respectively). Moreover, the GDP density is relatively small (772 CNY/km^2 in 2015). The population density is relatively small (96 people/km^2 in 2015). The vegetation coverage is relatively low (annual mean NDVI of 0.346), and the coefficient of variation is relatively large (mean value of 0.1926). The growth rate of the NDVI was 0.0049/a after 2000 when the GGP was implemented. (2) The analysis based on the GDM revealed the following: The individual factors that determine the changes in the spatial distribution of the NDVI are the climatic and environmental factors, such as precipitation ($q = 0.1476$), soil type ($q = 0.1373$), vegetation type ($q = 0.1353$), temperature ($q = 0.0500$), and geomorphic type ($q = 0.0417$). The main interactive factors are precipitation and soil type ($q = 0.2438$), and soil type and land-use type ($q = 0.2379$). The suitable precipitation for the NDVI ranged from 450 to 550 mm, the temperature range was smaller than 0 °C, the altitude ranged from 2963 to 4914 m, the slope ranged from 20° to 25°, the slope aspect was no slope aspect, the geomorphic type was large undulating mountain, the soil type was anthropogenic soil, the vegetation type was cultivated vegetation, the GDP density ranged from 500 to 1000 CNY/km^2, the population density ranged from 200 to 500 people/km^2, and the land-use type was farmland.

Observations were also made regarding the semi-humid area to the southeast of the 400 mm isohyet. (1) The analysis of the spatial–temporal evolution revealed that both the precipitation and the temperature are relatively high, with annual mean values of 531 mm and 8.3 °C, respectively. The geomorphic types are mainly hills and medium undulating mountains (accounting for 19.4 and 13.3%, respectively). The soil types are mainly primary soil and semi-eluvial soil (accounting for 30.9 and 14.9%, respectively). The main vegetation types are cultivated vegetation and grassland (accounting for 37.8 and 9.2%, respectively). The farmland and grassland are the main land-use types (accounting for 25.4 and 24.8% in 2015, respectively). Moreover, the GDP density is relatively large (834 CNY/km^2 in 2015). The population density is relatively large (212 people/km^2 in 2015). The vegetation coverage is relatively high (annual mean NDVI of 0.619), and the coefficient of variation is relatively small (mean value of 0.1165). The growth rate of the NDVI was 0.0093/a after the implementation of the GGP in 2000. (2) The analysis based on the GDM revealed the following. The individual factors that determine the changes in the spatial distribution of the NDVI are the anthropogenic factors, such as GDP density ($q = 0.1583$), land-use type ($q = 0.0761$), and population density ($q = 0.0520$). The main interactive factors are GDP density and geomorphic type ($q = 0.2470$), and GDP density and soil type ($q = 0.2300$). The suitable precipitation range for the NDVI was greater than 650 mm, the temperature range was smaller than 0 °C, the altitude ranged from 2963 to 4914 m, the slope ranged from 15° to 20°, the slope aspect was no slope aspect, the geomorphic type was large undulating mountain, the soil type was eluvial soil, the vegetation type was broad-leaved forest, the

GDP density ranged from 3000 to 5000 CNY/km^2, the population density ranged from 200 to 500 people/km^2, and the land-use type was woodland.

The conclusions above have important implications for ecological conservation and restoration in the different regions. The impacts of the climatic, environmental, and anthropogenic factors need to be comprehensively considered when formulating and implementing policies. The NDVI is sensitive to the climatic and environmental factors related to water changes in the semi-arid area where there is relatively little water. Therefore, appropriate vegetation types should be selected, regional vegetation structures should be optimized, and high water consumption vegetation should be gradually replaced by low water consumption vegetation such as grass and drought-tolerant plants. In the semi-humid area, the regional water conditions basically meet the needs of vegetation growth because of the suitable climatic and natural conditions. Moreover, human activities have a relatively large disturbance effect on the NDVI. Thus, more positive human activities should be encouraged, including Grain for Green, afforestation, urban greening, and agricultural modernization projects. Furthermore, negative human activities should be controlled, including disordered urban expansion, population expansion, and overgrazing.

4.2. Connections and Distinctions with Other Studies

(1) Vegetation NDVI: In this study, it was found that the NDVI on the entire LP has increased significantly after 2000 when the GGP was implemented, consistent with the findings of [26,37,38].

(2) Climatic factors: In this study, it was found that among all factors, precipitation has the greatest influence on vegetation growth in the semi-arid area of the LP, consistent with the findings of [39–41]. In the semi-humid area, precipitation is not the decisive factor controlling vegetation growth because of the relatively good water conditions. Several studies have suggested that temperature is also a key factor affecting the vegetation growth. In general, when the temperature is low, the physiological activity of the vegetation is low. An increase in temperature promotes photosynthesis and vegetation growth, but as the temperature increases further, the growth of vegetation is inhibited by the accelerated evaporation and soil drying [42,43]. It was found in this study that conditions are suitable for vegetation growth when the temperature smaller than 0 °C on the LP. This is the case because the vegetation growth is influenced by many factors other than temperature. It can be seen from Figures 2–5 and 8 that the NDVI value is large in the southwest corner of semi-arid area of the LP, where the precipitation is large, the land-use types are mainly woodland and grassland, and the altitude ranges from 2963 m to 4914 m (the high altitude causes the low temperature). The finding is consistent with the results of [29]. It was also found that compared with other factors, temperature has relatively small explanatory power in terms of the spatial distribution and its change in the NDVI on the LP, which agrees with the findings and results of [39].

(3) Environmental factors: In this study, it was found that the environmental factors such as topographic types have an impact on the vegetation growth and restoration, consistent with the findings and results of [44]. The study revealed that the explanatory power of interactions between the factors would be increased for the spatial distribution and its change in the NDVI, which is consistent with the practical situation. For instance, the soil moisture is lower for steeper slopes, inhibiting vegetation growth. In this study, it was found that the NDVI value is large when the slope is between 15° and 20°, while it decreases when the slope becomes greater.

(4) Human factors: Certain studies have suggested that the regional land-use types have changed since the implementation of the GGP [45,46], which is not consistent with the results of this study. The reason for this may be that the spatial resolutions of the remote sensing data used in these studies are different. In this study, it was found that the conservation and restoration of the original woodland and grassland have been strengthened since the implementation of the GGP on the LP, resulting in a significant increase in the NDVI throughout the entire region. Additionally, many studies have used the NDVI to

predict the grain yield because of the significant correlation between the grain yield and the NDVI [47,48]. The LP is a dry farming area, which has been affected by the scarcity of precipitation for a long time, resulting in a low grain yield [49]. With the agricultural technology developing recently, the optimization and adjustment of the crop planting structure has resulted in a continuous improvement in the grain yield on the LP. Therefore, the NDVI has increased significantly [19].

4.3. Possible Future Work

There is further research that can be carried out, including the following. (1) The factors affecting the spatial–temporal evolution of the NDVI are extremely complex. Due to data limitations, certain factors are not currently considered, such as agricultural fertilization, irrigation area, and CO_2 concentration [50,51], but can be further improved with the acquisition of new data in the future. (2) The interaction detector of the GDM only considers the interactions between two factors. There are interactive effects involving more than two factors. The combined impacts of multiple factors would provide insights on the evolution of vegetation, which can be considered in the future by introducing other methods, such as the analytical hierarchy process (AHP) and principal component analysis (PCA) [52]. (3) It is necessary to perform further collection of data with a relatively higher resolution through field investigations and monitoring to determine the practical situation of the surface undulations in the study area. (4) The cumulative and time-lag effects of various factors and indicators (e.g., standardized precipitation and evapotranspiration index, SPEI) should also be evaluated in the future [53]. (5) It is possible for the GDM to identify the impacts of driving forces on the evolution of vegetation from the perspective of spatial analysis and mathematical statistics. In order to further explore the evolutionary mechanism of vegetation, some methods from the field of landscape studies can be introduced in the future [54].

5. Conclusions

The main conclusions of this paper include the following.

(1) The spatial–temporal evolution characteristics of the NDVI on the entire LP, and in the semi-arid/humid areas during 2000–2015 were analyzed via the linear regression, coefficient of variation, and transfer matrix models.

- The proportion of the total area with an NDVI of greater than 0.6 increased significantly, and the proportions of relatively low vegetation (NDVI of 0.2 to 0.4) converted to medium vegetation (NDVI of 0.4 to 0.6), medium vegetation converted to relatively high vegetation (NDVI of 0.6 to 0.8), and relatively high vegetation converted to high vegetation (NDVI of 0.8 to 1.0) accounted for the largest proportions.
- The annual mean value of the NDVI on the LP was 0.529, decreasing from the southeastern semi-humid area (0.619) to the northwestern semi-arid area (0.346). The mean value of the NDVI coefficient of variation was 0.1406 on the LP, increasing from the southeastern semi-humid area (0.1165) to the northwestern semi-arid area (0.1926).
- The NDVI on the LP exhibited an upward trend. The annual growth rate of the NDVI in the entire LP was 0.0079, and the growth rate in the semi-humid area (0.0093) was higher than in the semi-arid (0.0049) area after the GGP was implemented.
- The area of the change in land-use types on the LP was not significant. Overall, a positive impact on the NDVI was found by the changes in the land-use type. The largest increments of the NDVI were due to grassland, farmland, and woodland, and these land-use types did not change.

(2) The GDM was adopted to quantitatively identify the impact of multiple factors on the spatial–temporal evolution of the NDVI on the entire LP, and in the semi-arid/humid areas.

- Using the factor detector, it was found that in the semi-arid area, the climatic and environmental factors were the decisive factors influencing the spatial distribution changes of the NDVI during 2000–2015, including precipitation, soil type, and vegetation type. The impacts of anthropogenic factors, such as the GDP density, land-use type, and population density, were more significant in the semi-humid area.

- Using the interaction detector, it was found that the explanatory power of interactions between factors were greater than their individual effects, exhibiting two types of nonlinear enhancement and bi-enhancement. For the semi-arid area, 83.6% of the interactive factor combinations exhibited nonlinear enhancement and 16.4% exhibited bi-enhancement. The interactive effect between precipitation and soil type was the strongest. For the semi-humid area, 61.8% of the interactive factor combinations exhibited nonlinear enhancement and 38.2% exhibited bi-enhancement. The interactive effect between GDP density and geomorphic type was the strongest.
- Using the risk and ecological detectors, the ranges or types of various factors that are suitable for vegetation growth and the differences of significance between factors for the NDVI on the LP were determined.

The conclusions of this study have important implications for policy makers and administrative managers in terms of the formulation and implementation of ecological conservation and restoration strategies in the different regions. In the semi-arid area, appropriate vegetation types should be selected, regional vegetation structures should be optimized, and high water consumption vegetation should be gradually replaced by low water consumption vegetation such as grass and drought-tolerant plants. In the semi-humid area, more positive human activities should be encouraged, including Grain for Green, afforestation, urban greening, and agricultural modernization projects. Furthermore, negative human activities should be controlled, including disordered urban expansion, population expansion, and overgrazing.

Author Contributions: Conceptualization, Data curation, Investigation, Methodology, Formal analysis, Writing—original draft: Y.D.; Conceptualization, Methodology, Formal analysis, Writing—review and editing, Supervision, Funding acquisition: D.Y.; Validation, Writing—review and editing: X.L. (Xiang Li); Project administration, Funding acquisition. J.H.; Supervision, Funding acquisition: W.S.; Validation, Writing—review and editing: X.L. (Xuecao Li); Supervision, Funding acquisition: H.W. All authors have read and agreed to the published version of the manuscript.

Funding: This research was supported by the National Natural Science Foundation of China (42171024, 51609122, 41701507), the National Key Research and Development Program of China (2018YFE0122700), the Open Research Fund Program of State Key Laboratory of Hydroscience and Engineering (sklhse-2020-A-07), and the Joint Open Research Fund Program of State key Laboratory of Hydroscience and Engineering and Tsinghua-Ningxia Yinchuan Joint Institute of Internet of Waters on Digital Water Governance (sklhse-2021-Iow06). Su Wei would like to thank the 2115 Talent Development Program of China Agricultural University.

Data Availability Statement: The shape file for the boundary of the LP, the SPOT NDVI data, and the influencing factors (including geomorphic type, soil type, vegetation type, socio-economic data, and land-use type) data were downloaded from the Resource and Environment Science and Data Center (RESDC), Chinese Academy of Sciences (available from http://www.resdc.cn, accessed on 6 October 2021). The DEM data were downloaded from the Shuttle Radar Topography Mission (available from http://srtm.csi.cgiar.org/, accessed on 6 October 2021). The precipitation and temperature data were from the China Meteorological Data Service Center (available from http://data.cma.cn/site/index.html, accessed on 6 October 2021).

Conflicts of Interest: The authors declare no conflict of interest.

References

1. Yao, R.; Cao, J.; Wang, L.; Zhang, W.; Wu, X. Urbanization Effects on Vegetation Cover in Major African Cities during 2001–2017. *Int. J. Appl. Earth Obs. Geoinf.* **2019**, *75*, 44–53. [CrossRef]
2. Hua, W.; Chen, H.; Zhou, L.; Xie, Z.; Qin, M.; Li, X.; Ma, H.; Huang, Q.; Sun, S. Observational Quantification of Climatic and Human Influences on Vegetation Greening in China. *Remote Sens.* **2017**, *9*, 425. [CrossRef]
3. He, B.; Chen, A.; Jiang, W.; Chen, Z. The Response of Vegetation Growth to Shifts in Trend of Temperature in China. *J. Geogr. Sci.* **2017**, *27*, 801–816. [CrossRef]

4.   Yang, Y.; Dou, Y.; Cheng, H.; An, S. Plant Functional Diversity Drives Carbon Storage Following Vegetation Restoration in Loess Plateau, China. *J. Environ. Manag.* **2019**, *246*, 668–678. [CrossRef] [PubMed]

5.   Liu, L.; Yang, X.; Zhou, H.; Liu, S.; Zhou, L.; Li, X.; Yang, J.; Han, X.; Wu, J. Evaluating the Utility of Solar-Induced Chlorophyll Fluorescence for Drought Monitoring by Comparison with NDVI Derived from Wheat Canopy. *Sci. Total Environ.* **2018**, *625*, 1208–1217. [CrossRef]

6.   Zhao, L.; Dai, A.; Dong, B. Changes in Global Vegetation Activity and Its Driving Factors during 1982–2013. *Agric. For. Meteorol.* **2018**, *249*, 198–209. [CrossRef]

7.   Huang, J.; Gómez-Dans, J.L.; Huang, H.; Ma, H.; Wu, Q.; Lewis, P.E.; Liang, S.; Chen, Z.; Xue, J.-H.; Wu, Y.; et al. Assimilation of Remote Sensing into Crop Growth Models: Current Status and Perspectives. *Agric. For. Meteorol.* **2019**, *276–277*, 107609. [CrossRef]

8.   Du, Z.; Zhang, X.; Xu, X.; Zhang, H.; Wu, Z.; Pang, J. Quantifying Influences of Physiographic Factors on Temperate Dryland Vegetation, Northwest China. *Sci. Rep.* **2017**, *7*, 40092. [CrossRef]

9.   Piao, S.; Mohammat, A.; Fang, J.; Cai, Q.; Feng, J. NDVI-Based Increase in Growth of Temperate Grasslands and Its Responses to Climate Changes in China. *Glob. Environ. Chang.* **2006**, *16*, 340–348. [CrossRef]

10.   Mao, J.; Shi, X.; Thornton, P.E.; Hoffman, F.M.; Zhu, Z.; Myneni, R.B. Global Latitudinal-Asymmetric Vegetation Growth Trends and Their Driving Mechanisms: 1982–2009. *Remote Sens.* **2013**, *5*, 1484–1497. [CrossRef]

11.   Qu, S.; Wang, L.; Lin, A.; Yu, D.; Yuan, M.; Li, C. Distinguishing the Impacts of Climate Change and Anthropogenic Factors on Vegetation Dynamics in the Yangtze River Basin, China. *Ecol. Indic.* **2020**, *108*, 105724. [CrossRef]

12.   Yao, R.; Wang, L.; Huang, X.; Chen, J.; Li, J.; Niu, Z. Less Sensitive of Urban Surface to Climate Variability than Rural in Northern China. *Sci. Total Environ.* **2018**, *628–629*, 650–660. [CrossRef] [PubMed]

13.   Zhang, Y.; Peng, C.; Li, W.; Tian, L.; Zhu, Q.; Chen, H.; Fang, X.; Zhang, G.; Liu, G.; Mu, X.; et al. Multiple Afforestation Programs Accelerate the Greenness in the 'Three North' Region of China from 1982 to 2013. *Ecol. Indic.* **2016**, *61*, 404–412. [CrossRef]

14.   Wang, S.; Fu, B.; Piao, S.; Lü, Y.; Ciais, P.; Feng, X.; Wang, Y. Reduced Sediment Transport in the Yellow River Due to Anthropogenic Changes. *Nat. Geosci.* **2016**, *9*, 38–41. [CrossRef]

15.   Zhao, G.; Mu, X.; Wen, Z.; Wang, F.; Gao, P. Soil Erosion, Conservation, and Eco-Environment Changes in the Loess Plateau of China. *Land Degrad. Dev.* **2013**, *24*, 499–510. [CrossRef]

16.   Gao, Y.; Liu, Z.; Li, R.; Shi, Z. Long-Term Impact of China's Returning Farmland to Forest Program on Rural Economic Development. *Sustainability* **2020**, *12*, 1492. [CrossRef]

17.   Li, G.; Sun, S.; Han, J.; Yan, J.; Liu, W.; Wei, Y.; Lu, N.; Sun, Y. Impacts of Chinese Grain for Green Program and Climate Change on Vegetation in the Loess Plateau during 1982–2015. *Sci. Total Environ.* **2019**, *660*, 177–187. [CrossRef] [PubMed]

18.   Wang, Z.; Yao, W.; Tang, Q.; Liu, L.; Xiao, P.; Kong, X.; Zhang, P.; Shi, F.; Wang, Y. Continuous Change Detection of Forest/Grassland and Cropland in the Loess Plateau of China Using All Available Landsat Data. *Remote Sens.* **2018**, *10*, 1775. [CrossRef]

19.   Wu, X.; Wang, S.; Fu, B.; Feng, X.; Chen, Y. Socio-Ecological Changes on the Loess Plateau of China after Grain to Green Program. *Sci. Total Environ.* **2019**, *678*, 565–573. [CrossRef]

20.   Hao, H.; Li, Y.; Zhang, H.; Zhai, R.; Liu, H. Spatiotemporal Variations of Vegetation and Its Determinants in the National Key Ecological Function Area on Loess Plateau between 2000 and 2015. *Ecol. Evol.* **2019**, *9*, 5810–5820. [CrossRef] [PubMed]

21.   Hu, Y.; Dao, R.; Hu, Y. Vegetation Change and Driving Factors: Contribution Analysis in the Loess Plateau of China during 2000–2015. *Sustainability* **2019**, *11*, 1320. [CrossRef]

22.   Yamori, W.; Hikosaka, K.; Way, D.A. Temperature Response of Photosynthesis in C3, C4, and CAM Plants: Temperature Acclimation and Temperature Adaptation. *Photosynth. Res.* **2014**, *119*, 101–117. [CrossRef]

23.   Hein, L.; de Ridder, N.; Hiernaux, P.; Leemans, R.; de Wit, A.; Schaepman, M. Desertification in the Sahel: Towards Better Accounting for Ecosystem Dynamics in the Interpretation of Remote Sensing Images. *J. Arid Environ.* **2011**, *75*, 1164–1172. [CrossRef]

24.   Liu, L.; Zhang, Y.; Bai, W.; Yan, J.; Ding, M.; Shen, Z.; Li, S.; Zheng, D. Characteristics of Grassland Degradation and Driving Forces in the Source Region of the Yellow River from 1985 to 2000. *J. Geogr. Sci.* **2006**, *16*, 131–142. [CrossRef]

25.   Ran, Q.; Hao, Y.; Xia, A.; Liu, W.; Hu, R.; Cui, X.; Xue, K.; Song, X.; Xu, C.; Ding, B.; et al. Quantitative Assessment of the Impact of Physical and Anthropogenic Factors on Vegetation Spatial-Temporal Variation in Northern Tibet. *Remote Sens.* **2019**, *11*, 1183. [CrossRef]

26.   Nie, T.; Dong, G.; Jiang, X.; Lei, Y. Spatio-Temporal Changes and Driving Forces of Vegetation Coverage on the Loess Plateau of Northern Shaanxi. *Remote Sens.* **2021**, *13*, 613. [CrossRef]

27.   Wang, J.-F.; Li, X.-H.; Christakos, G.; Liao, Y.-L.; Zhang, T.; Gu, X.; Zheng, X.-Y. Geographical Detectors-Based Health Risk Assessment and Its Application in the Neural Tube Defects Study of the Heshun Region, China. *Int. J. Geogr. Inf. Sci.* **2010**, *24*, 107–127. [CrossRef]

28.   Peng, W.; Kuang, T.; Tao, S. Quantifying Influences of Natural Factors on Vegetation NDVI Changes Based on Geographical Detector in Sichuan, Western China. *J. Clean. Prod.* **2019**, *233*, 353–367. [CrossRef]

29.   Meng, X.; Gao, X.; Li, S.; Lei, J. Spatial and Temporal Characteristics of Vegetation NDVI Changes and the Driving Forces in Mongolia during 1982–2015. *Remote Sens.* **2020**, *12*, 603. [CrossRef]

30. Zhu, L.; Meng, J.; Zhu, L. Applying Geodetector to Disentangle the Contributions of Natural and Anthropogenic Factors to NDVI Variations in the Middle Reaches of the Heihe River Basin. *Ecol. Indic.* **2020**, *117*, 106545. [CrossRef]
31. Guo, X.; Shao, Q. Spatial Pattern of Soil Erosion Drivers and the Contribution Rate of Human Activities on the Loess Plateau from 2000 to 2015: A Boundary Line from Northeast to Southwest. *Remote Sens.* **2019**, *11*, 2429. [CrossRef]
32. Zheng, J.; Yin, Y.; Li, B. A New Scheme for Climate Regionalization in China. *Acta Geogr. Sin.* **2010**, *65*, 3–12. [CrossRef]
33. Wang, J.-F.; Zhang, T.-L.; Fu, B.-J. A Measure of Spatial Stratified Heterogeneity. *Ecol. Indic.* **2016**, *67*, 250–256. [CrossRef]
34. Peng, W.; Wang, G.; Zhou, J.; Xu, X.; Luo, H.; Yang, C.; Zhao, J. Dynamic Monitoring of Fractional Vegetation Cover along Minjiang River from Wenchuan County to Dujiangyan City Using Multi-Temporal Landsat 5 and 8 Images. *Acta Ecol. Sin.* **2016**, *36*, 1975–1988. [CrossRef]
35. Hijmans, R.; Cameron, S.; Parra, J.; Jones, P.; Jarvis, A. Very High Resolution Interpolated Climate Surfaces of Global Land Areas. *Int. J. Climatol.* **2005**, *25*, 1965–1978. [CrossRef]
36. Zheng, K.; Ye, J.-S.; Jin, B.-C.; Zhang, F.; Wei, J.-Z.; Li, F.-M. Effects of Agriculture, Climate, and Policy on NDVI Change in a Semi-Arid River Basin of the Chinese Loess Plateau. *Arid Land Res. Manag.* **2019**, *33*, 321–338. [CrossRef]
37. Su, C.; Fu, B. Evolution of Ecosystem Services in the Chinese Loess Plateau under Climatic and Land Use Changes. *Glob. Planet. Chang.* **2013**, *101*, 119–128. [CrossRef]
38. Fu, B.; Liu, Y.; Lü, Y.; He, C.; Zeng, Y.; Wu, B. Assessing the Soil Erosion Control Service of Ecosystems Change in the Loess Plateau of China. *Ecol. Complex.* **2011**, *8*, 284–293. [CrossRef]
39. Naeem, S.; Zhang, Y.; Zhang, X.; Tian, J.; Abbas, S.; Luo, L.; Meresa, H.K. Both Climate and Socioeconomic Drivers Contribute to Vegetation Greening of the Loess Plateau. *Sci. Bull.* **2021**, *66*, 1160–1163. [CrossRef]
40. Li, Z.; Chen, Y.; Li, W.; Deng, H.; Fang, G. Potential Impacts of Climate Change on Vegetation Dynamics in Central Asia. *J. Geophys. Res. Atmos.* **2015**, *120*, 12345–12356. [CrossRef]
41. Zhao, W.; Hu, Z.; Guo, Q.; Wu, G.; Chen, R.; Li, S. Contributions of Climatic Factors to Interannual Variability of the Vegetation Index in Northern China Grasslands. *J. Clim.* **2020**, *33*, 175–183. [CrossRef]
42. Potter, C.; Klooster, S.; Genovese, V. Net Primary Production of Terrestrial Ecosystems from 2000 to 2009. *Clim. Chang.* **2012**, *115*, 365–378. [CrossRef]
43. Zhang, Q.; Qi, T.; Li, J.; Singh, V.P.; Wang, Z. Spatiotemporal Variations of Pan Evaporation in China during 1960–2005: Changing Patterns and Causes. *Int. J. Climatol.* **2015**, *35*, 903–912. [CrossRef]
44. He, J.; Shi, X.; Fu, Y. Identifying Vegetation Restoration Effectiveness and Driving Factors on Different Micro-Topographic Types of Hilly Loess Plateau: From the Perspective of Ecological Resilience. *J. Environ. Manag.* **2021**, *289*, 112562. [CrossRef] [PubMed]
45. Deng, L.; Shangguan, Z.; Sweeney, S. "Grain for Green" Driven Land Use Change and Carbon Sequestration on the Loess Plateau, China. *Sci. Rep.* **2014**, *4*, 7039. [CrossRef]
46. Zhu, Z.; Liu, L.; Chen, Z.; Zhang, J.; Verburg, P.H. Land-Use Change Simulation and Assessment of Driving Factors in the Loess Hilly Region—A Case Study as Pengyang County. *Environ. Monit. Assess.* **2010**, *164*, 133–142. [CrossRef]
47. Son, N.T.; Chen, C.F.; Chen, C.R.; Chang, L.Y.; Minh, V.Q. Monitoring Agricultural Drought in the Lower Mekong Basin Using MODIS NDVI and Land Surface Temperature Data. *Int. J. Appl. Earth Obs. Geoinf.* **2012**, *18*, 417–427. [CrossRef]
48. Fernandes, J.L.; Ebecken, N.F.F.; Esquerdo, J.C.D.M. Sugarcane Yield Prediction in Brazil Using NDVI Time Series and Neural Networks Ensemble. *Int. J. Remote Sens.* **2017**, *38*, 4631–4644. [CrossRef]
49. Li, F.-M.; Xiong, Y.-C.; Li, X.-G.; Zhang, F.; Guan, Y. Integrated Dryland Agriculture Sustainable Management in Northwest China. In *Innovations in Dryland Agriculture*; Farooq, M., Siddique, K.H.M., Eds.; Springer International Publishing: Cham, Switzerland, 2016; pp. 393–413, ISBN 978-3-319-47928-6.
50. Piao, S.; Yin, G.; Tan, J.; Cheng, L.; Huang, M.; Li, Y.; Liu, R.; Mao, J.; Myneni, R.B.; Peng, S.; et al. Detection and Attribution of Vegetation Greening Trend in China over the Last 30 Years. *Glob. Chang. Biol.* **2015**, *21*, 1601–1609. [CrossRef] [PubMed]
51. Xue, R.; Yang, Q.; Miao, F.; Wang, X.; Shen, Y.; Xue, R.; Yang, Q.; Miao, F.; Wang, X.; Shen, Y. Slope Aspect Influences Plant Biomass, Soil Properties and Microbial Composition in Alpine Meadow on the Qinghai-Tibetan Plateau. *J. Soil Sci. Plant Nutr.* **2018**, *18*, 1–12. [CrossRef]
52. Zhang, A.; Jia, G.; Ustin, S.L. Water Availability Surpasses Warmth in Controlling Global Vegetation Trends in Recent Decade: Revealed by Satellite Time Series. *Environ. Res. Lett.* **2021**, *16*, 074028. [CrossRef]
53. Zhao, A.; Yu, Q.; Feng, L.; Zhang, A.; Pei, T. Evaluating the Cumulative and Time-Lag Effects of Drought on Grassland Vegetation: A Case Study in the Chinese Loess Plateau. *J. Environ. Manag.* **2020**, *261*, 110214. [CrossRef] [PubMed]
54. Luo, Y.; Lü, Y.; Liu, L.; Liang, H.; Li, T.; Ren, Y. Spatiotemporal Scale and Integrative Methods Matter for Quantifying the Driving Forces of Land Cover Change. *Sci. Total Environ.* **2020**, *739*, 139622. [CrossRef] [PubMed]

Article

Assessing the Effects of Time Interpolation of NDVI Composites on Phenology Trend Estimation

Xueying Li [1,2,3], Wenquan Zhu [1,2,*], Zhiying Xie [1,2], Pei Zhan [1,2,4], Xin Huang [1,2], Lixin Sun [1,2] and Zheng Duan [3]

[1] State Key Laboratory of Remote Sensing Science, Jointly Sponsored by Beijing Normal University and Aerospace Information Research Institute of Chinese Academy of Sciences, Faculty of Geographical Science, Beijing Normal University, Beijing 100875, China; xueying.li@nateko.lu.se (X.L.); xiezy@mail.bnu.edu.cn (Z.X.); peizhan@nuist.edu.cn (P.Z.); huangxin@mail.bnu.edu.cn (X.H.); 201821051190@mail.bnu.edu.cn (L.S.)

[2] Beijing Engineering Research Center for Global Land Remote Sensing Products, Faculty of Geographical Science, Beijing Normal University, Beijing 100875, China

[3] Department of Physical Geography and Ecosystem Science, Lund University, 22362 Lund, Sweden; zheng.duan@nateko.lu.se

[4] School of Applied Meteorology, Nanjing University of Information Science & Technology, Nanjing 210044, China

* Correspondence: zhuwq75@bnu.edu.cn

Citation: Li, X.; Zhu, W.; Xie, Z.; Zhan, P.; Huang, X.; Sun, L.; Duan, Z. Assessing the Effects of Time Interpolation of NDVI Composites on Phenology Trend Estimation. *Remote Sens.* **2021**, *13*, 5018. https://doi.org/10.3390/rs13245018

Academic Editor: Jianxi Huang

Received: 31 October 2021
Accepted: 7 December 2021
Published: 10 December 2021

Publisher's Note: MDPI stays neutral with regard to jurisdictional claims in published maps and institutional affiliations.

Abstract: The accurate evaluation of shifts in vegetation phenology is essential for understanding of vegetation responses to climate change. Remote-sensing vegetation index (VI) products with multi-day scales have been widely used for phenology trend estimation. VI composites should be interpolated into a daily scale for extracting phenological metrics, which may not fully capture daily vegetation growth, and how this process affects phenology trend estimation remains unclear. In this study, we chose 120 sites over four vegetation types in the mid-high latitudes of the northern hemisphere, and then a Moderate Resolution Imaging Spectroradiometer (MODIS) MCD43A4 daily surface reflectance data was used to generate a daily normalized difference vegetation index (NDVI) dataset in addition to an 8-day and a 16-day NDVI composite datasets from 2001 to 2019. Five different time interpolation methods (piecewise logistic function, asymmetric Gaussian function, polynomial curve function, linear interpolation, and spline interpolation) and three phenology extraction methods were applied to extract data from the start of the growing season and the end of the growing season. We compared the trends estimated from daily NDVI data with those from NDVI composites among (1) different interpolation methods; (2) different vegetation types; and (3) different combinations of time interpolation methods and phenology extraction methods. We also analyzed the differences between the trends estimated from the 8-day and 16-day composite datasets. Our results indicated that none of the interpolation methods had significant effects on trend estimation over all sites, but the discrepancies caused by time interpolation could not be ignored. Among vegetation types with apparent seasonal changes such as deciduous broadleaf forest, time interpolation had significant effects on phenology trend estimation but almost had no significant effects among vegetation types with weak seasonal changes such as evergreen needleleaf forests. In addition, trends that were estimated based on the same interpolation method but different extraction methods were not consistent in showing significant (insignificant) differences, implying that the selection of extraction methods also affected trend estimation. Compared with other vegetation types, there were generally fewer discrepancies between trends estimated from the 8-day and 16-day dataset in evergreen needleleaf forest and open shrubland, which indicated that the dataset with a lower temporal resolution (16-day) can be applied. These findings could be conducive for analyzing the uncertainties of monitoring vegetation phenology changes.

Keywords: vegetation phenology; phenology trend; NDVI composites; time interpolation

1. Introduction

The vegetation phenology refers to the physiological and reproductive phenomenon of vegetation in an annual cycle, which is a robust and sensitive indicator of climate change [1–4]. Shifts in vegetation phenology can regulate interactions between vegetation and climate change by influencing the structure and functions of the terrestrial ecosystem [5–7]. The availability of accurate vegetation phenology shifts has significant implications for promoting the understanding of vegetation responses to climate change [8] and improving terrestrial ecosystem process models [9] and prediction skills in crop yield production [10].

Shifts in vegetation phenology across regional and global scales were frequently derived by using vegetation indices (VIs) from satellite remote sensing data at various spatial and temporal resolutions [11–19]. The accuracy of phenology trend estimation can be influenced by multiple variables such as geographical regions [20–22], vegetation types [23–25], and vegetation indexes [26–28] but mostly depends on the selection of remote sensing products, denoising methods, phenology extraction methods, and the different combinations of these factors [20,29]. Previous studies indicated discrepancies between phenology trends estimated by different remote sensing products [30–35]. For example, Peng et al. [36] investigated the shifts of spring green-up onset dates in six regularly updated land surface phenology products from Moderate Resolution Imaging Spectroradiometer (MODIS) and Advanced Very High Resolution Radiometer (AVHRR). Similar interannual shifts of green-up onset dates among all products only occurred in local regions while discrepancies were distributed across the contiguous United States. In Western Arctic Russia, the start of growing season (SOS) and the end of growing season (EOS) during 2000–2010 based on MODIS and SPOT-Vegetation datasets showed similar trends, but all were significantly different from the trend based on AVHRR data [37]. In addition, Zeng et al. [38] estimated the SOS trend over the northern high-latitude region and noticed that SOS continuously advanced from 2000 to 2010 by using MODIS data, but no advancing trends were shown in the AVHRR Global Inventory Modeling and Mapping Studies (GIMMS) time series. Such patterns have also been documented in Tibetan alpine grassland, where MODIS Normalized Difference Vegetation Index (NDVI) data captured the advancement of SOS throughout 2000–2014, but a delaying trend of the GIMMS NDVI estimated SOS was observed [39]. Discrepancies between phenology trends based on different denoising methods or extraction methods varied in research areas, research periods, and vegetation types [40–43]. For example, Zhu et al. [8] applied several commonly utilized vegetation phenology extraction methods on MOD09A1 (8-day) and MOD13A2 (16-day) datasets and found no significant differences between SOS or EOS trends derived from asymmetric Gaussian function, double logistic function, and the piecewise logistic function method. White et al. [20] compared 10 extraction methods for estimating the shifts in the start-of-spring dates based on the GIMMS NDVI dataset in North America; the results strongly suggested either no or very geographically limited trends towards earlier spring arrival. Wu et al. [19] applied six phenology extraction methods including the first-order, second-order, and third-order derivative; amplitude threshold; relative changing rate; and curvature change rate for deriving SOS and EOS from AVHRR. In the northern hemisphere, the SOS trends retrieved vary across methods from 1982 to 2018, while only the EOS trend estimated by the relative changing rate method was significantly advanced. In the southern hemisphere, EOS based on all methods demonstrated insignificant trends. Compared with denoising methods or extraction methods, the selection of datasets might be of a higher priority in vegetation dynamics monitoring [8,33].

Atmosphere conditions, such as cloud, dust, and other aerosols, can adversely affect the quality of satellite remote sensing VI data. To this regard, the maximum value composite method [44] was mostly used [45–49] and composited VI time-series data by retaining the maximum NDVI within a specific interval of days. Current VI composite products such as 15-day GIMMS NDVI 3 g data; MODIS 8-day (MOD09A1), 16-day (MOD13Q1), and 30-day (MOD13A3) data; and 10-day SPOT VGT S10 data were widely applied among regional [32,50,51] and global scales [12,52]. VI composite products should be interpolated into daily scales for extracting phenological metrics (i.e., SOS and EOS). However, the process of time interpolation may not fully capture real daily vegetation growth, especially during greening and senescence stages (where the curve is changing fast) [53], which may further affect the estimation of phenology trends. Current time interpolation methods include linear interpolation [54], cubic spline interpolation [55], and curve-fitting methods. Curve fitting methods smooth the noise while fitting the curve into a continuous daily scale line, including asymmetric Gaussian function fitting [56], fast Fourier transform [20], and double logistic function fitting [57], but these methods were mostly applied as denoising measures in previous studies [20,56,58,59]. For investigating the effects of data temporal resolution on phenology extraction and trend estimation, some studies compared phenology metrics or trends derived from daily NDVI data to those derived from multi-day NDVI composites [25,27,60]. However, the different performances of multi-day NDVI composites compared with daily NDVI data are directly caused by the process of time interpolation, and how this process affects phenology trend estimation remains unclear.

In this study, we used MCD43A4 daily surface reflectance data to construct a single-year daily (reference) NDVI dataset denoised by the Savitzky–Golay filter, and then single-year 8-day and single-year 16-day NDVI composite datasets were further generated. Four typical vegetation types, five time-interpolation methods, and three phenology extraction methods were chosen for estimating phenology trends. The main goals are to comprehensively investigate the effects of time interpolation on phenology trend estimation in the mid-high latitudes of the northern hemisphere from 2001 to 2019 among (1) different interpolation methods; (2) different vegetation types; and (3) different combinations of time interpolation methods and phenology extraction methods. In addition, we also analyzed the differences between the trends estimated from the 8-day and 16-day composite data, which would provide instructions on selecting relatively coarse temporal resolution (i.e., 16 day) data for phenology dynamics monitoring, as they are easier for collecting and storing.

2. Data and Methods

2.1. Study Area and Sites

We selected the mid-latitude and high-latitude area as our study region because the vegetation seasonal changes here are evident (noticeable amplitudes in NDVI curves), rendering extracting accurate phenological metrics possible [18]. In addition, NDVI datasets here are least contaminated by solar zenith angle effects [50,61]. Four typical widely distributed vegetation types (deciduous broadleaf forest (DBF), evergreen needleleaf forest (ENF), grassland (GRA), and open shrubland (OSH)) in the mid-high latitudes of the northern hemisphere (23.5°–70°N) were chosen as the study area (Figure 1). Vegetations around long-running experiment sites are usually well protected; thus, we selected sites with long-term (at least 10 years) observations from Fluxnet (https://fluxnet.fluxdata.org/sites/site-list-and-pages/, accessed on 20 March 2020; https://ameriflux.lbl.gov/, accessed on 23 March 2020; http://www.europe-fluxdata.eu/, accessed on 23 March 2020) and Phenocam (https://phenocam.sr.unh.edu/, accessed on 5 April 2020).

First, we filtered sites with the MCD12Q1 Land Cover Type product. If the vegetation type marked at each site was the same as the type in all 3×3 pixels (1500 m $\times$ 1500 m at 500 m resolution) centered on the site location, then the site was retained; otherwise, it was removed [62]. Second, in order to eliminate the influence of bare soil, sparse vegetation, and artificial vegetation (such as crops) on VI curves, sites meeting the following criteria in all years (2001–2019) were selected for further analysis [33,50,63–65]: (1) the mean NDVI during June–September should be higher than 0.10; (2) the annual maximum NDVI should occur during July–September; (3) the mean NDVI during July–September should be 1.2 times higher than the mean NDVI during November–March; and (4) the NDVI curve has a single growth cycle annually. Finally, according to the central limit theorem [66], data statistics will be close to normally distributed if the sample size is greater than or equal to 30. For each vegetation type, 30 sites meeting all criteria above were selected from higher to lower latitudes, which were 72 Fluxnet sites and 48 Phenocam sites in total (Figure 1).

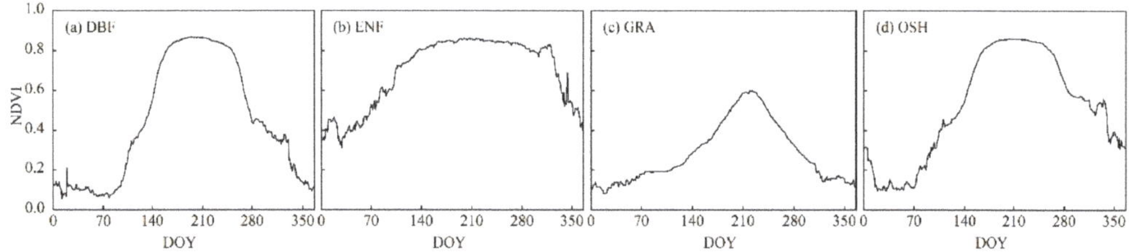

Figure 1. Sites and the distribution of vegetation types. DBF, ENF, GRA, and OSH are deciduous broadleaf forest, evergreen needleleaf forest, grassland, and open shrubland, respectively.

Figure 2 showed the typical NDVI curve of each vegetation type from the representative Fluxnet sites. The curve of DBF has the most apparent seasonal change (seasonal change is defined by the amplitude of the NDVI curve in Bradley et al. [67]). The curves of GRA and OSH have relatively weaker seasonal changes, and the curve of ENF has the weakest seasonal change.

Figure 2. Typical NDVI curves of four vegetation types in the study area. DBF, ENF, GRA, and OSH are deciduous broadleaf forest, evergreen needleleaf forest, grassland, and open shrubland, respectively; data of (**a–d**) are from Fluxnet sites CA-TP3, CZ-BK1, US-Wkg, and US-IB2, respectively; all curves are denoised by the Savitzky–Golay filtering method.

2.2. Data and Pre-Processing

MODIS provided an 8-day surface reflectance data (MOD09A1) and a 16-day NDVI composite (MOD13A1) data with the same georeference and spatial resolution (500 m) as the MCD43A4 product. However, there exist discrepancies in data generation. In MOD09A1, the selection of pixels within the 8-day composite period is based on the minimum channel 3 (blue) value, while MOD13A1 chooses the highest NDVI value within two 8-day composite periods. In addition, the deviations appear in data gap filling. MOD13A1 uses the climate modeling grid (CMG) average vegetation index product database for gap filling, which cannot be applied to the MOD09A1 NBAR dataset. All these discrepancies may increase bias during data pre-processing. Therefore, we chose to construct the daily NDVI data and NDVI composites based on the daily surface reflectance data MCD43A4.

Firstly, we chose daily surface reflectance data for red and near-infrared ranges from MCD43A4 product with 500 m spatial resolution during 2001–2019 (https://modis.ornl.gov/globalsubset/, accessed on 20 March 2020). Surface reflectance data was then pre-processed in the following four steps to obtain single-year daily NDVI data and finally generated single-year 8-day and single-year 16-day composite data.

(1) Calculating daily NDVI during 2001–2019

Daily NDVI from 2001 to 2019 was calculated by the mean nadir BRDF (bidirectional reflectance distribution function) adjusted reflectance (NBAR) values taken over 3×3 pixels in each site (Equation (1)). The days with unqualified data (labeled in "F") were skipped from the calculation:

$$\text{NDVI} = \frac{R_{NIR} - R_{Red}}{R_{NIR} + R_{Red}} \qquad (1)$$

where R_{NIR} is the mean of 3×3 pixels of near-infrared band surface reflectance; and R_{Red} is the mean of 3×3 pixels of red band surface reflectance.

(2) Constructing single-year daily NDVI data

For the missing daily NDVI values caused by NBAR data loss among a few sites (data labeled in "F"), a linear interpolation method was applied by using NDVI values of the same day in the nearest years (before and after) to fill up. If there were no qualified NDVI values among the nearest years, the multi-year (2001–2019) mean NDVI value of this day was then used for filling up the missing value.

(3) Denoising single-year daily NDVI data

Savitzky–Golay filter [68] is a simplified least squares fit convolution that can be applied for smoothing VI curves of a set of consecutive values [58]. The filter was proved to perform well by minimizing noises (e.g., cloud-contaminated NDVI values) effectively [69,70]. It was chosen to smooth the daily NDVI data in our study (equation and parameters are shown in Table 1).

Table 1. Data preprocessing methods and parameter settings.

Method	Equation	Parameter
Savitzky-Golay filter	$Y_j^* = \sum_{i=-m}^{i=m} C_i Y_{j+i}/N$	Y^* is the resultant NDVI value; Y is the original NDVI value; j is the running index of the original ordinate data; m is the half-width of the smoothing window (filter); C_i is the coefficient for the ith NDVI value of the filter; N is the amount of convoluting integers; the half-width of the smoothing window is set to 1/4 of the year length (90 days); the smoothing polynomial degree is set to 4 [58].
Maximum value composite	$y_{new} = MAX(y_1 + y_2 + \ldots + y_n)$	y_{new} is the resultant NDVI value; y_n is the original NDVI value; n is the days for compositing.

(4) Constructing single-year composite NDVI data

The maximum value composite method [44] was chosen to generate single-year 8-day and single-year 16-day composite data (Equation and parameters are shown in Table 1).

2.3. Methods

2.3.1. Time Interpolation

Five commonly used functions were chosen for interpolating the 8-day and 16-day composite data (equations and parameters were shown in Table 2): (1) piecewise logistic function fitting (PL). PL fits NDVI curve to a logistic function of time with no requirements of data pre-smoothing or threshold defining [71]. The function for NDVI data with a single growth cycle is shown in Table 2. (2) asymmetric Gaussian function fitting (AG). AG based on nonlinear least squares fits to the NDVI curves [56]. It is especially suited for describing the shape of the scaled VI curves in overlapping intervals around maxima and minima. (3) polynomial curve function fitting (PCF). PCF uses the least-square regression to analyze the relationship between NDVI data and the corresponding Julian day [50]. It effectively smooths the curve noises, and the degree of polynomial function is flexible according to the shape of the NDVI curve. (4) linear interpolation (Linear). Linear interpolation starts at the beginning of the NDVI (start point) curve and linearly constructs the missing value with the current start point and the next nearest point [54]. The function (Table 2) can also be understood as a weighted average. (5) cubic spline interpolation (Spline). Spline interpolates the data with piecewise cubic polynomials, and it allows the NDVI curve to pass through two specified endpoints with specified derivatives at each endpoint [55]. Spline is popularly used as it reduces both computational requirements and numerical instabilities arising with higher degree curves.

Table 2. Time interpolation methods and parameter settings.

Method	Equation	Parameter
Piecewise logistic function fitting (PL)	$y(t) = \frac{c}{1+e^{a+bt}} + d$	$y(t)$ is the resultant NDVI value at time t; t is the Julian days; a and b are fitting parameters; c is the amplitude of the NDVI curve; d is the minimum NDVI value [71].
Asymmetric Gaussian function fitting (AG)	$y(t) = w\mathrm{NDVI} + (m\mathrm{NDVI} - w\mathrm{NDVI}) \times g(t)$ $(t; a_1, a_2 \cdots a_5) = \begin{cases} \exp\left[-\left(\frac{t-a_1}{a_2}\right)^{a_3}\right], & \text{if } t > a_1 \\ \exp\left[-\left(\frac{a_1-t}{a_4}\right)^{a_5}\right], & \text{if } t < a_1 \end{cases}$	$y(t)$ is the resultant NDVI value at time t; $g(t)$ is the original NDVI value; $w\mathrm{NDVI}$ and $m\mathrm{NDVI}$ are the minimum and maximum NDVI value of the fitting part; a_1 is the position of the maximum or minimum value with respect to time t; a_2 (a_4) and a_3 (a_5) are the width and flatness of the right (left) half of the function [56].
Polynomial curve fitting (PCF)	$y(t) = \alpha_0 + \alpha_1 \times t^1 + \alpha_2 \times t^2 + \alpha_3 \times t^3 + \cdots + \alpha_n \times t^n$	$y(t)$ is the resultant NDVI value at time t; t is the Julian days; α_0-α_n are fitting parameters; n is the degree of smoothing polynomial; the smoothing polynomial degree is set to 6 [50].
Linear interpolation (Linear)	$y(t) = \frac{t-t_1}{t_0-t_1} y_0 + \frac{t-t_0}{t_1-t_0} y_1$	$y(t)$ is the resultant NDVI value at time t; t_0 and t_1 are the nearest day of year (DOY) of the missing value; y_0 and y_1 are the nearest NDVI of the missing value; t is the DOY of the interpolating point between t_0 and t_1.
Cubic spline interpolation (Spline)	$y_i(t) = a_i + b_i(t - t_i) + c_i(t - t_i)^2 + d_i(t - t_i)^3$	$y_i(t)$ is the resultant NDVI value at time t in the ith period; t is the interpolating point between t_i and t_{i+1}; a-d are function parameters decided by the DOY and NDVI matrix calculation results in the ith period and the $(I+1)$th period.

2.3.2. Phenology Extraction

For extracting phenological metrics, we chose three commonly used extraction methods (equations and parameters are shown in Table 3): (1) dynamic threshold (DT) method. In DT method, SOS and EOS are defined as the point in time at which the NDVI value increases and decreases to a specific level of seasonal amplitude [56]. Here, we defined the level percentage as 10%, 20%, and 30%, respectively [1,72,73]. (2) maximum rate of change (MRC) method. MRC defines the timing of the greatest NDVI change as the maximum (the left part of the curve, from the starting point to the peak) and minimum (the right part of the curve, from the peak to the ending point) values of NDVI ratio to determine the onset dates of the start and end of a growing season [50]. (3) change rate of the curvature (RCC) method. RCC defines the onset of senescence and dormancy dates as the point in time at which the rate of change in curvature in the NDVI curve exhibits local minimum or maximum values [71].

Table 3. Phenology extraction methods and parameter settings.

Method	Equation	Parameter
Dynamic threshold (DT)	$thd = \frac{NDVI(t) - NDVI_{min}}{NDVI_{max} - NDVI_{min}}$	$NDVI(t)$ is the original NDVI value at time t; $NDVI_{max}$ is the maximum value of the entire curve; $NDVI_{min}$ is the minimum value of the left/right curve (divided by the maximum NDVI); thd is the output ratio, ranging from 0–1 [56].
Maximum rate of change (MRC)	$NDVI_{ratio}(t) = \frac{NDVI(t+1) - NDVI(t)}{NDVI(t)}$	$NDVI(t)$ is the original NDVI value at time t; $NDVI(t+1)$ is the original NDVI value at time $t+1$; $NDVI_{ratio}(t)$ is the NDVI ratio at time t [50].
Change rate of curvature (RCC)	$NDVI(t) = \frac{y(t)''}{\left(1 + y(t)'^2\right)^{3/2}}$	$NDVI(t)$ is the rate of change of curve at time t; $y(t)'$ and $y(t)''$ are the first and the second derivative of curve at time t [71].

2.3.3. Phenology Trend Estimation

Extreme values caused by weather and human interference could affect phenology trend estimation; thus, outliers of extracted phenological metrics were removed in each site based on the 30-day rule proposed by Schaber and Badeck [74]. The NDVI values were considered as outliers if the estimated residuals of the linear regression model were larger than or equal to 30 days (Equation (2)), i.e., where $|e_{ij}| \geq 30$:

$$x_{ij} = m + a_i + b_j + e_{ij} \tag{2}$$

where x_{ij} is the NDVI data of year i on site j; m is a general mean (usually set to zero for finding a well-defined solution); a_i is the effect of year i (2001–2019); and b_j is the effect of site j ($j = 1, \ldots, 120$).

Then, the trend was calculated by linear regression (Equation (3)):

$$y = ax + b \tag{3}$$

where y is SOS or EOS for 2001–2019; x is the year for 2001–2019; b is the intercept; and a is the SOS or EOS trend for 2001–2019.

2.3.4. Statistical Analysis

The paired sample t-test was used to test if there existed statistically significant differences between each of the two experimental results (Table 4). A Kolmogorov–Smirnov (K-S) test was performed in advance to verify that all results obeyed normal distribution (p values are shown in Table S1). Pairs of experimental results being tested for statistically significant differences included the following: (1) phenology trends from the daily NDVI data and NDVI composites (8-day and 16-day) among five different interpolation methods; (2) phenology trends from the daily NDVI data and NDVI composites (8-day and 16-day) among different combinations of five interpolation methods and three extraction methods (the amount of the combinations is 50 in total for each vegetation type in SOS (EOS) trend estimation); and (3) phenology trends from the 8-day NDVI composite data and 16-day NDVI composite data. The level of $p < 0.05$ indicated significant difference.

Table 4. Statistically significant differences between different experiment results.

Number	Temporal Resolution	Time Interpolation Methods	Phenology Extraction Methods
(1)	1 d vs. 8 d, 1 d vs. 16 d	PL, AG, PCF, Linear, Spline	Mean of DT, MRC, and RCC
(2)	1 d vs. 8 d, 1 d vs. 16 d	PL, AG, PCF, Linear, Spline	DT, MRC, RCC
(3)	8 d vs. 16 d	PL, AG, PCF, Linear, Spline	Mean of DT, MRC, and RCC

PL, AG, PCF, Linear, and Spline are piecewise logistic function fitting, asymmetric Gaussian function fitting, polynomial curve fitting, linear interpolation, and cubic spline interpolation, respectively; DT, MRC, and RCC are dynamic threshold, maximum rate of change, and change rate of curvature, respectively; the experiment results of bold variables are tested by the paired sample t-test.

3. Results

3.1. Comparisons between Trends from Daily NDVI Data and NDVI Composites Based on Different Time Interpolation Methods

Trends estimated from NDVI composites were slightly different from the reference (daily) trend, but none of the interpolation methods had significant effects on trend estimation (Figure 3). The mean SOS trend of daily NDVI data was 0.07 d/year, for which its delaying rate was lower than all mean SOS trends from 8-day NDVI composite data (Figure 3a), and the SOS trends were 0.09 d/year (PL), 0.12 d/year (AG), 0.09 d/year (PCF and Linear), and 0.08 d/year (Spline), respectively. For 16-day NDVI composite data, the delaying rate of mean SOS trends based on AG (0.09 d/year) was higher than the daily SOS trend, but the SOS trends based on PL (0.04 d/year) and PCF (0.02 d/year) were lower. Linear (−0.06 d/year) and Spline (−0.08 d/year) yielded advanced mean SOS trends. For EOS trend estimation, minor differences existed between trends from NDVI composites and the daily NDVI data. The mean EOS trend of daily NDVI data was 0.03 d/year, and the advanced EOS trends were estimated based on Linear from 8-day NDVI composite data (−0.01 d/year), 16-day composite data (−0.05 d/year), and Spline from 16-day NDVI composite data (−0.01 d/year). (Figure 3b). Other trends all showed slightly delayed trends, which were 0.05 d/year (PL and AG), 0.10 d/year (PCF), and 0.01 d/year (Spline) from 8-day NDVI composite data, respectively. For 16-day NDVI composite data, the delaying rate of mean EOS trends were 0.02 d/year (PL), 0.08 d/year (AG), and 0.05 d/year (PCF), respectively.

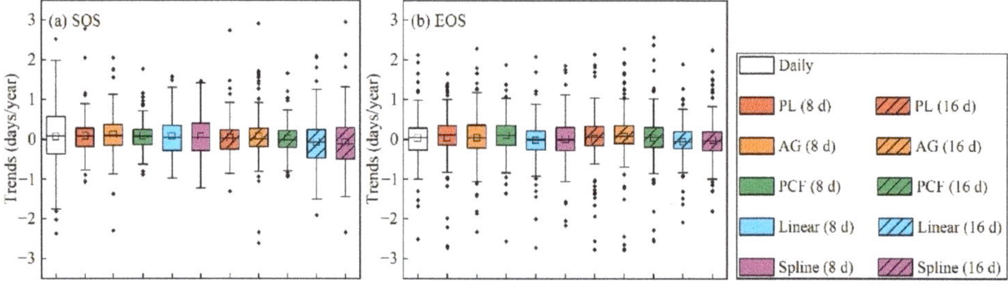

Figure 3. Comparisons between phenology trends from daily NDVI data and NDVI composites based on different time interpolation methods over all sites. (**a**) SOS trends comparisons, and (**b**) EOS trends comparisons. Phenology trends of daily NDVI data are unfilled, and phenology trends of NDVI composites are filled in colors; the bottom and top areas of boxes are the 25th and 75th percentiles; the lines through the boxes are the medians; the boxes designate the mean value; the diamonds beyond the ends of the whiskers are outliers; SOS and EOS are the start of growing season and the end of growing season; DBF, ENF, GRA, and OSH are deciduous broadleaf forest, evergreen needleleaf forest, grassland, and open shrubland, respectively; PL, AG, PCF, Linear, and Spline are piecewise logistic function fitting, asymmetric Gaussian function fitting, polynomial curve fitting, linear interpolation, and cubic spline interpolation, respectively.

3.2. Comparisons between Trends from Daily NDVI Data and NDVI Composites among Different Vegetation Types

For vegetations with apparent seasonal changes such as DBF, almost all time interpolation methods had significant effects on trend estimation (Figure 4a,e). For vegetations with weak seasonal changes such as ENF, almost no time interpolation methods had significant effects on trend estimation (Figure 4b,f). In DBF, there were significant differences between mean SOS trends estimated based on all interpolation methods and the mean SOS trend based on daily NDVI data (0.51 d/year) (Figure 4a). With the exception of the mean EOS trend (0.11 d/year) based on Spline from 8-day NDVI composite data, there were significant differences between the rest of mean EOS trends from NDVI composites and the mean EOS trend from daily NDVI data (0.40 d/year) (Figure 4e). In ENF, there was significant difference only between the mean SOS trend based on AG from 8-day NDVI composite data (0.30 d/year) and the mean SOS trend from daily NDVI data (0.04 d/year) (Figure 4b). In GRA, the mean EOS trends based on AG from 8-day NDVI composite data (0.29 d/year) and 16-day NDVI composite data (0.25 d/year) were all significantly different from the mean EOS trend from daily NDVI data (−0.07 d/year) (Figure 4g). In OSH, the mean SOS trends based on AG, PCF, and Linear from 8-day NDVI composite data were 0.05 d/year, −0.03 d/year, and −0.04 d/year, respectively; the SOS trends based on AG from 16-day NDVI composite data were 0.18 d/year, which were all significantly different from the mean trend from daily NDVI data (−0.17 d/year) (Figure 4d). The EOS trend based on PL from 8-day NDVI composite data (0.22 d/year) was significantly different from the mean EOS trend from daily NDVI data (−0.05 d/year) (Figure 4h).

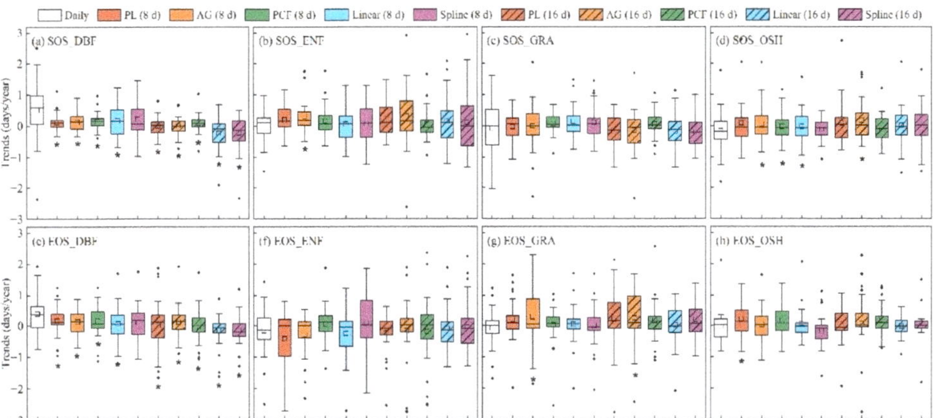

Figure 4. Comparisons between phenology trends from daily NDVI data and NDVI composites based on different time interpolation methods among different vegetation types. (**a**) SOS trends comparisons in DBF, (**b**) SOS trends comparisons in ENF, (**c**) SOS trends comparisons in GRA, (**d**) SOS trends comparisons in OSH, (**e**) EOS trends comparisons in DBF, (**f**) EOS trends comparisons in ENF, (**g**) EOS trends comparisons in GRA, and (**h**) EOS trends comparisons in OSH. Phenology trends of daily NDVI data are unfilled, and phenology trends of NDVI composites are filled in colors; the bottom and top areas of boxes are the 25th and 75th percentiles; the lines through the boxes are the medians; the boxes designate the mean value; the diamonds beyond the ends of the whiskers are outliers; SOS and EOS are the start of growing season and the end of growing season; DBF, ENF, GRA, and OSH are deciduous broadleaf forest, evergreen needleleaf forest, grassland, and open shrubland, respectively; PL, AG, PCF, Linear, and Spline are piecewise logistic function fitting, asymmetric Gaussian function fitting, polynomial curve fitting, linear interpolation, and cubic spline interpolation, respectively; * below the box indicates that there is significant difference ($p < 0.05$) between the phenology trend of the daily NDVI data and the trend estimated based on this time interpolation method.

3.3. Comparisons between Trends from Daily NDVI Data and NDVI Composites Based on Different Combinations of Time Interpolation Methods and Phenology Extraction Methods

There were 50 combinations of interpolation methods and extraction methods for each vegetation type in SOS (EOS) trend estimation, and the number of combinations for which its trends had significant differences from the trend of daily NDVI data is the largest in DBF among all vegetation types (Figure 5). In DBF, there were significant differences between SOS trends from 35 combinations and the daily SOS trend, among 30 of which included extraction methods of DT 10%, DT 20%, and DT 30% (Figure 5a). Significant differences were found between the EOS trends from 28 combinations and the daily EOS trend (Figure 5e). For 16-day NDVI composite data, significant differences were shown between the daily EOS trend and EOS trends from the combinations of PCF, Linear, Spline, and all extraction methods but only from the combinations of PCF, Linear, Spline and DT 10%, DT 20%, and DT 30% for 8-day NDVI composite data. In ENF, SOS trends from three combinations showed significant differences compared with the daily SOS trend, which shared DT 10% as the only extraction method. (Figure 5b). EOS trends from two combinations of Spline and MRC, AG, and RCC based on 8-day NDVI composite data were significantly different from the daily EOS trend (Figure 5f). In GRA, the SOS trend from only one combination was found to have significant difference from the daily SOS trend (Figure 5c), which was PCF and DT 10% based on 8-day NDVI composite data. Moreover, there were three EOS trends from only three combinations that were significantly different from the daily EOS trend, which were all based on 16-day NDVI composite data (Spline and DT 20%, PL and DT 10%, and AG and MRC, respectively) (Figure 5g). In OSH, SOS trends from 10 combinations showed significant differences compared with the daily SOS trend (Figure 5d), while EOS trends from seven combinations were significantly different from the daily EOS trend (Figure 5h), and no combinations included DT 10%.

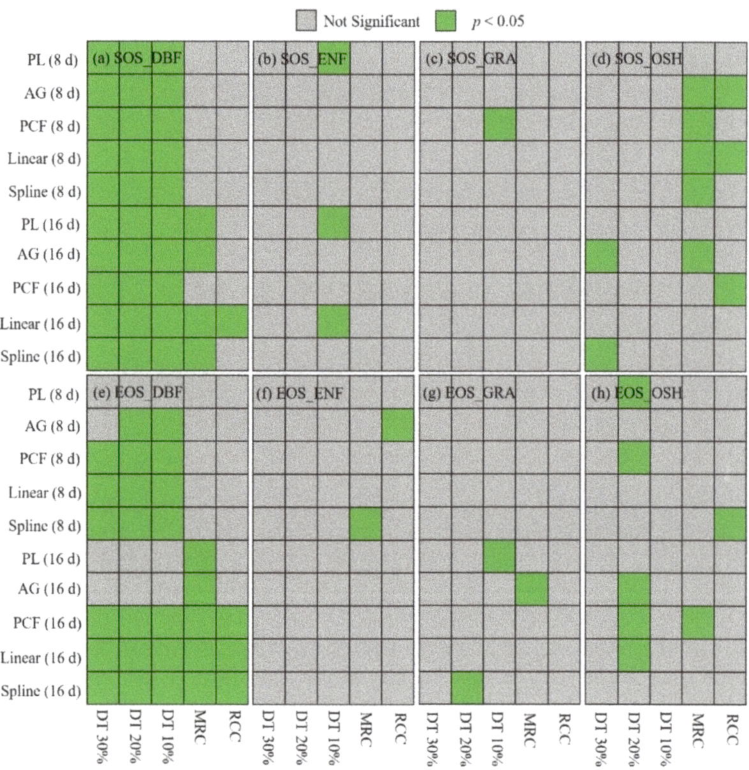

Figure 5. Comparisons between phenology trends from daily NDVI data and NDVI composites based on combinations of different time interpolation methods and extraction methods. (**a**) SOS trends comparisons in DBF, (**b**) SOS trends comparisons in ENF, (**c**) SOS trends comparisons in GRA, (**d**) SOS trends comparisons in OSH, (**e**) EOS trends comparisons in DBF, (**f**) EOS trends comparisons in ENF, (**g**) EOS trends comparisons in GRA, and (**h**) EOS trends comparisons in OSH. SOS and EOS are the start of growing season and the end of growing season; DBF, ENF, GRA, and OSH are deciduous broadleaf forest, evergreen needleleaf forest, grassland, and open shrubland, respectively; PL, AG, PCF, Linear, and Spline are piecewise logistic function fitting, asymmetric Gaussian function fitting, polynomial curve fitting, linear interpolation, and cubic spline interpolation, respectively; DT, MRC, and RCC are dynamic threshold, maximum rate of change, and change rate of curvature, respectively; grey boxes indicate that there are no significant differences ($p > 0.05$) between phenology trends from NDVI composites and daily NDVI data; green boxes indicate there are significant differences ($p < 0.05$) between phenology trends from NDVI composites and daily NDVI data.

3.4. Comparisons between Trends from the 8-Day and the 16-Day NDVI Composite Data

There were significant differences between phenology trends from the 8-day and the 16-day NDVI composite data (Figure 6) only in DBF and GRA. In DBF, significant differences occurred for all interpolation methods except PCF (Figure 6a). For PL, AG, Linear, and Spline, the mean SOS trends from the 8-day and 16-day NDVI composite data were 0.12 d/year and 0.00 d/year; 0.13 d/year and 0.01 d/year; 0.17 d/year and −0.24 d/year; and 0.22d /year and −0.23 d/year, respectively. There were also significant differences between the mean EOS trends from the 8-day and 16-day NDVI composite data among Linear and Spline, which were 0.07 d/year and −0.13 d/year (Linear), 0.11 d/year, and −0.14 d/year (Spline), respectively (Figure 6e). In GRA, the differences between the mean SOS trend from the 8-day and the 16-day NDVI composite data were significant among Linear and Spline, which were 0.14 d/year and −0.18 d/year (Linear), 0.11 d/year, and −0.20 d/year (Spline), respectively (Figure 6c). In addition, for Spline, the mean EOS trend from 8-day NDVI composite data was −0.09 d/year, which was significantly different from the 16-day mean EOS trend (0.19 d/year) (Figure 6g).

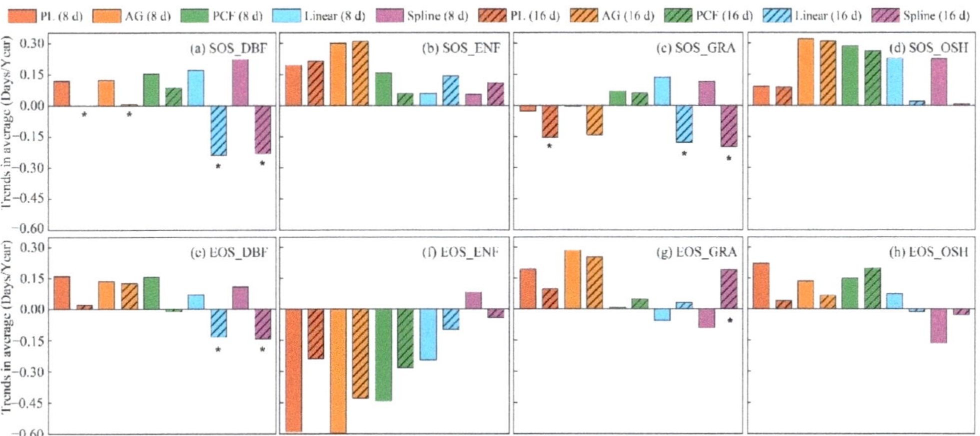

Figure 6. Comparisons between mean phenology trends from the 8-day and 16-day NDVI composite data among different interpolation methods. (**a**) SOS trends comparisons in DBF, (**b**) SOS trends comparisons in ENF, (**c**) SOS trends comparisons in GRA, (**d**) SOS trends comparisons in OSH, (**e**) EOS trends comparisons in DBF, (**f**) EOS trends comparisons in ENF, (**g**) EOS trends comparisons in GRA, and (**h**) EOS trends comparisons in OSH. SOS and EOS are the start of growing season and the end of growing season; DBF, ENF, GRA, and OSH are deciduous broadleaf forest, evergreen needleleaf forest, grassland, and open shrubland, respectively; PL, AG, PCF, Linear, and Spline are piecewise logistic function fitting, asymmetric Gaussian function fitting, polynomial curve fitting, linear interpolation, and cubic spline interpolation, respectively; * below the 16-day NDVI composite data indicates there is significant difference ($p < 0.05$) between the mean phenology trends from the 8-day and from the 16-day NDVI composite data.

4. Discussion

4.1. Effects of Time Interpolation on Trend Estimation among Different Interpolation Methods

Even though differences between the mean trends estimated from NDVI composites and from the reference (daily) data were insignificant, the discrepancies caused by time interpolation could not be ignored. The mean SOS trends based on Linear and Spline from 16-day NDVI composite data were slightly advanced while the mean trend based on daily data was delayed. The mean EOS trends based on Linear from the 8-day NDVI composite data, along with the Linear and Spline from 16-day NDVI composite data, all showed the advancing rates, which were inconsistent with the mean EOS trend based on daily data (showing the delaying rate). Therefore, it might be incomprehensive to evaluate the effects of multiple interpolation methods only by analyzing the mean trends of all sites. We further calculated the root-mean-square error (RMSE) between the trend estimated from the reference (daily) data and the trend from each interpolation method (Figures S1 and S2). The RMSE values of SOS and EOS trends based on all interpolation methods ranged from 0.35–0.52 d/year and 0.39–0.47 d/year (Figures S1 and S2), which were overall similar between each method, and the piecewise logistic function fitting (8 d) performed slightly better with the lowest RMSE values. For each interpolation method, the ratio of sites for which its absolute values of trends were lower than the corresponding RMSE value ranged from 56 to 77% for SOS trends and 58–71% for EOS trends (Table S2), which implied that the process of time interpolation on NDVI composites might even change the trend direction over half of all sites. RMSE is sensitive to outliers; thus, our calculation results might overestimate the effects of time interpolation on trend estimation, but the uncertainties caused by time interpolation should be considered.

4.2. Effects of Time Interpolation on Trend Estimation among Different Vegetation Types

For vegetation types with apparent seasonal changes such as DBF, almost all the time interpolation methods had significant effects on phenology trend estimation. However, for vegetation types with weaker seasonal changes such as ENF, time interpolation methods had almost no significant effects on trend estimation. Figures 7 and 8 showed an example of phenology extraction results and trends from the daily and 8-day NDVI composite data in DBF and ENF, respectively. During 90–150 in Julian day, changes of daily NDVI values and 8-day NDVI composite values in DBF ranged from 0.33 to 0.28 (Figure 7c), while it only ranged from 0.11 to 0.13 in ENF (Figure 7f). Meanwhile, during 270–330 in Julian day, changes of daily NDVI values and 8-day NDVI composite values in DBF ranged from 0.23 to 0.29 (Figure 8c), but it ranged from 0.03 to 0.07 in ENF (Figure 8f). The 8-day NDVI composite data values changed fast in DBF especially during greening and senescence stages (Figure 7b), making it hard for time interpolation to capture the detailed NDVI changes of each Julian day, which increased errors in the extraction of SOS (EOS) annually and in trend estimation (Figure 7a). Compared with DBF, values of NDVI composites changed slightly in ENF (Figure 8b), which resulted in a higher fidelity after time interpolation compared with daily NDVI data. Therefore, the annual extraction results of SOS (EOS) and their trends had higher accuracies (Figure 8a). The same pattern of results was found in our analysis of the 16-day NDVI composite data. We suggest that remote sensing data of daily temporal resolution should be used for estimating phenology trends in vegetation types especially with apparent seasonal changes. For vegetation with weaker seasonal changes, using NDVI composites (i.e., 8-day or 16-day) would have weaker effects on trend estimation.

4.3. Effects of Time Interpolation on Trend Estimation among Different Combinations of Time Interpolation Methods and Phenology Extraction Methods

The selection of phenology extraction methods should be fully considered based on study areas, vegetation types, satellite products, and interpolation methods. For vegetation types with apparent seasonal changes such as DBF, even though most time interpolation methods had significant effects on phenology trend estimation, the phenology trends from few specific combinations (i.e., polynomial curve function fitting and maximum rate of change based on the 16-day NDVI composite data in SOS (Figure 5a), asymmetric Gaussian function fitting, and dynamic threshold 30% based on the 8-day NDVI composite data in EOS (Figure 5e)) still showed no significant differences compared with the trends from daily NDVI data. In addition, for vegetation types with weaker seasonal changes such as ENF, there still existed phenology trends from specific combinations that had significant differences with trends from the daily NDVI data. Previous studies also indicated that different combinations could result in different accuracies of trend estimation [33,75,76]. According to our results, the maximum rate of change and the change rate of curvature method could be used for estimating phenology trends of DBF based on 8-day composite NDVI data, while a dynamic threshold of 20% and 30% had a better performance on phenology trend estimation for ENF. The dynamic threshold of 30% and the change rate of curvature method were suitable for GRA, and a dynamic threshold of 10% had a better result in OSH trend estimation.

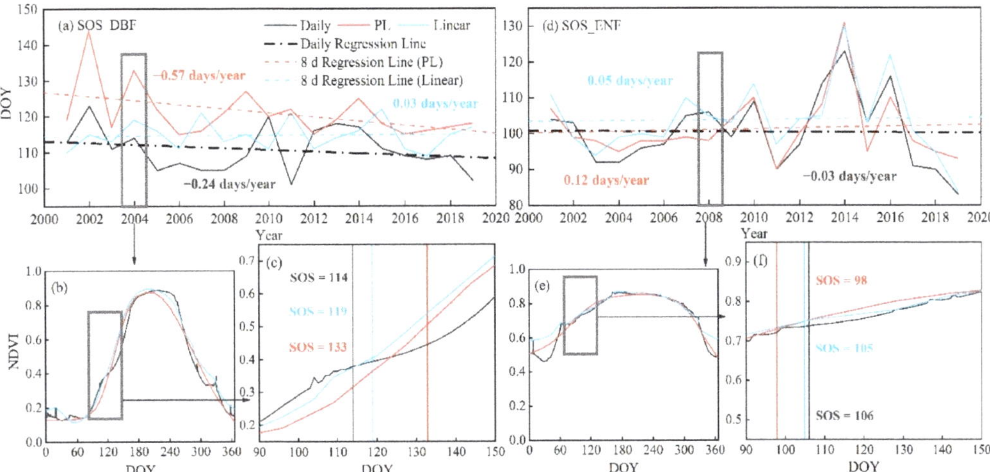

Figure 7. Extraction results and trends of the start of growing season (SOS) from the daily and the 8-day NDVI composite data in deciduous broadleaf forest (DBF) and evergreen needleleaf forest (ENF). (**a**) SOS trends of DBF from 2001 to 2019, (**b**) SOS trends of DBF in 2004, (**c**) SOS estimation results of DBF in 2004, (**d**) SOS trends of ENF from 2001 to 2019, (**e**) SOS trends of ENF in 2008, and (**f**) SOS estimation results of ENF in 2008. Piecewise logistic function fitting and linear interpolation are chosen as interpolation method examples; the dynamic threshold 30% is chosen as the extraction method.

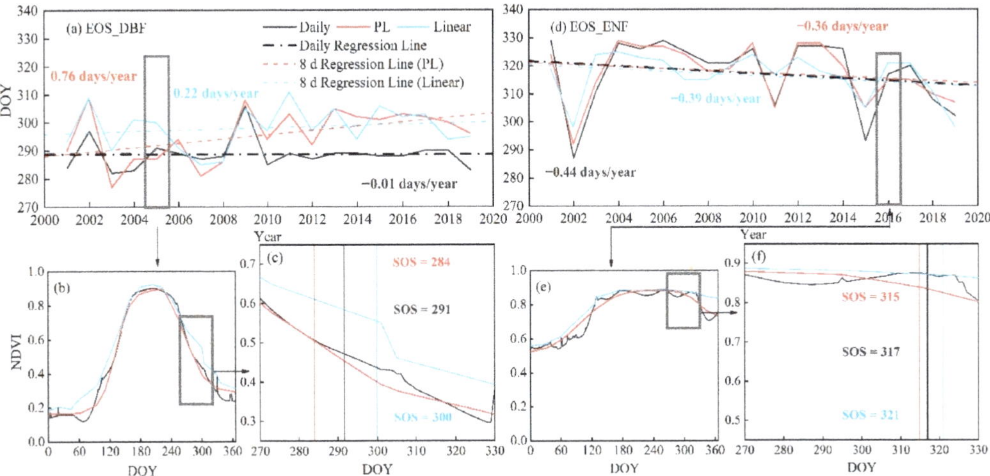

Figure 8. Extraction results and trends of the end of growing season (EOS) from the daily and 8-day NDVI composite data in deciduous broadleaf forest (DBF) and evergreen needleleaf forest (ENF). (**a**) EOS trends of DBF from 2001 to 2019, (**b**) EOS trends of DBF in 2005, (**c**) EOS estimation results of DBF in 2005, (**d**) EOS trends of ENF from 2001 to 2019, (**e**) EOS trends of ENF in 2016, and (**f**) EOS estimation results of ENF in 2016. Piecewise logistic function fitting and linear interpolation are chosen as interpolation method examples; a dynamic threshold of 30% is chosen as the extraction method.

4.4. Effects of Time Interpolation on Trend Estimation among Data with Different Temporal Resolutions

There were no significant differences between trends derived from the 8-day and 16-day NDVI composite data in ENF and OSH, and significant differences existed in DBF and GRA only among a few interpolation methods. Zhu et al. [8] used asymmetric Gaussian function fitting and piecewise logistic function fitting for estimating SOS trends of Tibetan Plateau alpine meadow from 2000 to 2015, and they found that the trends estimated from the fine temporal resolution (8-day) NDVI data and the coarse temporal resolution (16-day) NDVI data (MOD13A2) had no significant differences, which were in agreement with our research results (Figure 6c,g). Kross et al. [60] also observed that shifts in SOS were not sensitive to temporal resolution (4–28 days) among Canadian deciduous broadleaf sites. According to Figures S1 and S2, the RMSE values of trend estimated from the reference (daily) data and from 8-day NDVI composites were overall lower than those from 16-day NDVI composites. For DBF and GRA, especially among interpolation methods that caused significant differences between trends from the 8-day and 16-day NDVI composite data, we suggest NDVI composites of lower temporal resolution (8-day) for trend estimation when daily-scale datasets were not available. For ENF and OSH, there were no significant differences between trends estimated from the 8-day and 16-day NDVI composite data, which implied fewer discrepancies when applying coarse temporal resolution (16-day) data especially based on the AG and PCF method (Figure 6b,f,d,h) in which the discrepancies were relatively low. However, we only compared the mean trends between the 8-day and 16-day NDVI composite data among all sites; thus, our general conclusions may not apply for every site accurately. The selection of NDVI composites should still be fully considered based on the specific research area, vegetation type, and the data preprocessing method.

4.5. Limitations

We observed the delaying SOS mean trends and advancing EOS mean trends among daily-scale NDVI data and NDVI composites, which indicated opposite phenology trends (advancing SOS trends and delaying EOS trends) compared with former research [6,50,52,77,78]. We provide three explanations. First, the results demonstrated above (Figures 3 and 4) were the mean values of SOS and EOS trends, which cannot fully represent the trend of each site with different geographical locations and heterogeneous landscapes. Second, the opposite trends of phenology were also reported by various authors at continental scales over the northern high latitudes due to differences in data sources and scales [8,38], winter or spring warming [73,79], and the mixed impacts of increased spring–fall temperature and fall precipitation [80]. Finally, satellite-based phenological metrics may mainly reflect the spring phenology of early-unfolding (flowering) plant species, indicating that satellite-based phenology trends may follow the trends of ground-measured early plants. Fu et al. [3] found that most of these species showed a delayed trend in spring through the species-specific trend analysis, which confirmed that the delay of the SOS trends monitored by the satellite datasets truly exists. The uncertainties of the opposite phenology trends and their environmental/ecological consequences among different biome zones, study period, and remote sensing sensors still require deeper investigations.

In order to eliminate noise in NDVI time-series curves, we reconstructed the daily NDVI data by using the Savitzky–Golay filter as reference data. However, it still cannot completely simulate NDVI curves in a real natural state, which may cause uncertainties in trend estimation. In addition, similar studies replicated at additional locations, among various satellite products and vegetation types, are also needed for more comprehensive and reliable evaluation on the effects of time interpolation on phenology trend estimation. Due to the mismatch in observation scale (plant scale and pixel scale) and content (the definition of phenological events), we did not use the ground observations as reference data, but comparative studies between using remote-sensing tools and using high-accuracy ground-based measurements still constitute a common and direct method for assessing remote sensing approaches in predicting phenological events. In order to validate vegetation phenology products properly, ground observations from individual species, canopy cameras, or flux towers should be upscaled temporally and spatially for matching satellite pixels over various ecosystems and geographical regions.

5. Conclusions

In this study, we used MODIS MCD43A4 daily surface reflectance data to construct a daily NDVI time-series dataset as the reference data and then generated an 8-day and a 16-day NDVI composite dataset among 120 sites in the mid-high latitudes of the northern hemisphere during 2001–2019. The NDVI composites were used to comprehensively investigate the effects of time interpolation on trend estimation among (1) five time-interpolation methods; (2) four vegetation types; and (3) the combinations of five time-interpolation methods and three extraction methods. We also analyzed the differences of trends estimated between the 8-day and 16-day dataset.

Four main conclusions were drawn from our study. First, none of the interpolation methods had significant effects on trend estimation over all sites, but the discrepancies between trends estimated from NDVI daily data and from NDVI composites could not be ignored. For each interpolation method, the RMSE value of multi-day scale trends was higher than the absolute values of these trends among most of the sites (56–77% of all sites for SOS trends and 58–71% of all sites for EOS trends). Even the effects were insignificant, the process of time interpolation might still change trend direction compared with the trend from the NDVI daily data. Second, time interpolation had significant effects on phenology trend estimation among vegetation types with apparent seasonal changes, but had almost no significant effects among vegetation types with weak seasonal changes. In order to minimize estimation bias, we strongly suggest remote sensing datasets with a daily or high temporal resolution to be applied for estimating phenology trends in vegetation types especially sensitive to season changes. Third, the selection of extraction methods should be fully considered. Trends estimated based on the same interpolation method but different extraction methods were not consistent in showing significant (insignificant) differences with the trend estimated from the daily data, implying that the selection of extraction methods also affected trend estimation. The maximum rate of change and the change rate of curvature method could be used in deciduous broadleaf forest based on 8-day composite NDVI data, while the dynamic threshold of 20% and 30% had better performances for evergreen needleleaf forest. The dynamic threshold of 30% and the change rate of curvature were suitable for grassland, and the dynamic threshold of 10% had a better result in open shrubland. Lastly, for deciduous broadleaf forest and grassland, especially among interpolation methods that caused significant differences between trends from the 8-day and 16-day NDVI composite data, we suggest NDVI composites with a lower temporal resolution (8-day) for trend estimation when daily-scale datasets were not available. For evergreen needleleaf forest and open shrubland, there were fewer discrepancies between trends from 8-day and 16-day NDVI composite data, which implied the availability of using a coarse temporal resolution (16-day), especially based on the asymmetric Gaussian function and the polynomial curve function. In order to further enhance the comprehensive evaluation about the effects of time interpolation on phenology trend estimation, future studies should be carried out at additional locations and among various satellite products and vegetation types.

Supplementary Materials: The following are available online at https://www.mdpi.com/article/10.3390/rs13245018/s1, Table S1: p values of experimental results in Kolmogorov–Smirnov (K-S) test, Figure S1: Comparisons of the start of growing season (SOS) trends estimated from daily data and from each interpolation method, Figure S2: Comparisons of the end of growing season (EOS) trends estimated from the daily data and from each interpolation method, Table S2: The ratio of sites for which its absolute values of trends were lower than the corresponding root-mean-square error (RMSE) value among each interpolation method.

Author Contributions: X.L.: conceptualization, initial analysis, and writing—original draft preparation. W.Z.: conceptualization, funding acquisition, data curation, and writing—review and editing. Z.X.: data curation and writing—review and editing. P.Z.: data curation and writing—review and editing. X.H.: writing—review and editing. L.S.: writing—review and editing. Z.D.: writing—review and editing. All authors have read and agreed to the published version of the manuscript.

Funding: This study was funded by the National Natural Science Foundation of China (No. 41771047) and the National Key Research and Development Program of China (Grant No. 2020YFA0608504).

Conflicts of Interest: The authors declare that they have no known competing financial interests or personal relationships that could have appeared to influence the work reported in this paper.

References

1. White, M.A.; Thornton, P.E.; Running, S.W. A continental phenology model for monitoring vegetation responses to interannual climatic variability. *Glob. Biogeochem. Cycle* **1997**, *11*, 217–234. [CrossRef]
2. Richardson, A.D.; Bailey, A.S.; Denny, E.G.; Martin, C.W.; O'Keefe, J. Phenology of a northern hardwood forest canopy. *Glob. Chang. Biol.* **2006**, *12*, 1174–1188. [CrossRef]
3. Fu, Y.S.H.; Piao, S.L.; Op de Beeck, M.; Cong, N.; Zhao, H.F.; Zhang, Y.; Menzel, A.; Janssens, I.A. Recent spring phenology shifts in western Central Europe based on multiscale observations. *Glob. Ecol. Biogeogr.* **2014**, *23*, 1255–1263. [CrossRef]
4. Yang, X.Q.; Wu, J.P.; Chen, X.Z.; Ciais, P.; Maignan, F.; Yuan, W.P.; Piao, S.L.; Yang, S.; Gong, F.X.; Su, Y.X.; et al. A comprehensive framework for seasonal controls of leaf abscission and productivity in evergreen broadleaved tropical and subtropical forests. *Innovation* **2021**, *2*, 1–7. [CrossRef]
5. Piao, S.L.; Ciais, P.; Friedlingstein, P.; Peylin, P.; Reichstein, M.; Luyssaert, S.; Margolis, H.; Fang, J.Y.; Barr, A.; Chen, A.P.; et al. Net carbon dioxide losses of northern ecosystems in response to autumn warming. *Nature* **2008**, *451*, 49–52. [CrossRef]
6. Peñuelas, J.; Rutishauser, T.; Filella, I. Phenology Feedbacks on Climate Change. *Science* **2009**, *324*, 887–888. [CrossRef]
7. Dragoni, D.; Schmid, H.P.; Wayson, C.A.; Potter, H.; Grimmond, C.S.B.; Randolph, J.C. Evidence of increased net ecosystem productivity associated with a longer vegetated season in a deciduous forest in south-central Indiana, USA. *Glob. Chang. Biol.* **2011**, *17*, 886–897. [CrossRef]
8. Zhu, Y.X.; Zhang, Y.J.; Zu, J.X.; Wang, Z.P.; Huang, K.; Cong, N.; Tang, Z. Effects of data temporal resolution on phenology extractions from the alpine grasslands of the Tibetan Plateau. *Ecol. Indic.* **2019**, *104*, 365–377. [CrossRef]
9. Gu, F.X. Parameterization of Leaf Phenology for the Terrestrial Ecosystem Models. *Prog. Geogr.* **2006**, *25*, 68–75.
10. Walther, G.R.; Post, E.; Convey, P.; Menzel, A.; Parmesan, C.; Beebee, T.J.C.; Fromentin, J.M.; Hoegh-Guldberg, O.; Bairlein, F. Ecological responses to recent climate change. *Nature* **2002**, *416*, 389–395. [CrossRef]
11. Justice, C.O.; Townshend, J.R.G.; Holben, B.N.; Tucker, C.J. Analysis of the Phenology of Global Vegetation Using Meteorological Satellite Data. *Int. J. Remote Sens.* **1985**, *6*, 1271–1318. [CrossRef]
12. Myneni, R.B.; Keeling, C.D.; Tucker, C.J.; Asrar, G.; Nemani, R.R. Increased plant growth in the northern high latitudes from 1981 to 1991. *Nature* **1997**, *386*, 698–702. [CrossRef]
13. White, M.A.; Hoffman, F.; Hargrove, W.W.; Nemani, R.R. A global framework for monitoring phenological responses to climate change. *Geophys. Res. Lett.* **2005**, *32*, 1–4. [CrossRef]
14. Zhang, X.Y.; Friedl, M.A.; Schaaf, C.B. Global vegetation phenology from Moderate Resolution Imaging Spectroradiometer (MODIS): Evaluation of global patterns and comparison with in situ measurements. *J. Geophys. Res.* **2006**, *111*, 1–14. [CrossRef]
15. Zhu, W.Q.; Tian, H.Q.; Xu, X.F.; Pan, Y.Z.; Chen, G.S.; Lin, W.P. Extension of the growing season due to delayed autumn over mid and high latitudes in North America during 1982–2006. *Glob. Ecol. Biogeogr.* **2012**, *21*, 260–271. [CrossRef]
16. Yang, Y.T.; Guan, H.D.; Shen, M.G.; Liang, W.; Jiang, L. Changes in autumn vegetation dormancy onset date and the climate controls across temperate ecosystems in China from 1982 to 2010. *Glob. Chang. Biol.* **2015**, *21*, 652–665. [CrossRef]
17. Zhang, X.Y.; Liu, L.L.; Liu, Y.; Jayavelu, S.; Wang, J.M.; Moon, M.; Henebry, G.M.; Friedl, M.A.; Schaaf, C.B. Generation and evaluation of the VIIRS land surface phenology product. *Remote Sens. Environ.* **2018**, *216*, 212–229. [CrossRef]
18. Peng, D.L.; Wang, Y.; Xian, G.; Huete, A.R.; Huang, W.J.; Shen, M.G.; Wang, F.M.; Yu, L.; Liu, L.Y.; Xie, Q.Y.; et al. Investigation of land surface phenology detections in shrublands using multiple scale satellite data. *Remote Sens. Environ.* **2021**, *252*, 112133. [CrossRef]
19. Wu, W.; Sun, Y.; Xiao, K.; Xin, Q.C. Development of a global annual land surface phenology dataset for 1982–2018 from the AVHRR data by implementing multiple phenology retrieving methods. *Int. J. Appl. Earth Obs. Geoinf.* **2021**, *103*, 102487. [CrossRef]
20. White, M.A.; de Beurs, K.M.; Didan, K.; Inouye, D.W.; Richardson, A.D.; Jensen, O.P.; O'Keefe, J.; Zhang, G.; Nemani, R.R.; van Leeuwen, W.J.D.; et al. Intercomparison, interpretation, and assessment of spring phenology in North America estimated from remote sensing for 1982–2006. *Glob. Chang. Biol.* **2009**, *15*, 2335–2359. [CrossRef]
21. Hill, M.J.; Roman, M.O.; Schaaf, C.B.; Hutley, L.; Brannstrom, C.; Etter, A.; Hanan, N.P. Characterizing vegetation cover in global savannas with an annual foliage clumping index derived from the MODIS BRDF product. *Remote Sens. Environ.* **2011**, *115*, 2008–2024. [CrossRef]

22. Wang, Y.T.; Hou, X.Y.; Wang, M.J.; Wu, L.; Ying, L.L.; Feng, Y.Y. Topographic controls on vegetation index in a hilly landscape: A case study in the Jiaodong Peninsula, eastern China. *Environ. Earth Sci.* **2013**, *70*, 625–634. [CrossRef]

23. Soudani, K.; Hmimina, G.; Delpierre, N.; Pontailler, J.Y.; Aubinet, M.; Bonal, D.; Caquet, B.; de Grandcourt, A.; Burban, B.; Flechard, C.; et al. Ground-based Network of NDVI measurements for tracking temporal dynamics of canopy structure and vegetation phenology in different biomes. *Remote Sens. Environ.* **2012**, *123*, 234–245. [CrossRef]

24. Hartfield, K.A.; Marsh, S.E.; Kirk, C.D.; Carriere, Y. Contemporary and historical classification of crop types in Arizona. *Int. J. Remote Sens.* **2013**, *34*, 6024–6036. [CrossRef]

25. Hmimina, G.; Dufrene, E.; Pontailler, J.Y.; Delpierre, N.; Aubinet, M.; Caquet, B.; de Grandcourt, A.; Burban, B.; Flechard, C.; Granier, A.; et al. Evaluation of the potential of MODIS satellite data to predict vegetation phenology in different biomes: An investigation using ground-based NDVI measurements. *Remote Sens. Environ.* **2013**, *132*, 145–158. [CrossRef]

26. Ding, C.; Liu, X.N.; Huang, F.; Li, Y.; Zou, X.Y. Onset of drying and dormancy in relation to water dynamics of semi-arid grasslands from MODIS NDWI. *Agric. For. Meteorol.* **2017**, *234*, 22–30. [CrossRef]

27. Testa, S.; Soudani, K.; Boschetti, L.; Mondino, E.B. MODIS-derived EVI, NDVI and WDRVI time series to estimate phenological metrics in French deciduous forests. *Int. J. Appl. Earth Obs. Geoinf.* **2018**, *64*, 132–144. [CrossRef]

28. Huang, K.; Zhang, Y.J.; Tagesson, T.; Brandt, M.; Wang, L.H.; Chen, N.; Zu, J.X.; Jin, H.X.; Cai, Z.Z.; Tong, X.W.; et al. The confounding effect of snow cover on assessing spring phenology from space: A new look at trends on the Tibetan Plateau. *Sci. Total Environ.* **2021**, *756*, 144011. [CrossRef]

29. Fan, D.Q.; Zhao, X.S.; Zhu, W.Q.; Zheng, Z.T. Review of influencing factors of accuracy of plant phenology monitoring based on remote sensing data. *Prog. Geogr.* **2016**, *35*, 304–319.

30. Brown, M.E.; Pinzon, J.E.; Didan, K.; Morisette, J.T.; Tucker, C.J. Evaluation of the consistency of long-term NDVI time series derived from AVHRR, SPOT-Vegetation, SeaWiFS, MODIS, and Landsat ETM+ sensors. *IEEE Trans. Geosci. Electron.* **2006**, *44*, 1787–1793. [CrossRef]

31. Garrity, S.R.; Bohrer, G.; Maurer, K.D.; Mueller, K.L.; Vogel, C.S.; Curtis, P.S. A comparison of multiple phenology data sources for estimating seasonal transitions in deciduous forest carbon exchange. *Agric. For. Meteorol.* **2011**, *151*, 1741–1752. [CrossRef]

32. Zhang, G.L.; Zhang, Y.J.; Dong, J.W.; Xiao, X.M. Green-up dates in the Tibetan Plateau have continuously advanced from 1982 to 2011. *Proc. Natl. Acad. Sci. USA* **2013**, *110*, 4309–4314. [CrossRef] [PubMed]

33. Shen, M.G.; Zhang, G.X.; Cong, N.; Wang, S.P.; Kong, W.D.; Piao, S.L. Increasing altitudinal gradient of spring vegetation phenology during the last decade on the Qinghai-Tibetan Plateau. *Agric. For. Meteorol.* **2014**, *189*, 71–80. [CrossRef]

34. Donnelly, A.; Liu, L.L.; Zhang, X.Y.; Wingler, A. Autumn leaf phenology: Discrepancies between in situ observations and satellite data at urban and rural sites. *Int. J. Remote Sens.* **2018**, *39*, 8129–8150. [CrossRef]

35. Donnelly, A.; Yu, R.; Liu, L.L. Comparing in situ spring phenology and satellite-derived start of season at rural and urban sites in Ireland. *Int. J. Remote Sens.* **2021**, *42*, 7821–7841. [CrossRef]

36. Peng, D.L.; Zhang, X.Y.; Wu, C.Y.; Huang, W.J.; Gonsamo, A.; Huete, A.R.; Didan, K.; Tan, B.; Liu, X.J.; Zhang, B. Intercomparison and evaluation of spring phenology products using National Phenology Network and AmeriFlux observations in the contiguous United States. *Agric. For. Meteorol.* **2017**, *242*, 33–46. [CrossRef]

37. Zeng, H.Q.; Jia, G.S.; Forbes, B.C. Shifts in Arctic phenology in response to climate and anthropogenic factors as detected from multiple satellite time series. *Environ. Res.* **2013**, *8*, 035036. [CrossRef]

38. Zeng, H.Q.; Jia, G.S.; Epstein, H. Recent changes in phenology over the northern high latitudes detected from multi-satellite data. *Res. Lett.* **2011**, *6*, 045508. [CrossRef]

39. Wang, H.; Liu, H.Y.; Huang, N.; Bi, J.; Ma, X.L.; Ma, Z.Y.; Shangguan, Z.J.; Zhao, H.F.; Feng, Q.S.; Liang, T.G.; et al. Satellite-derived NDVI underestimates the advancement of alpine vegetation growth over the past three decades. *Ecology* **2021**, *102*, e03518. [CrossRef] [PubMed]

40. Beurs, K.M.d.; Henebry, G.M. Spatio-temporal statistical methods for modeling land surface phenology. In *Phenological Research*; Hudson, I., Keatley, M., Eds.; Springer: Dordrecht, The Netherlands, 2010; pp. 177–208.

41. White, K.; Pontius, J.; Schaberg, P. Remote sensing of spring phenology in northeastern forests: A comparison of methods, field metrics and sources of uncertainty. *Remote Sens. Environ.* **2014**, *148*, 97–107. [CrossRef]

42. Cong, N.; Shen, M.G.; Piao, S.L. Spatial variations in responses of vegetation autumn phenology to climate change on the Tibetan Plateau. *J. Plant Ecol.* **2017**, *10*, 744–752. [CrossRef]

43. Shen, R.Q.; Lu, H.B.; Yuan, W.P.; Chen, X.Z.; He, B. Regional evaluation of satellite-based methods for identifying end of vegetation growing season. *Remote Sens. Ecol. Conserv.* **2021**, *175*, 88–98. [CrossRef]

44. Holben, B.N. Characteristics of Maximum-Value Composite Images from Temporal Avhrr Data. *Int. J. Remote Sens.* **1986**, *7*, 1417–1434. [CrossRef]

45. Viovy, N.; Arino, O.; Belward, A.S. The best index slope extraction (BISE)-a method for reducing noise in NDVI yime-series. *Int. J. Remote Sens.* **1992**, *13*, 1585–1590. [CrossRef]

46. Taddei, R. Maximum value interpolated (MVI): A maximum value composite method improvement in vegetation index profiles analysis. *Int. J. Remote Sens.* **1997**, *18*, 2365–2370. [CrossRef]

47. Ma, M.G.; Veroustraete, F. Reconstructing pathfinder AVHRR land NDVI time-series data for the Northwest of China. *Adv. Space Res.* **2006**, *37*, 835–840. [CrossRef]

48. Julien, Y.; Sobrino, J.A. Comparison of cloud-reconstruction methods for time series of composite NDVI data. *Remote Sens. Environ.* **2010**, *114*, 618–625. [CrossRef]

49. Fan, X.W.; Liu, Y.B.; Wu, G.P.; Zhao, X.S. Compositing the minimum NDVI for daily water surface mapping. *Remote Sens.* **2020**, *12*, 700. [CrossRef]

50. Piao, S.L.; Fang, J.Y.; Zhou, L.M.; Ciais, P.; Zhu, B. Variations in satellite-derived phenology in China's temperate vegetation. *Glob. Chang. Biol.* **2006**, *12*, 672–685. [CrossRef]

51. Liu, Q.; Fu, Y.S.H.; Zeng, Z.Z.; Huang, M.T.; Li, X.R.; Piao, S.L. Temperature, precipitation, and insolation effects on autumn vegetation phenology in temperate China. *Glob. Chang. Biol.* **2016**, *22*, 644–655. [CrossRef] [PubMed]

52. Jeong, S.J.; Ho, C.H.; Gim, H.J.; Brown, M.E. Phenology shifts at start vs. end of growing season in temperate vegetation over the Northern Hemisphere for the period 1982–2008. *Glob. Chang. Biol.* **2011**, *17*, 2385–2399. [CrossRef]

53. Liang, L.A.; Schwartz, M.D.; Fei, S.L. Validating satellite phenology through intensive ground observation and landscape scaling in a mixed seasonal forest. *Remote Sens. Environ.* **2011**, *115*, 143–157. [CrossRef]

54. Meijering, E. A chronology of interpolation: From ancient astronomy to modern signal and image processing. *Proc. IEEE* **2002**, *90*, 319–342. [CrossRef]

55. Wolberg, G.; Alfy, I. Monotonic cubic spline interpolation. *Comput. Graph. Int.* **1999**, 188–195.

56. Jönsson, P.; Eklundh, L. Seasonality extraction by function fitting to time-series of satellite sensor data. *IEEE Trans. Geosci. Remote Sensing.* **2002**, *40*, 1824–1832. [CrossRef]

57. Beck, P.S.A.; Atzberger, C.; Hogda, K.A.; Johansen, B.; Skidmore, A.K. Improved monitoring of vegetation dynamics at very high latitudes: A new method using MODIS NDVI. *Remote Sens. Environ.* **2006**, *100*, 321–334. [CrossRef]

58. Chen, J.; Jönsson, P.; Tamura, M.; Gu, Z.H.; Matsushita, B.; Eklundh, L. A simple method for reconstructing a high-quality NDVI time-series data set based on the Savitzky-Golay filter. *Remote Sens. Environ.* **2004**, *91*, 332–344. [CrossRef]

59. Cong, N.; Piao, S.L.; Chen, A.P.; Wang, X.H.; Lin, X.; Chen, S.P.; Han, S.J.; Zhou, G.S.; Zhang, X.P. Spring vegetation green-up date in China inferred from SPOT NDVI data: A multiple model analysis. *Agric. For. Meteorol.* **2012**, *165*, 104–113. [CrossRef]

60. Kross, A.; Fernandes, R.; Seaquist, J.; Beaubien, E. The effect of the temporal resolution of NDVI data on season onset dates and trends across Canadian broadleaf forests. *Remote Sens. Environ.* **2011**, *115*, 1564–1575. [CrossRef]

61. Slayback, D.A.; Pinzon, J.E.; Los, S.O.; Tucker, C.J. Northern hemisphere photosynthetic trends 1982-99. *Glob. Chang. Biol.* **2003**, *9*, 1–15. [CrossRef]

62. Klosterman, S.T.; Hufkens, K.; Gray, J.M.; Melaas, E.; Sonnentag, O.; Lavine, I.; Mitchell, L.; Norman, R.; Friedl, M.A.; Richardson, A.D. Evaluating remote sensing of deciduous forest phenology at multiple spatial scales using PhenoCam imagery. *Biogeoences.* **2014**, *11*, 4305–4320. [CrossRef]

63. Zhou, L.; Kaufmann, R.K.; Tian, Y.; Myneni, R.B.; Tucker, C.J. Relation between interannual variations in satellite measures of northern forest greenness and climate between 1982 and 1999. *J. Geophys. Res.* **2003**, *108*, 1–11. [CrossRef]

64. Ding, M.J.; Zhang, Y.L.; Sun, X.M.; Liu, L.S.; Wang, Z.F.; Bai, W.Q. Spatiotemporal variation in alpine grassland phenology in the Qinghai-Tibetan Plateau from 1999 to 2009. *Chin. Sci. Bull.* **2013**, *58*, 396–405. [CrossRef]

65. Zu, J.X.; Zhang, Y.J.; Huang, K.; Liu, Y.J.; Chen, N.; Cong, N. Biological and climate factors co-regulated spatial-temporal dynamics of vegetation autumn phenology on the Tibetan Plateau. *Int. J. Appl. Earth Obs. Geoinf.* **2018**, *69*, 198–205. [CrossRef]

66. Kwak, S.G.; Kim, J.H. Central limit theorem: The cornerstone of modern statistics. *Korean J. Anesthesiol.* **2017**, *70*, 144–156. [CrossRef] [PubMed]

67. Bradley, B.A.; Jacob, R.W.; Hermance, J.F.; Mustard, J.F. A curve fitting procedure to derive inter-annual phenologies from time series of noisy satellite NDVI data. *Remote Sens. Environ.* **2007**, *106*, 137–145. [CrossRef]

68. Savitzky, A.; Golay, M.J.E. Smoothing and differentiation of data by simplified least squares procedures. *Anal. Chem.* **1964**, *36*, 1627–1639. [CrossRef]

69. Hird, J.N.; McDermid, G.J. Noise reduction of NDVI time series: An empirical comparison of selected techniques. *Remote Sens. Environ.* **2009**, *113*, 248–258. [CrossRef]

70. Chen, Y.; Cao, R.Y.; Chen, J.; Liu, L.C.; Matsushita, B. A practical approach to reconstruct high-quality Landsat NDVI time-series data by gap filling and the Savitzky-Golay filter. *ISPRS-J. Photogramm. Remote Sens.* **2021**, *180*, 174–190. [CrossRef]

71. Zhang, X.Y.; Friedl, M.A.; Schaaf, C.B.; Strahler, A.H.; Hodges, J.C.F.; Gao, F.; Reed, B.C.; Huete, A. Monitoring vegetation phenology using MODIS. *Remote Sens. Environ.* **2003**, *84*, 471–475. [CrossRef]

72. Heumann, B.W.; Seaquist, J.W.; Eklundh, L.; Jönsson, P. AVHRR derived phenological change in the Sahel and Soudan, Africa, 1982–2005. *Remote Sens. Environ.* **2007**, *108*, 385–392. [CrossRef]

73. Yu, H.Y.; Luedeling, E.; Xu, J.C. Winter and spring warming result in delayed spring phenology on the Tibetan Plateau. *Proc. Natl. Acad. Sci. USA* **2010**, *107*, 22151–22156. [CrossRef]

74. Schaber, J.; Badeck, F.W. Evaluation of methods for the combination of phenological time series and outlier detection. *Tree Physiol.* **2002**, *22*, 973–982. [CrossRef] [PubMed]

75. Zhang, X.X.; Cui, Y.P.; Liu, S.J.; Li, N.; Fu, Y.M. Evaluation of the accuracy of phenology extraction methods for natural vegetation based on remote sensing. *Chin. J. Ecol.* **2019**, *38*, 1589–1599.

76. Li, N.; Zhan, P.; Pan, Y.Z.; Zhu, X.F.; Li, M.Y.; Zhang, D.J. Comparison of Remote Sensing Time-Series Smoothing Methods for Grassland Spring Phenology Extraction on the Qinghai-Tibetan Plateau. *Remote Sens.* **2020**, *12*, 3383. [CrossRef]

77. Menzel, A.; Sparks, T.H.; Estrella, N.; Koch, E.; Aasa, A.; Ahas, R.; Alm-Kubler, K.; Bissolli, P.; Braslavska, O.; Briede, A.; et al. European phenological response to climate change matches the warming pattern. *Glob. Chang. Biol.* **2006**, *12*, 1969–1976. [CrossRef]
78. Steltzer, H.; Post, E. Seasons and Life Cycles. *Science* **2009**, *324*, 886–887. [CrossRef]
79. Jin, H.X.; Jonsson, A.M.; Olsson, C.; Lindstrom, J.; Jonsson, P.; Eklundh, L. New satellite-based estimates show significant trends in spring phenology and complex sensitivities to temperature and precipitation at northern European latitudes. *Int. J. Biometeorol.* **2019**, *63*, 763–775. [CrossRef]
80. Li, H.B.; Wang, C.Z.; Zhang, L.J.; Li, X.X.; Zang, S.Y. Satellite monitoring of boreal forest phenology and its climatic responses in Eurasia. *Int. J. Remote Sens.* **2017**, *38*, 5446–5463. [CrossRef]

Article

Effects of Environmental Factors on the Changes in MODIS NPP along DEM in Global Terrestrial Ecosystems over the Last Two Decades

Zhaoqi Wang [1,*], Hong Wang [1], Tongfang Wang [1], Lina Wang [1], Xiaotao Huang [2,3], Kai Zheng [1] and Xiang Liu [1]

[1] State Key Laboratory of Plateau Ecology and Agriculture, Qinghai University, Xining 810016, China; wanghong@qhu.edu.cn (H.W.); wangtongfang@qhu.edu.cn (T.W.); wanglina@qhu.edu.cn (L.W.); zhengkai2019@qhu.edu.cn (K.Z.); liuxiang12@mails.ucas.ac.cn (X.L.)

[2] Key Laboratory of Restoration Ecology for Cold Regions Laboratory in Qinghai, Northwest Institute of Plateau Biology, Chinese Academy of Sciences, Xining 810008, China; xthuang@nwipb.cas.cn

[3] Key Laboratory of Adaptation and Evolution of Plateau Biota, Chinese Academy of Sciences, Xining 810008, China

* Correspondence: wangzhaoqi@qhu.edu.cn; Tel.: +86-15910851420

Abstract: Global warming has exerted widespread impacts on the terrestrial ecosystem in the past three decades. Vegetation is an important part of the terrestrial ecosystem, and its net primary productivity (NPP) is an important variable in the exchange of materials and energy in the terrestrial ecosystem. However, the effect of climate variation on the spatial pattern of zonal distribution of NPP has remained unclear over the past two decades. Therefore, we analyzed the spatiotemporal patterns and trends of MODIS NPP and environmental factors (temperature, radiation, and soil moisture) derived from three sets of reanalysis data. The moving window method and digital elevation model (DEM) were used to explore their changes along elevation gradients. Finally, we explored the effect of environmental factors on the changes in NPP and its elevation distribution patterns. Results showed that nearly 60% of the global area exhibited an increase in NPP with increasing elevation. Soil moisture has the largest uncertainty either in the spatial pattern or inter-annual variation, while temperature has the smallest uncertainty among the three environmental factors. The uncertainty of environmental factors is also reflected in its impact on the elevation distribution of NPP, and temperature is still the main dominating environmental factor. Our research results imply that the carbon sequestration capability of vegetation is becoming increasingly prominent in high-elevation regions. However, the quantitative evaluation of its carbon sink (source) functions needs further research under global warming.

Keywords: net primary productivity (NPP); global warming; digital elevation model (DEM); uncertainty

Citation: Wang, Z.; Wang, H.; Wang, T.; Wang, L.; Huang, X.; Zheng, K.; Liu, X. Effects of Environmental Factors on the Changes in MODIS NPP along DEM in Global Terrestrial Ecosystems over the Last Two Decades. *Remote Sens.* **2022**, *14*, 713. https://doi.org/10.3390/rs14030713

Academic Editors: Xuejia Wang, Tinghai Ou, Wenxin Zhang and Youhua Ran

Received: 16 December 2021
Accepted: 31 January 2022
Published: 2 February 2022

1. Introduction

Since the nineteenth century, the global near-surface temperature has continued to increase according to the Fifth Assessment Report of the Intergovernmental Panel on Climate Change (IPCC). The temperature in the last 10 years after 2000 resulted in the hottest 10 years in history, and 1983–2012 was the hottest 30 years in the past 800 years. The widespread impact of global warming has caused a series of negative ecological consequences, such as drought [1,2], melting [3], rising sea levels [4], and frequent extreme climates [5,6]. The accelerating global warming has become a major challenge that is restricting the sustainable development of human society [7–9].

Vegetation is an important part of the terrestrial ecosystem and plays a crucial role in sequestering carbon and mitigating climate change [10]. The vegetation ecosystem is found to be more vulnerable and sensitive to climate change than the other ecosystems. As a variable that reflects the efficiency of vegetation fixation and conversion of light energy, net primary productivity (NPP) is widely used in the monitoring of vegetation dynamics [11,12]. It is related to life activities such as vegetation growth, development, and reproduction, and it also provides an indispensable material basis for the life activities of other biological members in the entire ecosystem. Most of the studies focus on the spatiotemporal changes in NPP [13,14] and the impact of climate and human activities on NPP [15,16], modeling organic carbon storage with NPP as input data, and changes in carbon sources and sinks of the terrestrial ecosystem [12,17–19]. Studies on the changes in NPP with increasing elevation gradients (EG) are mainly conducted in local regions, and all come from instantaneous surveys [20–24]. Therefore, we still lack the knowledge and a full picture of the changes in NPP along EG at the global scale.

Typically, NPP declines with increasing elevation, which can largely be explained by the decreasing temperature as elevation increases [20,22]. However, the impact of temperature on the NPP of vegetation may be more significant in high-elevation areas with the intensification of global warming [25,26], because the warming rate in high-elevation areas is often greater than that in low-elevation areas [25,27]. Thus far, the warming rate has been intensifying at the global scale for decades [26,28–30], which is likely to cause a completely different spatial pattern of NPP on EG [31]. A few studies have focused on determining the spatial pattern of changes in vegetation greenness along EG in recent years [29,32,33], and their results indicated that the signal of the vegetation greenness increases with increasing elevation is found at the global scale and regional scale. However, vegetation greenness is not completely equal to vegetation productivity, and it does not directly participate in the process of the carbon cycle. This situation raises a scientific question as to whether the elevation pattern of NPP has changed under the influence of global warming.

Considering the importance of the effect of environmental variation on vegetation NPP, and the defects and gaps in current scientific research, we analyzed the inter-annual variation of MODIS NPP product (MOD17A2HGF, version 6.1) and environmental factors (air temperature, solar radiation, and soil moisture that derived from the reanalysis data of ERA5, MERRA2, and NCEP2). We calculated the changes in NPP, temperature, radiation, and soil moisture along EG by using the digital elevation model (DEM), which was also used to verify whether the elevation pattern of NPP has been altered under environmental variation. Furthermore, we analyzed the effect of environmental factors on the spatiotemporal changes and elevation distribution of the NPP from 2001 to 2020. We hope that the results of this study can provide references for the evaluation of terrestrial ecosystem carbon source and sink functions and the improvement and development of carbon cycle models. This study will contribute to our understanding of the impact of environmental factors on the elevational distribution of NPP in recent decades and will help us formulate strategies for mitigating climate change. We also expect that the outcomes of this study can provide references for the evaluation of terrestrial ecosystem carbon sink (source) functions and the improvement of current carbon cycle models.

2. Materials and Methods

2.1. Datasets

2.1.1. NPP Data

The NPP data used in this study come from the MODIS (Moderate Resolution Imaging Spectroradiometer) remote sensing product of MOD17A3HGF Version 6.1 [34], which will be generated at the end of each year when the entire yearly 8-day MOD15A2H is available. MOD17A3HGF has two data fields named Npp_500m and Npp_QC_500m, which represent NPP data and their quality control (QC). The poor-quality inputs were cleaned from the 8-day leaf area index and the fraction of photosynthetically active radiation based on the QC label for every pixel. The data type of Npp_QC_500m is an unsigned 1-byte integer (uint8) with a valid range from 0 to 100 (Units = 100%). Finally, we use the QC file to select pixels with good quality to participate in the analysis. The annual NPP, with a spatial resolution of 500 m, is derived from the sum of all 8 days of NPP in a certain year from 2001 to 2020. We further converted the coordinate system to World Geodetic System 1984 and resampled the spatial resolution to 0.008° (≈1 km) by using the bicubic method to match the resolution of the DEM data.

2.1.2. DEM Data

The Shuttle Radar Topography Mission (SRTM) is a joint project between the National Geospatial-Intelligence Agency and the National Aeronautics and Space Administration. The SRTM30_PLUS DEM dataset with a spatial resolution of 30-arc second (≈1 km) was developed by the Scripps Institute of Oceanography, University of California San Diego [35]. The coordinate system is the World Geodetic System 1984. The land data are based on the 1 km averages of topography derived from the United States Geological Survey (USGS) SRTM30 (30 arc-sec) grided DEM data product created with data from the NASA SRTM [36]. Global 30 Arc-Second Elevation (GTOPO30) data are used for high latitudes where SRTM data are not available. GTOPO30 is a global DEM with a horizontal grid spacing of 30 arc seconds (approximately 1 km) [37]. SRTM has been extensively used and provides a good representation of the topography [38], which performs better than Advanced Spaceborne Thermal Emission and Reflection Radiometer (ASTER) Global Digital Elevation Model (GDEM) (the spatial resolution is 500 m) in terms of micro-topography, hydrologic-network and structural information characterization [39,40]. Considering the high quality of the SRTM data and the computational efficiency of the data, we chose SRTM30_PLUS as the DEM data in this study.

2.1.3. Reanalysis Data

We selected three sets of reanalysis data (Table 1) to study the spatiotemporal changes of environmental factors (temperature, radiation, and soil moisture), their distribution along EG, and the uncertainties of the correlation between environmental factors and NPP. The three sets of reanalysis data are the fifth generation of European ReAnalysis (ERA5), the recent Modern-Era Retrospective Analysis for Research and Applications version 2 (MERRA2), and National Centers for Environmental Prediction Reanalysis 2 (NCEP2). These reanalysis data can completely cover the data period of MODIS NPP and contain three kinds of environmental factor data at the same time. In data processing, we convert the unit of temperature from K to °C by subtracting 273.15, and the unit of radiation was converted from W/m^2 to MJ/m^2 because the unit of absorbed photosynthetically active radiation is MJ/m^2 in the NPP simulation. Therefore, we can intuitively observe the magnitude of the spatial distribution of radiation. The soil moisture has different names in each reanalysis dataset, and we selected the variables with similar meanings and consistent units. Furthermore, we supplemented the monthly data with mean annual temperature (MAT), radiation (MAR), and soil moisture (MASM) raster data. Then, we extracted these environmental data from 2001 to 2020 for analysis. To keep the coordinate system of the reanalysis datasets consistent with the DEM, we set the resolution of the MERRA2 and NCEP2 data to 0.5° and 1.875°, respectively. After that, we upscaled the DEM data to match them, and then performed the moving window operation. We explain the reason for this in Section 2.2. However, trend and correlation analysis were performed at the resolution of 0.008° to keep the data resolution consistent throughout the main text, and also to be consistent with the NPP data.

2.2. Calculation of the Changes in NPP and Environmental Factors along EG

The changes in NPP along EG (NPP_{EG}) were calculated in three steps that are the same as those for the change of environmental factors along EG [29,32,33]. Step 1: We selected a 9×9 moving window to traverse the DEM and NPP (or environmental factors) raster data (Figure 1). The size of the moving window determines the amount of data used for analysis for each moving step. A previous study indicated that the difference caused by the window size is extremely small and recommended choosing a window size of 9×9 for calculation [29,33]. We also tested the window sizes of 5, 7, 9, and 11, and then expanded the size of the moving window to 19, 29, 39, and 49. We found that the changes in the areas of positive NPP_{EG} were 0.0039% from the window size of 5 to 11, while the changes in the areas were reached 1% from the window size of 19 to 49. Therefore, we followed the recommendation of previous studies and chose a window size of 9×9 for the study. Step 2: The data extracted by the moving window have two dimensions. Thus, we need to reduce the dimensions and arrange them in pairs for linear regression analysis. A 9×9 moving window contains 81 data after dimensionality reduction. Step 3: We assign the regression coefficient obtained by linear regression analysis to the center pixel of the moving window and traverse the entire raster data accordingly to obtain the distribution map of the change of the variable along EG. The statistical significance of the slope is tested by a *t*-test. It should be noted that we set the NPP value of the ocean to no data. Therefore, when the moving window contains pixels of the ocean, the center pixel of the moving window will be filled with no data. In addition, we upscaled the DEM data to match the original resolution of environmental factors derived from three reanalysis datasets, such that we can avoid the regression analysis was established between the same environmental factors (coarser grid) and different DEM data (fine grid) in a moving window.

Table 1. Summary of reanalysis data products used in this study.

Name	Category	Timespan	Spatial Resolution	Temporal Resolution	References	Data Acquisition
ERA5	Temperature (K)	1981–present	$0.1° \times 0.1°$	Monthly	[41]	https://cds.climate.copernicus.eu (accessed on 10 December 2021)
	Radiation (W/m^2)					
	Volumetric soil water (m^3 m^{-3})					
MERRA2	Temperature (K)	1980–present	$0.5° \times 0.625°$	Monthly	[42]	https://disc.gsfc.nasa.gov/ (accessed on 10 December 2021)
	Radiation (W/m^2)					
	Water root zone (m^3 m^{-3})					
NCEP2	Temperature (K)	1979–present	$1.875° \times 1.904°$	Monthly	[43]	https://psl.noaa.gov/ (accessed on 10 December 2021)
	Radiation (W/m^2)					
	Volumetric Soil Moisture (m^3 m^{-3})					

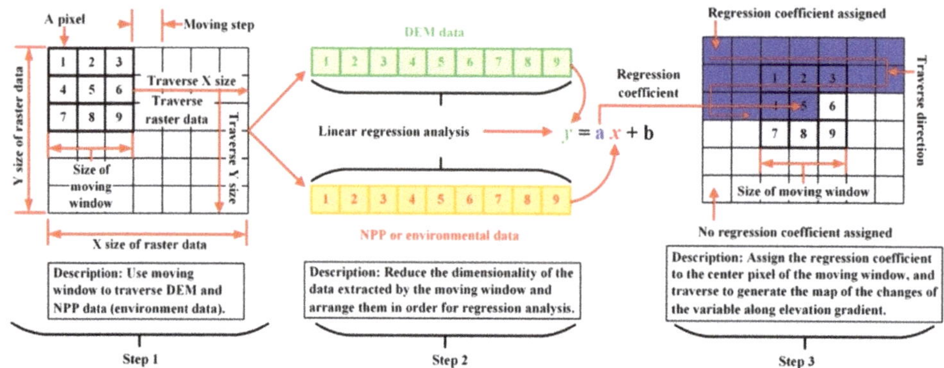

Figure 1. Schematic diagram of the calculation of the NPP or environmental factors changes along EG. We take a 3×3 window as an example.

2.3. Trend Analysis

We use the ordinary least squares regression method [15,44–46] to determine the variation of NPP and environmental factors in time series:

$$\text{Slope} = \frac{n \times \sum_{i=1}^{n} i \times (\text{VAR})_i - \left(\sum_{i=1}^{n} i\right)\left(\sum_{i=1}^{n} (\text{VAR})_i\right)}{n \times \sum_{i=1}^{n} i^2 - \left(\sum_{i=1}^{n} i\right)^2}, \tag{1}$$

where VAR can be NPP, MAT, MAR, MASM, and their changes along EG; i is the sequence number of the year (from 2001 to 2020); n represents the total number of years. The significance of the trend was determined by the *T*-test.

2.4. Correlation Analysis

We use Pearson correlation analysis to calculate the correlation between NPP and environmental factors (MAT, MAR, and MASM). The correlation coefficient (r) of the two variables can be calculated by Equation (2).

$$r_{xy} = \frac{\sum_{i=1}^{n} (x_i - \bar{x})(y_i - \bar{y})}{\sqrt{\sum_{i=1}^{n} (x_i - \bar{x})^2 \sum_{i=1}^{n} (y_i - \bar{y})^2}}, \tag{2}$$

The correlation coefficient between NPP and MAT was taken as an example. Variables xi and yi denote the NPP and MAT in year i, respectively; $\bar{x}$ and $\bar{y}$ are the mean values of NPP and MAT from 2001 to 2020, respectively.

3. Results

3.1. Trends of NPP and Environmental Factors

NPP exhibited an increasing trend (1.75 gC/m^2/yr) from 2001 to 2020 (Figure 2a), which was almost below the mean value for the first 10 years, and above the mean value for the next 10 years. The trends of the three kinds of reanalysis data showed that MAT was significantly increased. However, ERA5 was significantly decreased, while both MERRA2 and NCEP2 showed significant increasing trends. By contrast, the trends of MAR exhibited a decreasing trend, but only MERRA2 was statistically significant (Figure 2b–d). The spatial pattern of mean annual NPP is consistent with our perception that tropical forests have the highest NPP, while NPP is relatively low in alpine and arid regions (Figure 3a). The spatial pattern of the NPP trend showed that 70.63% of the global areas (16.21% were statistically significant) presented an increasing trend from 2001 to 2020, which were mainly found in central and western Canada, parts of central and northern China, central and southern South America, and central Africa (Figure 3b). By contrast, the regions with a decreasing trend of NPP occupied 29.37% of the global areas, and 3.02% of them were statistically significant, which were mainly distributed in northern South America.

The uncertainties of the trend of environmental factors were reflected in the spatial distribution (Figure 3.2). MAT showed a significant increasing trend in high−latitude regions of Asia and Europe, but opposite trends were found in central Africa (Figure 3.2a,d,g). The consistency of the positive trend of MAT accounted for 56.80% of the global area, whereas the areas with a negative trend accounted for 7.28%, mainly in northeastern North America, the Iranian plateau, and parts of the region across central India (Figure 3.2 j). Strong spatial heterogeneities are observed in the spatial pattern of MAR trends of the three sets of reanalysis data (Figure 3.2b,e,h). The inconsistent trends of MAR accounted for 61% of the global area. By contrast, the consistency of negative trends (27.55%) was greater than that of positive trends (11.65%). It is difficult to find areas where the MASM trend is consistent (Figure 3.2c,f,i), and MASM trends are inconsistent across most regions of the globe (73.84%) (Figure 3.2l). We found similar spatial patterns of MAT, MAR, and MASM (Figure 5). MAT is lower in high−elevation areas (Figure 5a,d,g). However, the cold climate of the Qinghai–Tibet Plateau is not reflected in NCEP2 MAT data (Figure 5g). Therefore, we speculated that the NCEP2 MAT probably fail to capture the temperature distribution.

The higher uncertainty of MAT includes western North America, the Andes in South America, Greenland, the Mongolian Plateau, and the Qinghai–Tibetan Plateau (Figure 5j). MAR in high latitudes is lower than that in other regions (Figure 5b,e,h), which is highly uncertain in Greenland and the Sahara Desert (Figure 5k). MASM has relatively high values in the high latitudes of the northern hemisphere, the eastern United States, southeastern China, and tropical forests (Figure 5c,f,i). By contrast, it is lower in the western United States, the Mongolian plateau, the Sahara Desert, southern Africa, and Australia. The higher uncertainty of MASM is mainly found in the high latitudes of the northern hemisphere, eastern China, and tropical forests (Figure 5l).

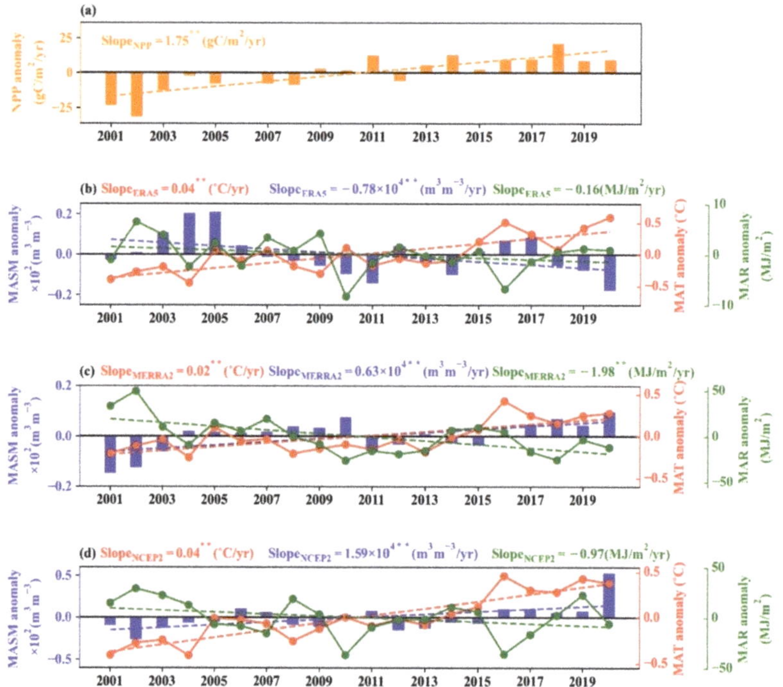

Figure 2. Trend of the NPP (**a**), MAT, MAR, and MASM (**b–d**) anomaly from 2001 to 2020. The dashed straight lines denote the trendlines in subfigure (**a**). Asterisks (******) denote that the slope is statistically significant at the 0.01 level.

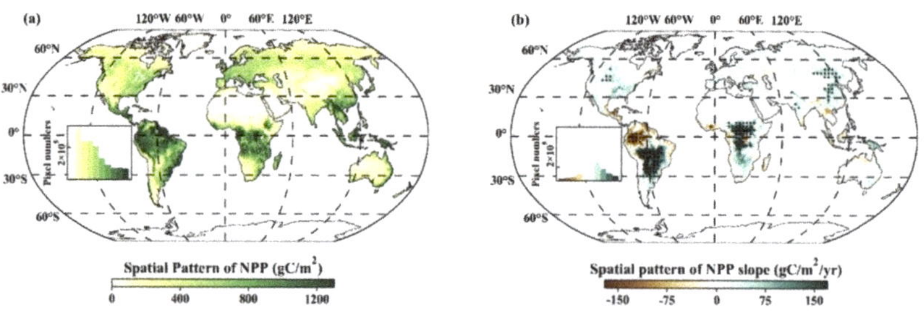

Figure 3. Spatial pattern of global NPP (**a**) and its trend (**b**) from 2001 to 2020. The frequency of the uncertainty value is in the left of each subfigure. The regions with black dots in (**b**) indicate that the trend is statistically significant at the 0.05 level.

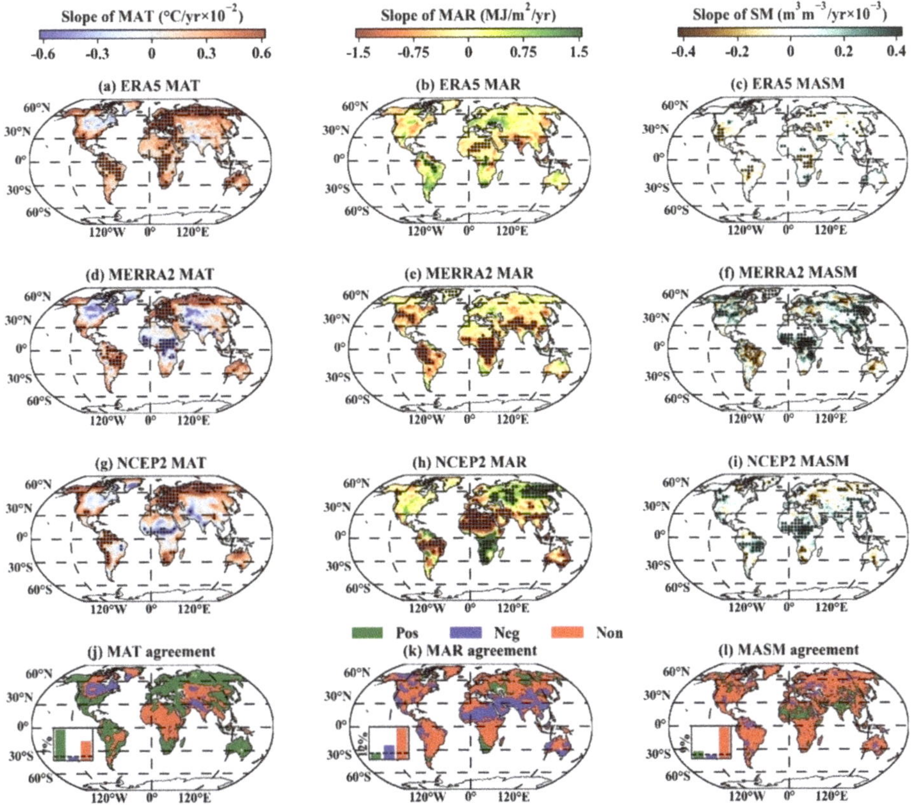

Figure 4. Spatial patterns of the trends of MAT (**a,d,g**), MAR (**b,e,h**), MASM (**c, f,i**) (derived from ERA5, MERRA2, and NCEP2), and their spatial consistencies (**j,k,l**) from 2001 to 2020. The frequency of the uncertainty value is in the left of subfigures (**j,k,l**). The regions with black dots indicate that the trend is statistically significant at the 0.05 level. "Pos" and "Neg" denote the regions with the positive and negative agreement, respectively. "Non" denotes that the regions have not reached an agreement (the same meaning thereafter).

3.2. Elevational Distribution of NPP and Environmental Factors

Changes in NPP and its slope along EG (NPP_{EG} and NPP_{EG}^{slope}) are shown in Figure 6. The positive NPP_{EG} accounted for 59.98% of the global area, and 31.32% of them reached a significance level of 0.05. The positive NPP_{EG} was scattered around the Great Lakes, the eastern foothills of the Andes, the eastern Brazilian plateau, the sub−Saharan African continent, the Indochina Peninsula, and eastern Australia. By contrast, the region with negative NPP_{EG} accounted for 40.02%, and the significantly negative region occupied 11.18%. These regions are mainly found in northeastern North America, southern South America, and Central Asia (Figure 6a). The NPP_{EG}^{slope} showed strong spatial heterogeneity during the study period (Figure 6b), and there is no obvious spatial distribution feature exists. The region with positive NPP_{EG}^{slope} accounted for 53.88%. By contrast, 46.12% of the global area had negative NPP_{EG}^{slope}.

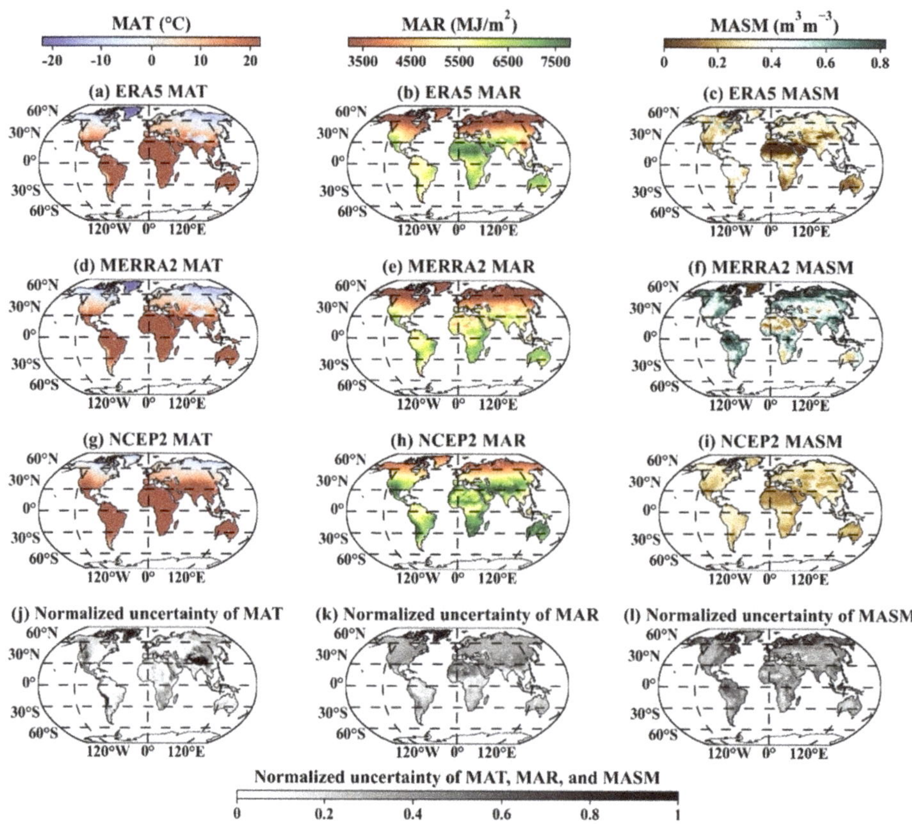

Figure 5. Spatial patterns of the MAT (**a,d,g**), MAR (**b,e,h**), MASM (**c,f,i**) (derived from ERA5, MERRA2, and NCEP2), and their spatial consistencies (**j,k,l**) from 2001 to 2020.

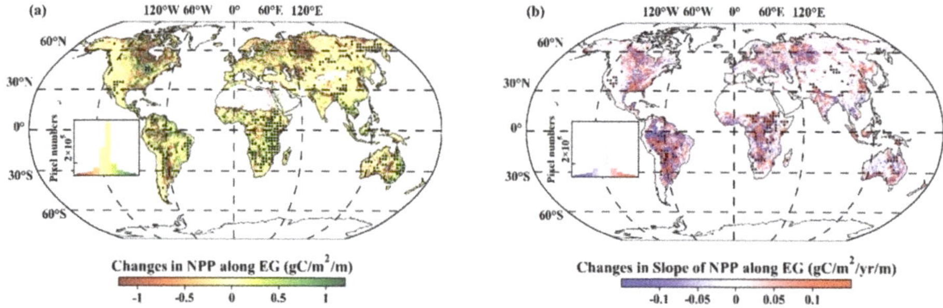

Figure 6. Spatial pattern of the changes in NPP (**a**) and its slope (**b**) along EG from 2001 to 2020. The value frequency is in the left of each subfigure. The regions with black dots indicate that the trend is statistically significant at the 0.05 level.

The spatial pattern of MAT, MAR, and MASM along EG (MAT_{EG}, MAR_{EG}, $MASM_{EG}$) is shown in Figure 7. MAT_{EG}^{ERA5} and MAT_{EG}^{ERA2} follow the natural law that temperature decreases with the increase in elevation, and the area of the two reached 84.36% and 83.67% (Figure 7a,d), with the statistically significant area even reaching up to 78.57% and

74.35%, respectively. By contrast, only 32.90% of the area of $\text{MAT}_{\text{EG}}^{\text{NCEP2}}$ conforms to the natural law, which indicates that the temperature data of NECP2 are likely to be wrong in elevational distribution (Figure 7g). Therefore, we excluded NCEP2 when calculating the spatial uncertainty of MAT. The results showed that the spatial consistency of $\text{MAT}_{\text{EG}}^{\text{ERA5}}$ and $\text{MAT}_{\text{MERRA2}}$ reached 78.84%. MAR_{EG} does not show obvious spatial distribution characteristics (Figure 7b,e,h), and the areas with positive MAR_{EG} (59.73%) were larger than those with negative MAR_{EG} (40.27%). The spatial consistency of the MAR_{EG} is 39.06%, which is mainly distributed in central and eastern North America, most of Europe, and central and western Russia. However, the inconsistent regions still accounted for most of the global area (Figure 7k). The areas with positive MASM_{EG} have strong spatial heterogeneity, although it occupied $(64.09 \pm 7.02)\%$ of the global area (Figure 7c,f,i), and the spatial inconsistency of MASM_{EG} is up to 62.73% (Figure 7l).

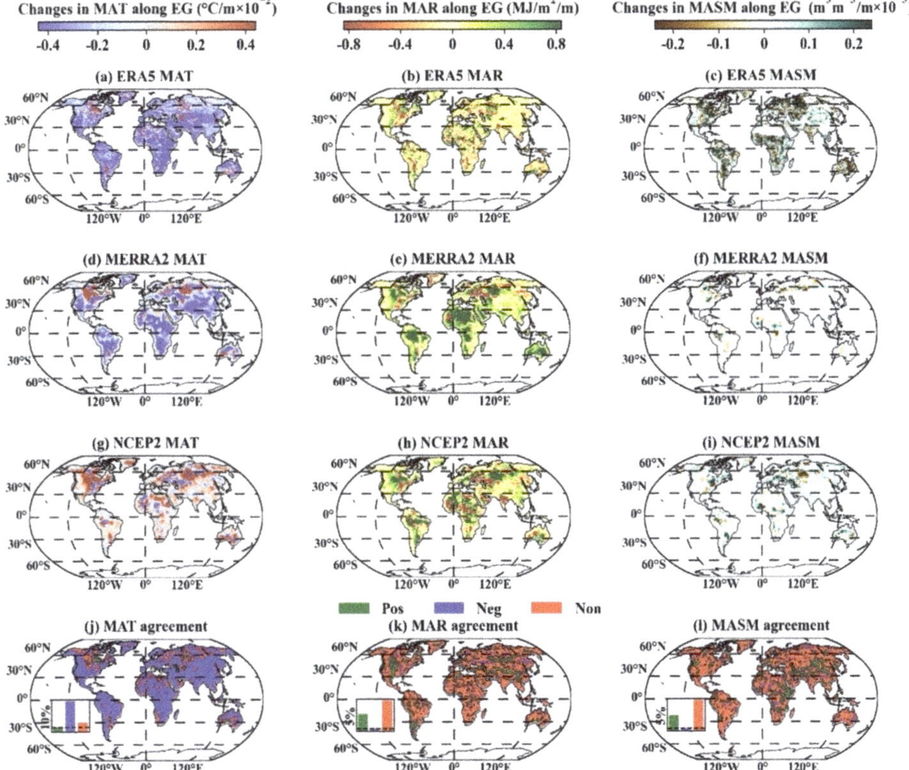

Figure 7. Spatial patterns of MAT (**a,d,g**), MAR (**b,e,h**), MASM (**c,f,i**) (derived from ERA5, MERRA2, and NCEP2) along EG and their spatial consistencies (**j,k,l**) from 2001 to 2020. The frequency of the uncertainty value is in the left of subfigures (**j,k,l**).

We further investigated the spatial pattern of changes in the slope of MAT, MAR, and MASM along EG ($\text{MAT}_{\text{EG}}^{\text{slope}}$, $\text{MAR}_{\text{EG}}^{\text{slope}}$ and $\text{MASM}_{\text{EG}}^{\text{slope}}$) (Figure 8). The positive and negative $\text{MAT}_{\text{EG}}^{\text{slope}}$ accounted for 52.03% and 47.97% of the global area, respectively (Figure 8a,d,g). The area with consistent changes of $\text{MAT}_{\text{EG}}^{\text{slope}}$ accounts for 56.28% of the global area, and the remaining regions are highly uncertain (Figure 8j). The areas with positive $\text{MAR}_{\text{EG}}^{\text{slope}}$ occupied $(45.50 \pm 3.00)\%$ of the global area, which is less than that of negative MAR_{EG}^{slope} $(54.50 \pm 3.00)\%$ (Figure 8b,e,h). Anyway, MAR_{EG}^{slope} is highly uncertain in most of the global area (68.11%) (Figure 8k). The areas of positive MASM_{EG}^{slope} (51.49%) are slightly more than that of negative MASM_{EG}^{slope} (48.51%) (Figure 8c,f,i). The inconsistent region covered up to 72.09% of the global area (Figure 8l).

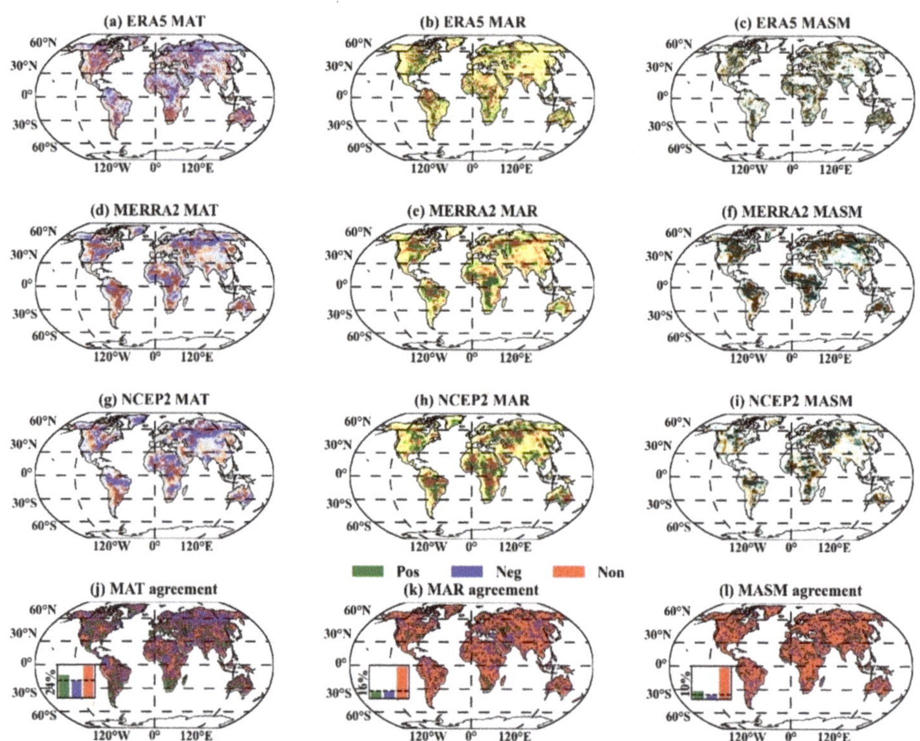

Figure 8. Spatial patterns of the slope of MAT (**a,d,g**), MAR (**b,e,h**), and MASM (**c,f,i**) (derived from ERA5, MERRA2, and NCEP2) along EG and their spatial consistencies (**j,k,l**) from 2001 to 2020. The frequency of the uncertainty value is in the left of subfigures (**j,k,l**).

3.3. Effect of Environmental Factors on the Elevational Distribution of NPP

Figure 9 illustrates spatial patterns of the dominating environmental factor on the changes in NPP from 2001 to 2020. ERA5 data showed that MAT and MAR seem to be the dominating factors on the changes in NPP in the north of 30°N, central and northern parts of South America, central Africa, and Southeast Asia, whereas MASM mainly affects the changes in NPP in the central and eastern United States, the eastern part of the Brazilian plateau, southern Africa, and Australia (Figure 9a). By contrast, the effects of environmental factors of MERRA2 on the changes in NPP have a clear spatial distribution pattern. MAT mainly affects the changes in NPP in the Qinghai–Tibet Plateau and parts of central South America. MAR is the dominant factor in the high latitudes of the northern hemisphere, southern South America, and Southeast Asia. MASM has a wider range of influence, including southern North America, Eurasia from 30°N to 60°N, the Indian peninsula, most of Africa, and Australia (Figure 9b). The MAT and MAR of NCEP2 are the dominant factors in the changes in NPP in the north of 30°N. MAR is also the main environmental factor that affects NPP changes in northwestern South America, central Africa, and Southeast Asia. The areas where MASM presented a dominant environmental factor include most of the United States, northern South America and most of Brazil, the southern edge of the Sahara

Desert and southern Africa, most of Eurasia from 30°N to 60°N, the Indian Peninsula, and Australia (Figure 9c). MASM has the highest spatial consistency (25.15%). These regions are mainly distributed in the United States, eastern and southern South America, southern Africa, Eastern Europe, Central Asia, parts of East Asia, India, and Australia, followed by MAR (9.74%) and MAT (7.70%), which can be found in high latitudes in the northern hemisphere (Figure 9d).

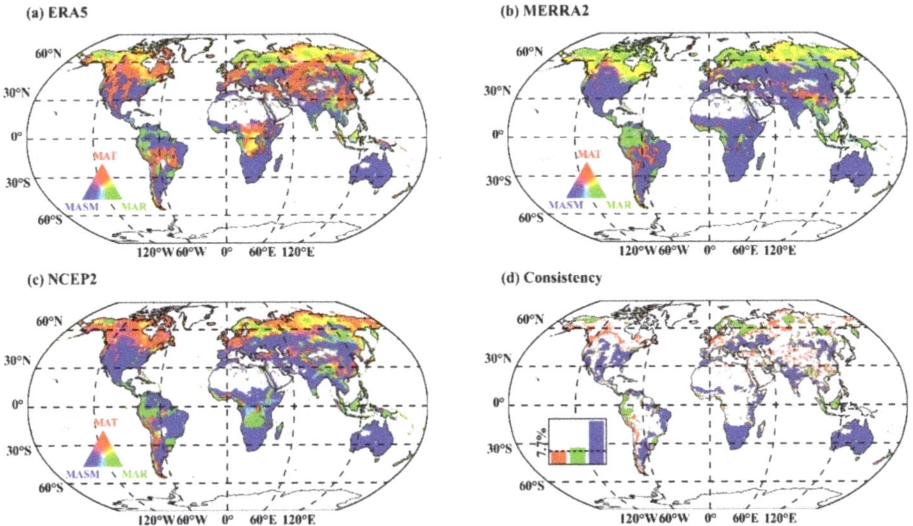

Figure 9. Spatial patterns of the effect of MAT, MAR, and MASM derived from ERA5 (**a**), MERRA2 (**b**), and NCEP2 (**c**) on the changes in NPP and their spatial consistencies (**d**) from 2001 to 2020.

We found that NPP has an obvious turning point at the elevation of 3050 m and divided the elevation into high and low gradients based on the turning point (Figure A1 in Appendix A). Then, we explored the impact of environmental factors on NPP in high− and low−elevation areas (Figure 10a). The three kinds of reanalysis datasets showed that MAT still maintained the highest R^2 in the three environmental factors. We further fitted the regression equation curve and uncertainty range of NPP and environmental factors (Figure 10). The effect of MAT on NPP is linear and non−linear at low and high elevations, respectively (Figure 10a). The R^2 between MAT and NPP exceeds 0.9 and is statistically significant at the 0.01 level. In general, the uncertainty of the effect of MAT on NPP gradually decreases as the temperature increases, and it decreases more at low elevations. The effect of MAR on NPP is nonlinear at both low and high elevations (Figure 10b). NPP increases as MAR increases at low elevations, whereas opposite trends were observed at high elevations. The uncertainty of the impact of MAR on NPP gradually decreases with the increase in MAR at low elevations. By contrast, the uncertainty does not decrease significantly as the MAR decreases at high elevations (Figure 10c). We found a linear and nonlinear effect of MASM on NPP at low and high elevations, respectively. NPP increases as MASM increases at low elevations. However, the upper and lower limits of uncertainty trend toward the opposite direction. The same situation occurs at high elevations that larger uncertainty of the effect of MASM on NPP, and MASM remains stable with the increase in NPP (Figure 10d). The R^2 of all fitted equations is statistically significant at both high and low elevations.

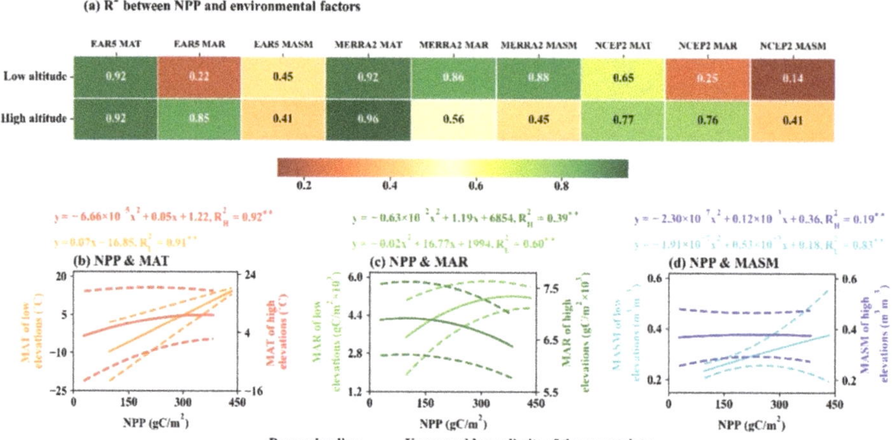

Figure 10. Regression curves, functions, and their R^2 between NPP and environmental factors. (**a**) represent R^2 between NPP and environmental factors. The regression curves and functions can be found in (**b–d**). Subscript L of R^2 means low elevations, and H means high elevations, and ** denotes the R^2 is significant at the 0.01 level.

4. Discussion

4.1. Elevational Patterns of Environmental Factors and NPP

In this article, we use the moving window method to explore the changes in NPP and environmental factors along EG at the global scale. Unlike field surveys, this method is based on remote sensing and DEM data, which can quickly and economically monitor the changes of variables along EG. In addition, the moving window method used in this article is intended to investigate the change of NPP and environmental factors at the relative EG within the moving window, which is different from the studies on the absolute EG conducted through a field survey. The accuracy of this method can be indirectly proved by the natural law that the temperature decreases with the increase in elevation. Nearly 80% of the regions showed a decrease in temperature with increasing elevation, and more than 70% of the regions are statistically significant. This result means that the moving window method can effectively monitor the change of the variable in the EG and has high accuracy. Undeniably some regions (although very few) have a local micro-climate, the temperature of which will not decrease with the increase in elevation. We can also evaluate the data quality of temperature based on the natural law of temperature–elevation and the moving window method. For example, NCEP2 fails to capture the elevation distribution of MAT because it is contrary to the natural law that the temperature decreases with the increase in elevation. Unfortunately, we are still unable to assess the data quality of radiation and soil moisture because the spatial distributions of the reanalysis data and their trends are highly uncertain, and there is no corresponding natural law to follow. Typically, NPP decreased with increasing elevation (negative NPP_{EG}) because of the limitation of low temperature. However, we found that the areas with positive NPP_{EG} accounted for 59.98% of the global area, which means the elevation pattern of NPP in our perception has changed. The spatial pattern of NPP slope demonstrated that, compared with NPP in low-elevation areas, NPP in high-elevation areas has a higher increase rate, and this phenomenon has become more common worldwide.

4.2. Uncertainty of Environmental Factors and Their Effect on NPP

We found that MAR and MASM had larger uncertainties than MAT in the inter-annual variations and the spatial distribution along EG. Such a large uncertainty makes it difficult to assess their effect on the changes in NPP and further forms a lower spatial correlation consistency. Factors affecting MAR mainly include cloud height, thickness, shape, and aerosols. [47]. The factors that affect soil moisture include soil texture [48], the data assimilation process [49], the sample size of observation data [50], and land use type [51].

We found that the effect of radiation on NPP is nonlinear regardless of whether it is in high- or low-elevation regions. The positive correlation between NPP and radiation in low-elevation is likely to be that, as the elevation increases, temperature and water inhibit the photosynthesis of plants, thereby reducing the use of photosynthetically active radiation. Moreover, the radiation available for plants decreases, thereby forming a non-linear positive correlation between NPP and radiation. However, the non-linear negative correlation between radiation and NPP in high-elevation regions suggests that radiation is not the main driving force for changes in NPP. We suppose that radiation cannot be fully utilized by plants in high-elevation areas, and temperature is still the main environmental factor that determines changes in NPP along EG. The effect of soil moisture on NPP shows a non-linear positive correlation in low elevation areas, but it is highly uncertain because of the large differences in the trend of soil moisture among ERA5, NCEP2, and MERRA2. Anyway, the positive correlation between NPP and soil moisture has also been confirmed in the arid region [52], whereas soil moisture in high-elevation areas is higher than that in low-elevation areas, it has not caused an increase in NPP. Therefore, we hold the view that soil moisture determines the lower limit of NPP, and the upper limit of NPP is determined by both soil moisture and temperature.

Overall, the temperature is still the dominating climate factor that determines the spatial patterns of NPP along EG, and it is generally positively correlated with NPP whether in high- or low-elevation areas. However, NPP is more sensitive to temperature in low-elevation areas, which indicates that temperatures are more conducive to vegetation photosynthesis in this region. By contrast, the temperature gradually decreases with the increase in high-elevation, and the photosynthesis of vegetation is also limited by low temperature. Thus, NPP appears to rise first and gradually stabilizes. The effect of temperature on NPP along EG has been confirmed in regional-scale studies [53,54].

4.3. Implications for Carbon Cycle

NPP is the source of energy in the carbon cycle of the terrestrial ecosystem. Our research shows that nearly 60% of the global area exhibits an increase in NPP with increasing elevation, which means that vegetation in high-elevation areas plays an increasingly prominent role in absorbing atmospheric CO_2 and mitigating climate change. However, this increase in NPP is caused by global climate change, especially the increase in temperature in high mountain areas [33], which improves the photosynthesis capacity of plants. However, when the temperature exceeds the optimal temperature of the plants, it will inhibit the photosynthesis of plants, and even cause the death of local species because they cannot adapt to the rapid warming. This kind of research on vegetation degradation caused by warming has been widely reported [12]. What we want to emphasize is that this phenomenon of increasing NPP with elevation may be beneficial to CO_2 fixation in the short term, and an uncontrolled increase in temperature will inevitably lead to vegetation degradation and even ecosystem collapse. The increase in temperature will also cause the increase in plant autotrophic respiration and soil heterotrophic respiration, and the CO_2 produced by the respiration process is directly discharged out of the vegetation–soil system. With the high degree of uncertainty in soil respiration, much uncertainty exists in the quantitative evaluation of the carbon source and sink functions of the ecosystem.

Environmental factors have a strong influence on the terrestrial carbon cycle. Temperature is the basic input data to establish its impact on the soil carbon cycle. In this study, the difference in the spatial distribution of ERA5 and MERRA2 temperature is very small. However, the algorithm for the effect of temperature on the decomposition of soil organic carbon (shown as a nonlinear positive correlation effect) has slight differences. The differences in the algorithms are reflected in space, mainly located in high latitude areas and the Qinghai–Tibet Plateau with great uncertainty. This condition means that the uncertainty of the temperature algorithm in the alpine region will greatly affect the decomposition of soil organic carbon in the carbon cycle, which in turn affects the spatial distribution of soil organic carbon storage in the alpine region, leading to distortions in the evaluation of the carbon source and sink functions of these regions.

Numerous studies have shown that soil moisture is crucial to the carbon cycle process of terrestrial ecosystems [55–57]. However, the three sets of reanalysis data show that the spatial distribution of soil moisture has great uncertainty. Similarly, there are large discrepancies in the algorithm for the effect of soil moisture on the decomposition of soil organic carbon. The spatial uncertainty of soil moisture involved tropical rain forests and the Sahara Desert [17]. The fundamental reason is that the understanding of the effect of soil moisture on the decomposition of soil organic carbon varies, especially as to whether high-humidity environments will inhibit the decomposition of soil organic matter. An anaerobic environment created by high humidity will inhibit the decomposition of soil organic matter in the carbon cycle model, which will lead to the accumulation of soil organic carbon (high soil organic carbon), and vice versa, leading to the decomposition of soil organic carbon (low soil organic carbon). The uncertainty of the original input data of soil moisture and the algorithm makes it difficult for us to evaluate its effect on the carbon cycle. Considering the importance of soil moisture in the carbon cycle of terrestrial ecosystems, we propose that future studies need to further strengthen the observation of soil moisture and improve its simulation accuracy to more accurately assess the carbon source and sink functions of terrestrial ecosystems.

5. Conclusions

This study analyzed the spatiotemporal changes of the MODIS NPP product and environmental factors (temperature, radiation, and soil moisture data derived from the reanalysis data ERA5, MERRA2, and NCEP2) and their distribution along EG. We also identified the spatial uncertainty of environmental factors and their effects on the elevation distribution of NPP. We found that nearly 60% of the global area presented an increase in NPP with increasing elevation, which implied that the elevation pattern of NPP has changed, and the carbon sequestration capacity of vegetation is increasing elevation. However, soil respiration is likely to increase as well. Quantitatively evaluating the carbon sink (source) function of vegetation remains to be further studied in high-elevation regions. The temperature of NCEP2 failed to capture the alpine environment of the Qinghai–Tibet Plateau, and it does not clearly show the natural law that the temperature decreases with the increase in elevation. Soil moisture has the largest areas of spatial consistency in affecting the spatiotemporal changes in NPP among the three environmental factors. However, its spatial pattern and variation are the most uncertain among the three environmental factors, even though it is essential to the carbon cycle of terrestrial ecosystems. NPP has obvious elevation differentiation with an elevation of 3060 m as the demarcation point, which divides the elevation into low and high. MAT is the main driving force that affects the elevation distribution of NPP, with its effect on NPP exhibiting a significant linear and nonlinear positive correlation at low and high elevations, respectively. The results of this study are expected to contribute to our understanding of the changes in NPP along EG and provide references for the development of terrestrial ecosystem carbon cycle models.

Author Contributions: Conceptualization, Z.W. and X.H.; data curation, Z.W., K.Z. and X.L.; Formal analysis, Z.W.; investigation, Z.W., H.W. and T.W.; methodology, Z.W. and L.W.; writing—original draft, Z.W.; writing—review and editing, Z.W. All authors have read and agreed to the published version of the manuscript.

Funding: This research was funded by the National Natural Science Foundation of China (Grant No. 32160278); the Open Project of State Key Laboratory of Plateau Ecology and Agriculture, Qinghai University (Grant No. 2020-ZZ-14, 2021-KF-08); the National Natural Science Foundation of China

(Grant No. NSFC32001188); the Natural Science Foundation of Qinghai Province of China (Grant No. 2021-ZJ-973Q); the National Natural Science Foundation of China (Grant No. 42067070).

Institutional Review Board Statement: Not applicable.

Informed Consent Statement: Not applicable.

Data Availability Statement: The data presented in this study are openly available in (NASA EOSDIS Land Processes DAAC) at (doi.org/10.5067/MODIS/MOD17A3HGF.061), reference number [33].

Conflicts of Interest: The authors declare no conflict of interest.

Appendix A

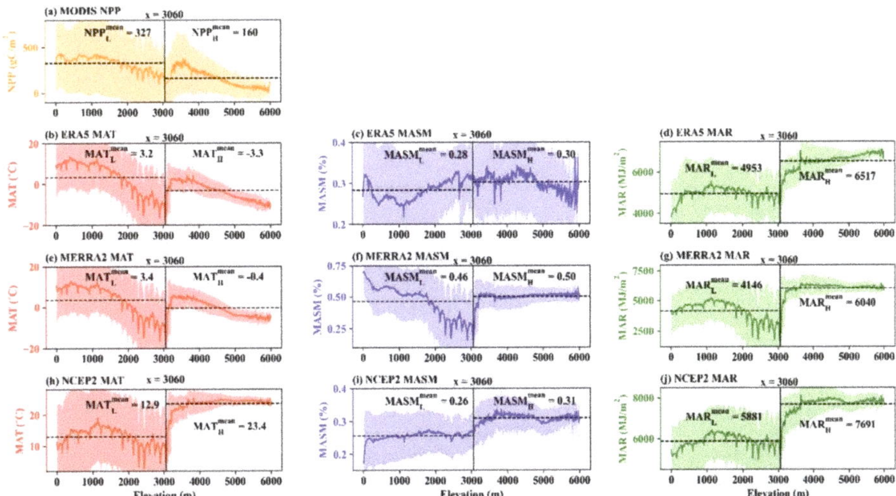

Figure A1. Changes in NPP and environmental factors along EG from 2001 to 2020. The elevation of 3060 m is used to distinguish high and low elevation (the vertical black solid line in Figure (**a–j**)). The dashed line represents the average value of the variables at high and low elevations. The shaded areas with different colors in each figure represent ±1 SD.

References

1. Xu, C.; McDowell, N.G.; Fisher, R.A.; Wei, L.; Sevanto, S.; Christoffersen, B.O.; Weng, E.; Middleton, R.S. Increasing impacts of extreme droughts on vegetation productivity under climate Change. *Nat. Clim. Chang.* **2019**, *9*, 948–953. [CrossRef]
2. Vicente-Serrano, S.M.; Quiring, S.M.; Peña-Gallardo, M.; Yuan, S.; Domínguez-Castro, F. A review of environmental droughts: Increased risk under global warming? *Earth Sci. Rev.* **2019**, *201*, 102953. [CrossRef]
3. Kraaijenbrink, P.D.A.; Bierkens, M.F.P.; Lutz, A.F.; Immerzeel, W.W. Impact of a global temperature rise of 1.5 degrees Celsius on Asia's glaciers. *Nature* **2017**, *549*, 257–260. [CrossRef] [PubMed]
4. Mengel, M.; Levermann, A.; Frieler, K.; Robinson, A.; Marzeion, B.; Winkelmann, R. Future sea level rise constrained by observations and long-term commitment. *Proc. Natl. Acad. Sci. USA* **2016**, *113*, 2597–2602. [CrossRef]
5. Cai, W.; Wang, G.; Dewitte, B.; Wu, L.; Santoso, A.; Takahashi, K.; Yang, Y.; Carréric, A.; McPhaden, M.J. Increased variability of eastern Pacific El Niño under greenhouse warming. *Nature* **2018**, *564*, 201–206. [CrossRef]
6. Wang, G.; Cai, W.; Gan, B.; Wu, L.; Santoso, A.; Lin, X.; Chen, Z.; McPhaden, M.J. Continued increase of extreme El Niño frequency long after 1.5 °C warming stabilization. *Nat. Clim. Chang.* **2017**, *7*, 568–572. [CrossRef]
7. Zhu, Z.; Piao, S.; Yan, T.; Ciais, P.; Bastos, A.; Zhang, X.; Wang, Z. The accelerating land carbon sink of the 2000s may not be driven predominantly by the warming hiatus. *Geophys. Res. Lett.* **2018**, *45*, 1402–1409. [CrossRef]
8. Liu, Y.; Piao, S.; Gasser, T.; Ciais, P.; Yang, H.; Wang, H.; Keenan, T.F.; Huang, M.; Wan, S.; Song, J.; et al. Field-experiment constraints on the enhancement of the terrestrial carbon sink by CO_2 fertilization. *Nat. Geosci.* **2019**, *12*, 809–814. [CrossRef]
9. IPCC. *Climate Change 2013—The Physical Science Basis: Working Group I Contribution to the Fifth Assessment Report of the Intergovernmental Panel on Climate Change*; Cambridge University Press: Cambridge, UK, 2014.

10. Alward, R.D.; Detling, J.K.; Milchunas, D.G. Grassland vegetation changes and nocturnal global warming. *Science* **1999**, *283*, 229. [CrossRef]
11. Cramer, W.; Kicklighter, D.W.; Bondeau, A.; Iii, B.M.; Churkina, G.; Nemry, B.; Ruimy, A.; Schloss, A.L. Comparing global models of terrestrial net primary productivity (NPP): Overview and key results. *Glob. Chang. Biol.* **1999**, *5*, 1–15. [CrossRef]
12. Wang, Z.; Zhang, Y.; Yang, Y.; Zhou, W.; Gang, C.; Zhang, Y.; Li, J.; An, R.; Wang, K.; Odeh, I.; et al. Quantitative assess the driving forces on the grassland degradation in the Qinghai–Tibet Plateau, in China. *Ecol. Inf.* **2016**, *33*, 32–44. [CrossRef]
13. Cuo, L.; Zhang, Y.; Xu, R.; Zhou, B. Decadal change and inter-annual variability of net primary productivity on the Tibetan Plateau. *Clim. Dyn.* **2021**, *56*, 1837–1857. [CrossRef]
14. Mao, F.; Du, H.; Li, X.; Ge, H.; Cui, L.; Zhou, G. Spatiotemporal dynamics of bamboo forest net primary productivity with climate variations in Southeast China. *Ecol. Indic.* **2020**, *116*, 106505. [CrossRef]
15. Zhang, Y.; Wang, Q.; Wang, Z.; Yang, Y.; Li, J. Impact of human activities and climate change on the grassland dynamics under different regime policies in the Mongolian Plateau. *Sci. Total Environ.* **2020**, *698*, 134304. [CrossRef] [PubMed]
16. Wang, Y.; Yue, H.; Peng, Q.; He, C.; Hong, S.; Bryan, B.A. Recent responses of grassland net primary productivity to climatic and anthropogenic factors in Kyrgyzstan. *Land Degrad. Dev.* **2020**, *31*, 2490–2506. [CrossRef]
17. Wang, Z.; Yang, Y.; Li, J.; Zhang, C.; Chen, Y.; Wang, K.; Odeh, I.; Qi, J. Simulation of terrestrial carbon equilibrium state by using a detachable carbon cycle scheme. *Ecol. Indic.* **2017**, *75*, 82–94. [CrossRef]
18. Wang, Z. Estimating of terrestrial carbon storage and its internal carbon exchange under equilibrium state. *Ecol. Model.* **2019**, *401*, 94–110. [CrossRef]
19. Wang, Z.; Chang, J.; Peng, S.; Piao, S.; Ciais, P.; Betts, R. Changes in productivity and carbon storage of grasslands in China under future global warming scenarios of 1.5 °C and 2 °C. *J. Plant Ecol.* **2019**, *12*, 804–814. [CrossRef]
20. Raich, J.W.; Russell, A.E.; Vitousek, P.M. Primary productivity and ecosystem development along an elevational gradient on Mauna Loa, Hawaii. *Ecology* **1997**, *78*, 707–721. [CrossRef]
21. Sierra Cornejo, N.; Leuschner, C.; Becker, J.N.; Hemp, A.; Schellenberger Costa, D.; Hertel, D. Climate implications on forest above- and belowground carbon allocation patterns along a tropical elevation gradient on Mt. Kilimanjaro (Tanzania). *Oecologia* **2021**, *195*, 797–812. [CrossRef]
22. Luo, T.; Pan, Y.; Ouyang, H.; Shi, P.; Luo, J.; Yu, Z.; Lu, Q. Leaf area index and net primary productivity along subtropical to alpine gradients in the Tibetan Plateau. *Glob. Ecol. Biogeogr.* **2004**, *13*, 345–358. [CrossRef]
23. Girardin, C.A.J.; Malhi, Y.; Aragão, L.E.O.C.; Mamani, M.; Huaraca Huasco, W.; Durand, L.; Feeley, K.J.; Rapp, J.; Silva-Espejo, J.E.; Silman, M.; et al. Net primary productivity allocation and cycling of carbon along a tropical forest elevational transect in the Peruvian Andes. *Glob. Chang. Biol.* **2010**, *16*, 3176–3192. [CrossRef]
24. Lin, H.; Zhang, Y. Evaluation of six methods to predict grassland net primary productivity along an altitudinal gradient in the Alxa Rangeland, Western Inner Mongolia, China. *Grassl. Sci.* **2013**, *59*, 100–110. [CrossRef]
25. Rangwala, I.; Miller, J.R. Climate change in mountains: A review of elevation-dependent warming and its possible causes. *Clim. Chang.* **2012**, *114*, 527–547. [CrossRef]
26. Bertrand, R.; Lenoir, J.; Piedallu, C.; Riofrío-Dillon, G.; de Ruffray, P.; Vidal, C.; Pierrat, J.-C.; Gégout, J.-C. Changes in plant community composition lag behind climate warming in lowland forests. *Nature* **2011**, *479*, 517–520. [CrossRef] [PubMed]
27. Qin, J.; Yang, K.; Liang, S.; Guo, X. The altitudinal dependence of recent rapid warming over the Tibetan Plateau. *Clim. Chang.* **2009**, *97*, 321–327. [CrossRef]
28. Pepin, N.; Bradley, R.; Diaz, H.; Baraër, M.; Caceres, E.; Forsythe, N.; Fowler, H.; Greenwood, G.; Hashmi, M.; Liu, X. Elevation-dependent warming in mountain regions of the world. *Nat. Clim. Chang.* **2015**, *5*, 424–430. [CrossRef]
29. Gao, M.; Piao, S.; Chen, A.; Yang, H.; Liu, Q.; Fu, Y.H.; Janssens, I.A. Divergent changes in the elevational gradient of vegetation activities over the last 30 years. *Nat. Commun.* **2019**, *10*, 2970. [CrossRef]
30. Pearson, R.G.; Phillips, S.J.; Loranty, M.M.; Beck, P.S.; Damoulas, T.; Knight, S.J.; Goetz, S.J. Shifts in Arctic vegetation and associated feedbacks under climate change. *Nat. Clim. Chang.* **2013**, *3*, 673–677. [CrossRef]
31. Shen, M.; Zhang, G.; Cong, N.; Wang, S.; Kong, W.; Piao, S. Increasing altitudinal gradient of spring vegetation phenology during the last decade on the Qinghai–Tibetan Plateau. *Agric. For. Meteorol.* **2014**, *189*, 71–80. [CrossRef]
32. Wang, Z.; Liu, X.; Wang, H.; Zheng, K.; Li, H.; Wang, G.; An, Z. Monitoring vegetation greenness in response to climate variation along the elevation gradient in the three-river source region of China. *ISPRS Int. J. Geo. Inf.* **2021**, *10*, 193. [CrossRef]
33. Wang, Z.; Cui, G.; Liu, X.; Zheng, K.; Lu, Z.; Li, H.; Wang, G.; An, Z. Greening of the Qinghai–Tibet plateau and its response to climate variations along elevation gradients. *Remote Sens.* **2021**, *13*, 3712. [CrossRef]
34. Running, S.W.; Zhao, M. MODIS/Terra Net Primary Production Gap-Filled Yearly L4 Global 500 m Sin Grid V061. 2021. Available online: https://lpdaac.usgs.gov/products/mod17a3hgfv006/ (accessed on 14 December 2021). [CrossRef]
35. Becker, J.J.; Sandwell, D.T.; Smith, W.H.F.; Braud, J.; Binder, B.; Depner, J.; Fabre, D.; Factor, J.; Ingalls, S.; Kim, S.H.; et al. Global bathymetry and elevation data at 30 arc seconds resolution: SRTM30_PLUS. *Mar. Geod.* **2009**, *32*, 355–371. [CrossRef]
36. Farr, T.G.; Rosen, P.A.; Caro, E.; Crippen, R.; Duren, R.; Hensley, S.; Kobrick, M.; Paller, M.; Rodriguez, E.; Roth, L.; et al. The Shuttle Radar Topography Mission. *Rev. Geophys.* **2007**, *45*, RG2004. [CrossRef]
37. Miliaresis, G.C.; Argialas, D.P. Segmentation of physiographic features from the global digital elevation model/GTOPO30. *Comput. Geosci.* **1999**, *25*, 715–728. [CrossRef]

38. Grohmann, C.H. Evaluation of TanDEM-X DEMs on selected Brazilian sites: Comparison with SRTM, ASTER GDEM and ALOS AW3D30. *Remote Sens. Environ.* **2018**, *212*, 121–133. [CrossRef]
39. Bannari, A.; Kadhem, G.; El-Battay, A.; Hameid, N. Comparison of SRTM-V4. 1 and ASTER-V2. 1 for accurate topographic attributes and hydrologic indices extraction in flooded areas. *J. Earth Sci. Eng.* **2018**, *8*, 8–30. [CrossRef]
40. Han, H.; Zeng, Q.; Jiao, J. Quality assessment of TanDEM-X DEMs, SRTM and ASTER GDEM on selected Chinese sites. *Remote Sens.* **2021**, *13*, 1304. [CrossRef]
41. Muñoz-Sabater, J.; Dutra, E.; Agustí-Panareda, A.; Albergel, C.; Arduini, G.; Balsamo, G.; Boussetta, S.; Choulga, M.; Harrigan, S.; Hersbach, H.; et al. ERA5-Land: A state-of-the-art global reanalysis dataset for land applications. *Earth Syst. Sci. Data* **2021**, *13*, 4349–4383. [CrossRef]
42. Gelaro, R.; McCarty, W.; Suárez, M.J.; Todling, R.; Molod, A.; Takacs, L.; Randles, C.A.; Darmenov, A.; Bosilovich, M.G.; Reichle, R.; et al. The Modern-Era Retrospective Analysis for Research and Applications, Version 2 (MERRA-2). *J. Clim.* **2017**, *30*, 5419–5454. [CrossRef]
43. Kanamitsu, M.; Ebisuzaki, W.; Woollen, J.; Yang, S.-K.; Hnilo, J.; Fiorino, M.; Potter, G. NCEP-DOE AMIP-II reanalysis (R-2). *Bull. Am. Meteorol. Soc.* **2002**, *83*, 1631–1644. [CrossRef]
44. Gang, C.; Wang, Z.; Chen, Y.; Yang, Y.; Li, J.; Cheng, J.; Qi, J.; Odeh, I. Drought-induced dynamics of carbon and water use efficiency of global grasslands from 2000 to 2011. *Ecol. Indic.* **2016**, *67*, 788–797. [CrossRef]
45. Zhang, Y.; Zhang, C.; Wang, Z.; Chen, Y.; Gang, C.; An, R.; Li, J. Vegetation dynamics and its driving forces from climate change and human activities in the Three-River Source Region, China from 1982 to 2012. *Sci. Total Environ.* **2016**, *563*, 210–220. [CrossRef] [PubMed]
46. Chen, J.; Yan, F.; Lu, Q. Spatiotemporal variation of vegetation on the Qinghai–Tibet plateau and the influence of climatic factors and human activities on vegetation trend (2000–2019). *Remote Sens.* **2020**, *12*, 3150. [CrossRef]
47. Wang, K.; Dickinson, R.E. Global atmospheric downward longwave radiation at the surface from ground-based observations, satellite retrievals, and reanalyses. *Rev. Geophys.* **2013**, *51*, 150–185. [CrossRef]
48. Yang, S.; Li, R.; Wu, T.; Hu, G.; Xiao, Y.; Du, Y.; Zhu, X.; Ni, J.; Ma, J.; Zhang, Y.; et al. Evaluation of reanalysis soil temperature and soil moisture products in permafrost regions on the Qinghai-Tibetan Plateau. *Geoderma* **2020**, *377*, 114583. [CrossRef]
49. Shangguan, W.; Dai, Y.; Liu, B.; Zhu, A.; Duan, Q.; Wu, L.; Ji, D.; Ye, A.; Yuan, H.; Zhang, Q.; et al. A China data set of soil properties for land surface modeling. *J. Adv. Modeling Earth Syst.* **2013**, *5*, 212–224. [CrossRef]
50. Du, Y.; Li, R.; Zhao, L.; Yang, C.; Wu, T.; Hu, G.; Xiao, Y.; Zhu, X.; Yang, S.; Ni, J.; et al. Evaluation of 11 soil thermal conductivity schemes for the permafrost region of the central Qinghai-Tibet Plateau. *Catena* **2020**, *193*, 104608. [CrossRef]
51. Zucco, G.; Brocca, L.; Moramarco, T.; Morbidelli, R. Influence of land use on soil moisture spatial–temporal variability and monitoring. *J. Hyd.* **2014**, *516*, 193–199. [CrossRef]
52. Yue, D.; Zhou, Y.; Guo, J.; Chao, Z.; Guo, X. Relationship between net primary productivity and soil water content in the Shule River Basin. *Catena* **2022**, *208*, 105770. [CrossRef]
53. Guan, X.; Shen, H.; Li, X.; Gan, W.; Zhang, L. Climate control on net primary productivity in the complicated mountainous area: A case study of Yunnan, China. *IEEE J. Sel. Top. Appl. Earth Obs. Remote Sens.* **2018**, *11*, 4637–4648. [CrossRef]
54. Xu, H.-J.; Zhao, C.-Y.; Wang, X.-P. Elevational differences in the net primary productivity response to climate constraints in a dryland mountain ecosystem of northwestern China. *Land Degrad. Dev.* **2020**, *31*, 2087–2103. [CrossRef]
55. Humphrey, V.; Berg, A.; Ciais, P.; Gentine, P.; Jung, M.; Reichstein, M.; Seneviratne, S.I.; Frankenberg, C. Soil moisture–atmosphere feedback dominates land carbon uptake variability. *Nature* **2021**, *592*, 65–69. [CrossRef] [PubMed]
56. Green, J.K.; Seneviratne, S.I.; Berg, A.M.; Findell, K.L.; Hagemann, S.; Lawrence, D.M.; Gentine, P. Large influence of soil moisture on long-term terrestrial carbon uptake. *Nature* **2019**, *565*, 476–479. [CrossRef] [PubMed]
57. Stocker, B.D.; Zscheischler, J.; Keenan, T.F.; Prentice, I.C.; Peñuelas, J.; Seneviratne, S.I. Quantifying soil moisture impacts on light use efficiency across biomes. *N. Phytol.* **2018**, *218*, 1430–1449. [CrossRef] [PubMed]

remote sensing

Article

The Forest Change Footprint of the Upper Indus Valley, from 1990 to 2020

Xinrong Yan [1,2] and Juanle Wang [1,2,3,4,*]

1   State Key Laboratory of Resources and Environmental Information System, Institute of Geographic Sciences and Natural Resources Research, Chinese Academy of Sciences, Beijing 100101, China; yanxr@lreis.ac.cn
2   University of Chinese Academy of Sciences, Beijing 100049, China
3   China-Pakistan Earth Science Research Center, Islamabad 45320, Pakistan
4   Jiangsu Center for Collaborative Innovation in Geographical Information Resource Development and Application, Nanjing 210023, China
*   Correspondence: wangjl@igsnrr.ac.cn; Tel.: +86-010-6488-8016

Abstract: The upper Indus Valley is the most important and vulnerable water tower in the South Asian subcontinent, which provides a vital water supply for 230 million people in the basin. Forests play an important role in water conservation in this region, and the security of upstream forests forms the foundation downstream water and food security. However, a big challenge is to effectively monitor the dynamics of the forest in this region. Thus, we used the LandTrendr spectral-temporal segmentation algorithm combined with 8203 scenes of multi-source remote sensing data to study the forest change footprint in the upper Indus Valley. The overall accuracy of LandTrendr extraction for forest disturbance and recovery was 86.01%, and the Kappa coefficient was 0.73. The results showed the following: (1) From 1990 to 2020, the area of forest recovery was 1.01% more than that of disturbance, 70% of disturbance occurred between 1990 and 2001, and 60% of recovery occurred between 1999 and 2012. (2) Although the overall trend of forest disturbance and recovery was balanced, there were significant differences in forest management status among the different regions. Nepal has the highest forest stability, India has the largest area of forest disturbance, and Pakistan and China have the largest areas of forest recovery. (3) India's Himachal Pradesh and Jammu and Kashmir are the two provinces with the largest disturbed areas, primarily due to grazing, fires, and commercial tree planting. Pakistan's North-West Frontier, Azad Kashmir, and China's Tibet Ali region were major contributors to the recovery, which was driven by afforestation policies in both countries. This study provides an important data base and monitoring method for planning land and forest use in Indus Valley countries, protecting fragile environments, and promoting policies for the Sustainable Development Goals.

Keywords: forest disturbance; forest recovery; footprint information; LandTrendr spectral-temporal segmentation algorithm; upper Indus Valley

---

Citation: Yan, X.; Wang, J. The Forest Change Footprint of the Upper Indus Valley, from 1990 to 2020. *Remote Sens.* **2022**, *14*, 744. https://doi.org/10.3390/rs14030744

Academic Editor: Wenxin Zhang

Received: 20 December 2021
Accepted: 3 February 2022
Published: 5 February 2022

Publisher's Note: MDPI stays neutral with regard to jurisdictional claims in published maps and institutional affiliations.

1. Introduction

Forests are the main component of terrestrial ecosystem and the largest "carbon pool" on land, which plays an important role in regulating climate and mitigating global warming [1]. Monitoring forest disturbance and recovery has received abundant attention over the past few decades, especially to identify the important role of forests in curbing climate warming and achieving "carbon neutrality" [2]. The United Nations Sustainable Development Goal (SDG) 15 calls for the protection, restoration, and promotion of the sustainable use of terrestrial ecosystems and sustainable management of forests as well as a series of assessment indicators [3]. Forest disturbance is the main factor that affects forest growth, structure, and function. This disturbance is largely due to the degradation and disappearance of forests caused by environmental changes such as drought, hurricanes, geological disasters, and various human activities [4,5]. Forest recovery involves a series of favorable conditions that can promote the growth of forests, including internal and external factors. In areas with more favorable climatic conditions, forest disturbances can be recovered through succession of forest communities or internal self-regulation of ecosystems [6]. In high latitude and ecologically fragile areas, the implementation of forest protection policies plays a positive role in forest recovery [7,8]. Using the Landsat pixel scale, forest disturbance was defined as partial or complete disappearance of the forest canopy. Forest recovery is defined as the inverse ratio of forest loss, which represents an increase in forest cover of the tree canopy [9].

The upper Indus Valley is one of the most critical water towers in the world [10], and ecosystem security in this region is directly linked to the welfare of 230 million individuals downstream. The Karakoram, Hindu Kush, Ladakh, and Himalayas in the upper reaches form diverse forest ecosystems, which are important carbon sinks and water conservation areas. The security of forest ecosystems is related to the security of water and agricultural resources in the Indus Valley basin. However, forest disturbances in this region have been a concern in recent decades. Studies have shown that the annual deforestation rate in this area reaches 2.2% [11]. A previous study by Rashid et al. (2017) showed that the evergreen broad-leaved forest in Kashmir was degraded, and aboveground biomass and carbon storage were lost, thereby adversely affecting carbon fixation in this region [12]. There is evidence that forests in the upper Indus Valley were destroyed, and severe forest disturbances devastated the stability of the original forest ecosystem, leading to catastrophic floods in 1992 and 2010. In response to these problems, the governments of relevant countries have issued several commercial logging bans, but this has not prevented the loss of upstream forest areas. The rate of deforestation and loss of total forest area in this region are still higher than the global average [11,13]. Thus, it is urgent to obtain information on the disturbance and recovery of forests in recent decades through various monitoring methods to provide a reference for the formulation of forest policies in this region.

Forest disturbance and recovery information can be obtained through many technical means such as regular field surveys of forest resources. China, Canada, and Finland have developed relatively complete forest resource survey techniques, but these require higher human and economic costs. The upper Indus Valley includes five countries, and it is difficult to organize forest surveys and obtain forest disturbance and recovery data for all countries. Remote sensing technology is currently considered the most effective and lowest cost method, providing spatiotemporal awareness of forest change on multiple scales [14,15]. The principle of remote sensing technology for monitoring disturbance and recovery is to track the change information of the vegetation canopy spectrum in a time-series image. When the band or index representing the vegetation canopy suddenly or gradually exceeds a certain threshold, the forest is considered disturbed or recovered [16]. Time-series images provide important, consistent, and continuous data sources for the acquisition of forest footprint information in forest-covered areas. Early forest change monitoring based on remote sensing mainly came from the comparative analysis of the interpretation results of two or more temporal remote sensing images. However, because of the limitation of the timeliness of interpretation of samples and other reasons, such methods generally span many years, such as 5 or 10 years, ignoring the short-interval footprint information in the process of forest change, and it is difficult to achieve continuous monitoring of forest change information from a large area.

In the upper Indus Valley, Joshi et al. (2014) interpreted images of the central Himalaya in 1979, 1999, and 2009 by means of artificial visual interpretation, in which forest stands are reduced and forest areas are affected by degradation and isolation [17]. To obtain information on forest loss in the Himalayas, Qamer et al. (2016) used remote sensing images during two time periods from 1990 to 2000 and 2000 to 2010 [18]. Owing to the limitation of computing power or samples, these methods did not describe the forest loss footprint information in detail, and the images of the three periods could not dynamically restore the process of forest change over 21 years. However, most of these were short-term and partial studies, making it difficult to create an overall assessment of the forest status of the entire upper Indus Valley. With the accumulation of remote sensing images and the enhancement of data computing capability, forest change monitoring has gradually developed into a time-series image combined with a change detection algorithm. Since the release of Earth big data computing platforms, such as Google Earth Engine (GEE), increased data computing power has enabled forest change monitoring to use higher spatial resolution data. [19]. In recent years, the big data cloud computing platform for Earth observation has increasingly integrated interference and recovery monitoring algorithms. This includes the vegetation change tracker [20], breaks for additive seasonal and trend [21], vegetation regeneration and disturbance estimates over time [22], continuous change detection and classification [23], and continuous monitoring of land disturbance [24]. These algorithms are widely used in forest event monitoring, such as forest fires, deforestation, and degradation, and are important for recognizing long-term changes in forests.

Based on this background, this research intends to combine Landsat time-series imagery and Earth observation big data cloud computing platform GEE to carry out forest disturbance and recovery monitoring research in the upper Indus Valley. The main objectives include the following: (1) combining the time-spectral segmentation algorithm LandTrendr to obtain the forest change footprint in the upper Indus Valley from 1990 to 2020; and (2) analysis of the results of forest disturbance and recovery to obtain the spatiotemporal cognition of the forest change footprint in the upper Indus Valley.

2. Materials and Methods

2.1. Study Area

The Indus River is a major river in Central Asia and an important source of agricultural irrigation that feeds more than 230 million people. The Indus Valley, ranging from 24° N to 37° N and from 66° E to 82° E (Figure 1), has a diverse geographical environment, covering an area of 618,580.35 km^2. It borders the Karakoram Mountains and Himalayas to the northeast, the Thar Desert in India to the southeast, the Hindu Kush Mountains in Afghanistan to the northwest, the Baluchistan Plateau to the southwest, and the Arabian Gulf to the south. The Indus Valley has a subtropical climate, characterized by a distinct monsoon climate. However, owing to the influence of the high mountains in the northeast, the climate is usually between dry and semi-dry, tropical, and subtropical. The year is divided into four seasons.

The upper Indus Valley is the intersection of many mountain ranges, and the complex geographical environment has led to the development of a variety of forest ecosystems. The natural vegetation mainly includes coniferous forests (subalpine forests, arid temperate forests, humid temperate forests, and subtropical pine forests), shrub forests (arid subtropical broad-leaved forests and arid tropical thorn forests), economic forests, and shelter forests. The forests are widely distributed between 500 and 5500 m elevation, and 90% of the forests are located above 1500 m.

According to a report from the Worldwide Fund for Nature, the population boom coupled with poverty and a lack of awareness have led to illegal and unsustainable logging, excessive logging for fuel and charcoal, and increased small-scale agriculture, which continue to reduce the amount of forest cover in the upper Indus Valley. In addition, forest fires, natural disasters, climate change, pests, and diseases further contribute to the degradation and disappearance of forests [25].

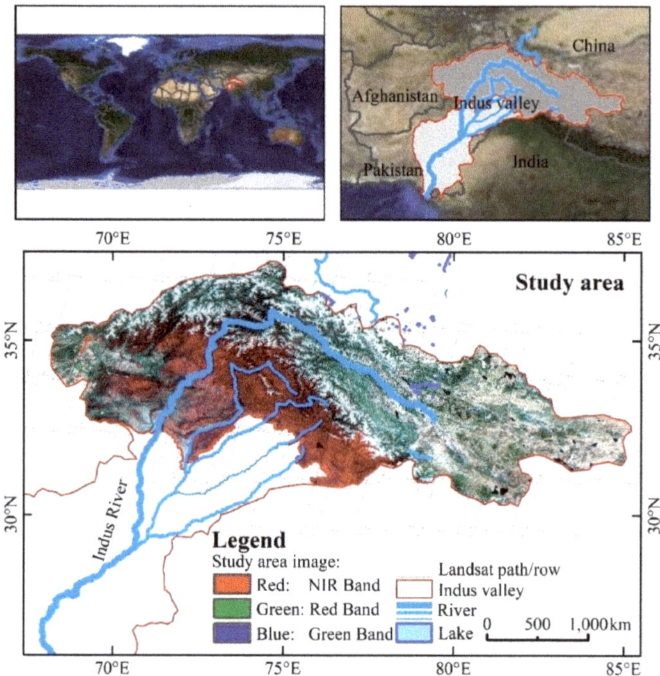

Figure 1. The study area is in the upper Valley of the Indus River. To highlight the forests, we used a Sentinel-2 image synthesized with false colors (the darker red represents the lusher vegetation).

2.2. Data Preprocessing

Table 1 shows the sources and descriptions of multi-source remote sensing data used in this study. Landsat time-series data were used from the United States Geological Survey (USGS), including thematic mapper (TM), enhanced thematic mapper (ETM), and operational land imager (OLI) sensors. The data have the advantages of having high resolution, a short revisit cycle, and easy access and processing. GEE was used as the main data processing platform, which provided access to the atmospheric correction surface reflectance data of all Landsat sensors. We connected all available remote sensing images of vegetation growth season on the GEE platform from July 15 to October 1, which can effectively avoid images caused by seasonal snow in high-altitude mountainous areas. Landsat 7 ETM+ scan line corrector failure (SLC-off) images with data gaps were removed. A total of 8203 remote sensing images were used in this study. In order to reduce the influence of the difference in reflection wavelength between ETM+ and OLI sensors on the results, the harmonic function from OLI to ETM+ was used to process the data consistently [26]. In the high mountains of South Asia, clouds, snow, ice, and mountain shadows have a serious impact on the results of forest change monitoring. We used the image quality assessment band (QA) and the CFMask algorithm to generate masks to remove these four types of elements, and the annual image was generated by the median synthesis method [27]. According to the statistics obtained from annual synthetic images, cloud-free observation has been achieved in 90% of the forest for more than 20 years, and the average cloud-free observation time for each pixel was 25.5 years from 1990 to 2010. The annual synthetic images meet the requirements for LandTrendr algorithm fitting.

Table 1. Datasets used in this study.

Name	Data Source	Data Description
Landsat time-series remote sensing image	USGS, filter and synthesize on GEE	A total of 8203 remote sensing images were used, path and row as shown in Figure 1.
Google Earth image	Google Earth Pro Software	Assist in determining thresholds for different levels of disturbance and recovery; collect samples for accuracy assessment.
Hansen Global Forest Change datasets v1.8 (2000–2020) (HGFC) [9]	University of Maryland	Results from time-series analysis of Landsat images in characterizing global forest extent and change from 2000 through 2020. The data are used for an accuracy assessment.

2.3. Forest Change Footprint Information Extraction

Figure 2 shows the workflow for obtaining forest change footprint information. This process mainly includes data collection and processing, mask construction, algorithm segmentation, different levels of disturbance and recovery classification, and accuracy assessment.

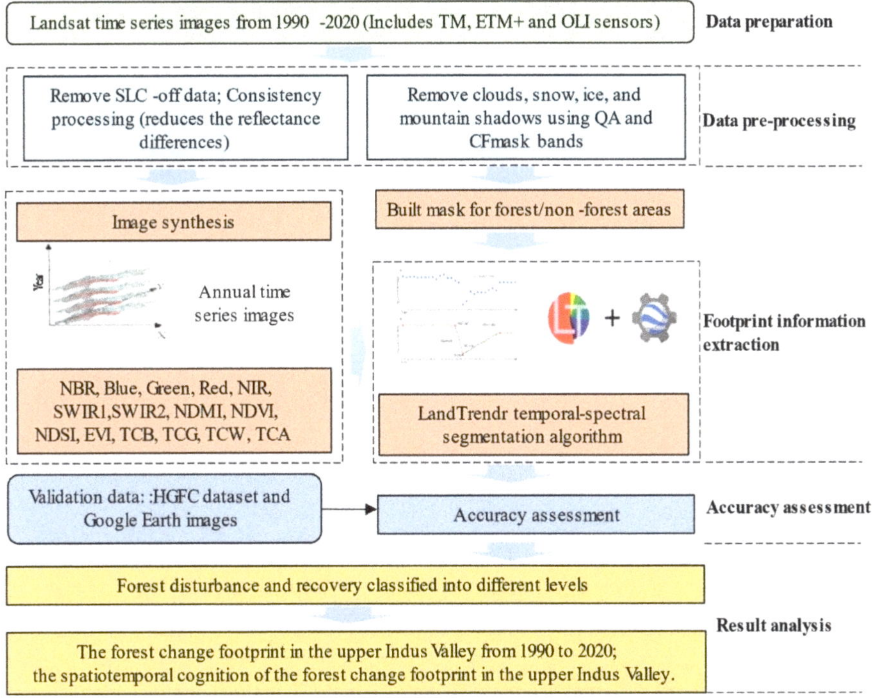

Figure 2. Flowchart of forest change footprint monitoring in the upper Indus Valley.

2.3.1. Built Mask for Forest/Non-Forest Areas

The forest mask can effectively avoid the influence of cultivated land and dense grassland on LandTrendr segmentation results. In this study, the forest area needed to be a maximum union of the forest distribution from 1990 to 2020, which cannot be achieved by using remote sensing images or datasets of a single time phase.

We obtained all available remote sensing images for July 1 to October 1 from 1990 to 2020, and based on the forest extract rule, a normalized difference vegetation index (NDVI) > 0.6 and short-waved infrared radiation (SWIR) < 0.1 generated the annual forest distribution area. Zhu et al. (2012) showed that the NDVI of the forest growth season is always above 0.6, but sometimes other vegetation types, such as crops, grass, and shrubs, could also have an overall forest NDVI value above 0.6. Since forests are generally dark in the SWIR band compared to other vegetation types, a surface reflectance threshold of 0.1 in the overall band 7 excludes other vegetation types that may have high NDVI values [14,28]. The maximum forest union from 1990 to 2020 showed that there was a total forest area of 46,192.25 km^2 in the study area, with a forest coverage fraction of 7.47%.

2.3.2. Detection Methods for Forest Disturbance and Recovery

LandTrendr is a set of spectral-temporal segmentation algorithms based on remote sensing image pixels that are useful for change detection in a time series of moderate-resolution satellite imagery. The time segmentation algorithm is considered an effective method for detecting forest disturbance and recovery. The trajectory of the generated spectral time-series data had almost no interannual signal noise [29]. The algorithm uses a time segmentation strategy based on regression and a point-to-point fitting spectral index as a time function, allowing for the capture of slow-evolving processes such as recovery and unexpected events [30]. Interactive Data Language (IDL) initially implemented LandTrendr, and later Google engineers ported LandTrendr to the GEE platform [19,31].

The GEE framework nearly eliminates the onerous data management and image preprocessing aspects of the IDL implementation. LandTrendr combined with GEE also simplifies tedious data management and image preprocessing by directly accessing geospatial datasets. Figure 3 shows the process of the LandTrendr algorithm for extracting forest disturbances. The discrete original value of the spectral index or band was divided into a series of straight lines and breakpoints. From the segmentation results, the start year, end year, duration, and magnitude of the spectral index or band of forest disturbance can be easily obtained.

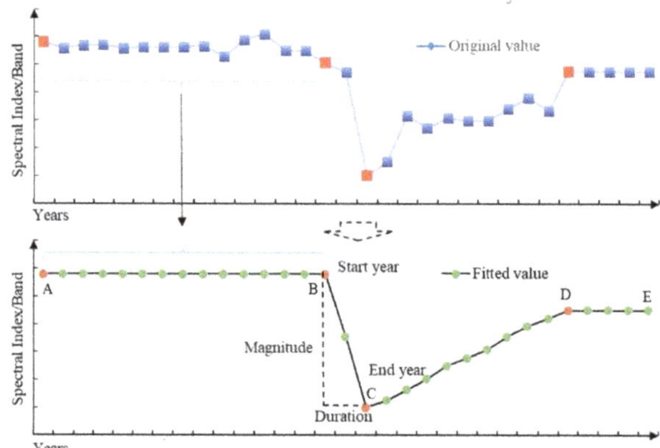

Figure 3. LandTrendr pixel time-series segmentation. Image data were reduced to a single band or spectral index and then divided into a series of straight-line segments by breakpoint (red points) identification. For example, this figure shows the segmentation process of the NDVI, AB, and CD, which represent the steady state of the forest, where BC is fitted as a disturbance event, and CD is an inverse process of BC and is identified as a recovery process.

The LandTrendr algorithm requires a series of surface reflectance bands and spectral indices as inputs when performing spectral-time segmentation. The first band or index was used for the main segmentation, and the other bands were used for fitting to supplement the best results. Cohen et al. (2018) showed that using the 13 bands and indices in Table 2 can effectively reduce the error rate of the LandTrendr segmentation [32]. Table 2 shows the bands and indices used in this study. After obtaining the segmentation results from LandTrendr, we filtered the segmentation results using a forest mask to remove the effects of dense grassland and cropland.

Table 2. Bands/indices used in this study.

Band/Index Name	Description	Calculate Method	Reference
Blue, Green, Red, NIR, SWIR1, SWIR2	Original bands from the TM, ETM+, and OLI.	The SLC-off data were removed, and the harmonization function [26] was used for consistency processing along with ETM+ and OLI data to obtain the annual sequence images of six bands.	/
Normalized burn ratio (NBR)	Normalized difference indices generated by TM, ETM+, and OLI sensors.	NBR = (NIR-SWIR2)/(NIR+SWIR2)	[33]
Normalized difference moisture index (NDMI)	Normalized difference indices generated by TM, ETM+, and OLI sensors.	NDMI = (NIR-SWIR1)/(NIR+SWIR1)	[34]
Normalized difference vegetation index (NDVI)	Normalized difference indices generated by TM, ETM+, and OLI sensors.	NDVI = (NIR-Red)/(NIR+Red)	[35]
Normalized difference snow index (NDSI)	Normalized difference indices generated by TM, ETM+, and OLI sensors.	NDSI = (Green-SWIR1)/(Green+SWIR1)	[36]
Enhanced vegetation index (EVI)	A vegetation index calculated from three bands of TM, ETM+, and OLI sensors.	EVI = 2.5 * ((NIR - Red)/(NIR + 6 * Red - 7.5 * Blue + 1))	[37]
Tasseled cap brightness (TCB)	It is derived from spectral data and the tasselled-cap transformation algorithm. The algorithm can compress spectral data into several bands of physical scene features with minimal information loss.	TCB = 0.2043(Blue) + 0.4158(Green) + 0.5524(Red) + 0.5741(NIR) + 0.3124(SWIR1) + 0.2303(SWIR2)	[38–40]
Tasseled cap greenness (TCG)		TCG = −0.1603(Blue) − 0.2819(Green) − 0.4934(Red)) + 0.7940(NIR) − 0.0002(SWIR1) − 0.1446(SWIR2)	
Tasseled cap wetness (TCW)		TCW = 0.0315(Blue) + 0.2021(Green) + 0.3102(Red)) + 0.1594(NIR) − 0.6806(SWIR1) − 0.6109(SWIR2)	
Tasseled cap wetness (TCA)		TCA = arctan (TCB/TCG)	

2.3.3. Classification of Forest Disturbance and Recovery Levels

Three levels of disturbance and recovery were classified to further study the extent of forest disturbance and recovery (Table 3). In the segmentation results of the LandTrendr algorithm, the "magnitude" band records the variation amplitude of disturbance and recovery in different years. Different levels of disturbance and recovery were obtained by reclassifying this band. Referring to previous study [41], we obtained the thresholds of disturbance and recovery using the visual interaction method between classification results and high-resolution remote sensing images from Google Earth.

Table 3. Different levels of disturbance and recovery [41].

Type	Level	Description	Thresholds
Disturbance	Serious	Land use type changes, deforestation, and forest fires caused complete changes in the surface; for example, the transition from forests to agricultural land and buildings.	$500 <$ magnitude
	Moderate	Due to different reasons such as selective logging, drought, or pests and diseases, the forest has been severely disturbed.	$350 <$ magnitude ≤ 500
	Light	Local changes in the forest, forest disturbances can be reflected in high-resolution images, such as plenter-thinning in the management process.	$200 <$ magnitude ≤ 350
Recovery	Strong	The transition from non-forest land use types to forest types is mainly through afforestation, and areas with better climate conditions can also be self-regulated or community succession through forest ecosystems.	<-500
	Moderate	The opposite process of moderate disturbance indicates the change of forest structure, such as the change from sparse forest to dense forest.	$-500 \leq$ magnitude < -350
	Light	Due to afforestation or self-recovery of forests, the density of forest structure gradually becomes higher, which can be observed in high-resolution images.	$-350 \leq$ magnitude < -200

Serious disturbance and strong recovery categories indicate distinct processes of forest change, such as clear-cutting or barren afforestation. Moderate disturbance, moderate recovery, light disturbance, and light recovery may be changes in forest growth trends, such as varying levels of forest degradation or the growth process from young forests to mature forests [41].

2.3.4. Accuracy Assessment

In this study, we validated the forest disturbance and recovery results using two data sources: HGFC datasets and Google Earth images. The validation data were classified into four categories: disturbance, recovery, and both disturbance and recovery (disturbance + recovery).

- HGFC was produced by the Global Land Analysis and Discovery Laboratory at the University of Maryland, in partnership with the Global Forest Watch, and provides annually updated global-scale forest loss data derived using Landsat time-series imagery (https://storage.googleapis.com/earthenginepartners-hansen/GFC-2020-v1.8/download.html, accessed on 10 December 2021)[9]. This dataset was generated based on multi-source remote sensing data, such as Landsat and MODIS from 2000 to 2020, combined with bagged decision tree classification methods. In contrast to the Hansen product, we extended the time to 1990 and used a change detection algorithm based on spectral-temporal segmentation, a near-automated change detection algorithm that has the advantage of requiring less input data and having a high-detection efficiency compared to the classification method. We downloaded and synthesized HGFC data from 2000 to 2020 for validation, including the disturbance and recovery bands.
- Validation samples were obtained through visual interpretation of high-resolution Google Earth historical images.

3. Results

3.1. Spatial and Temporal Patterns of Forest Disturbance and Recovery

Figure 4 shows the spatial distribution of forest disturbances (including serious, moderate, and light disturbances) in the upper Indus Valley from 1990 to 2020 using transitional colors. Forest disturbances are widely distributed in the middle of the study area, mainly in the southwestern Himalayas and the southern Hindu Kush. Large areas of forest disturbance occurred in the upper reaches of the three rivers of Jhelum, Chenab, and Ravi and on both sides of the Kashmir Valley, largely distributed around cities and farmlands, between 1995 and 2015. In the past 31 years, a total of 13,233.55 km^2 of forest was disturbed, accounting for 28.64% of the total forest area. The average disturbed area was 426.88 km^2 per year, indicating that 0.92% of forests were disturbed at different levels per year on average. The annual disturbed area showed a fluctuating downward trend. In 1990, the disturbed forest area was 1080.17 km^2, and reached a peak of 1410.23 km^2 in 1997. After 2009, the disturbed forest area was less than 300 km^2. The disturbance area from 1990 to 2001 accounted for 70% of the total disturbed area in 31 years, indicating that the disturbance mostly occurred in the first 10 years of the study period.

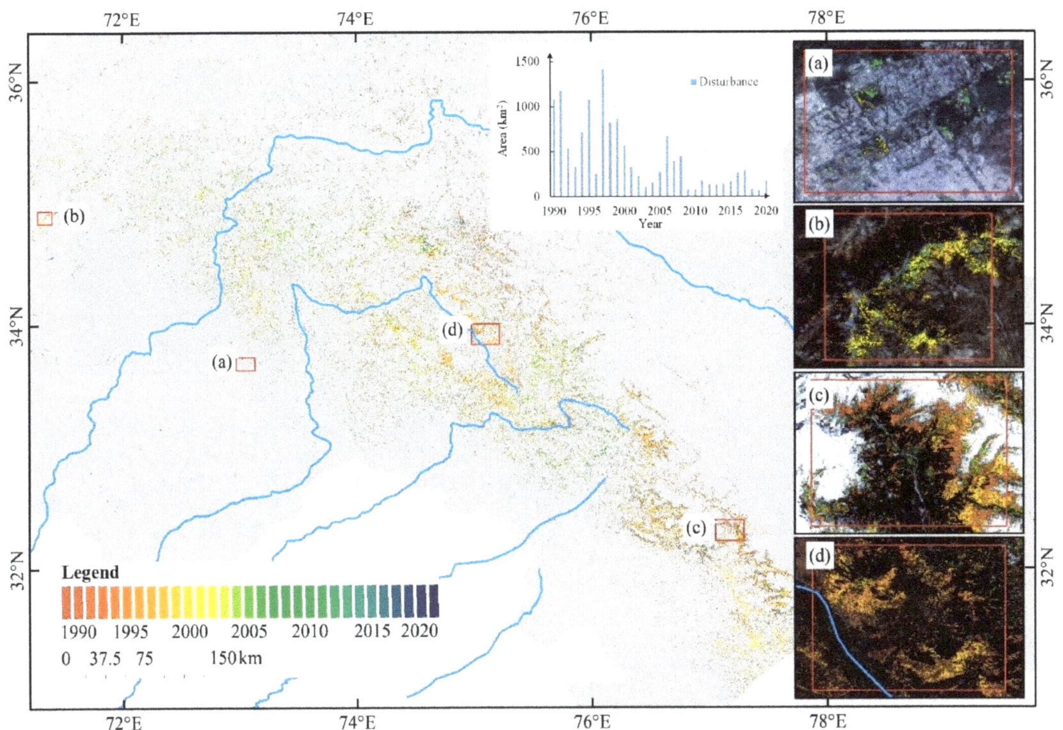

Figure 4. Spatial and temporal footprint distribution map of forest disturbance. Four selected subareas in the figure represent disturbance result examples in different areas, including (**a**) plain area, (**b**) plain and alpine transition area, (**c**) alpine area, and (**d**) river valley area.

Figure 5 shows the spatiotemporal information of forest recovery (including strong, moderate, and light recovery) in the upper Indus Valley from 1990 to 2020, and the recovery area is widely distributed in the middle region as well as the disturbed area. Similar to the distribution of disturbances, the restored areas were also concentrated in the southwestern Himalayas and the southern Hindu Kush. In the past 31 years, a total of 13,702.55 km^2 of forest was restored, accounting for 29.66% of the total forest area. The average annual recovery area was 442.01 km^2, and 0.96% of the forests were restored on average. The annual recovery area showed a hump distribution trend. At the beginning, the recovery forest area was 924 km^2 in 1990, and reached the peak of 1464.34 km^2 in 2000. The area recovered from 1999 to 2010 accounted for 54.78% of the total area of 31 years, indicating that the forest recovery was mainly concentrated in the middle part of the study period.

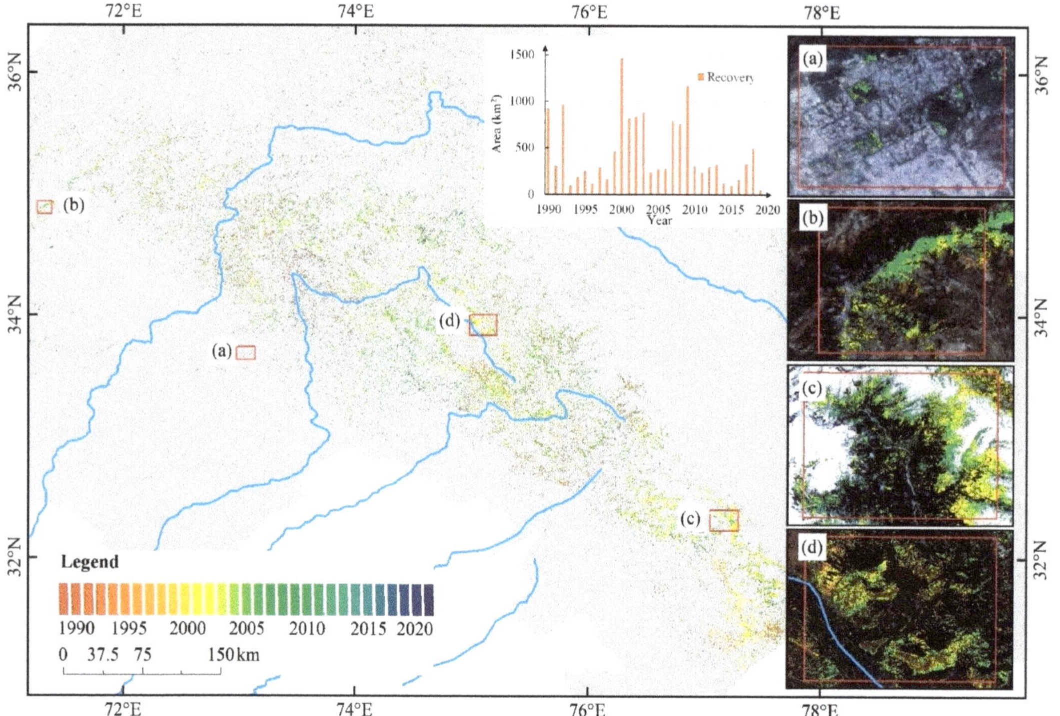

Figure 5. Spatial and temporal footprint distribution map of forest recovery. Four selected sub-areas in the figure represent recovery result examples in different areas, including (**a**) plain area, (**b**) plain and alpine transition area, (**c**) alpine area, and (**d**) river valley area.

From the results of the long-term analysis, it can be concluded that forest recovery and disturbance areas tend to be balanced in the upper Indus Valley. The recovery area was only 469 km^2 larger than the disturbance area, accounting for 1.01% of the forest area in the upper Indus Valley. Figure 6 shows the interannual variation characteristics of disturbance and recovery. In 1990–2000, 70% of forest disturbance occurred, followed by 60% forest recovery in 1999–2012. In terms of spatial distribution, 70% of disturbances in 1990–2000 and 60% of recovery in 1999–2012 were mostly spatially coincident. This indicates that the disturbed forests from 1990 to 2000 were restored in the following 10 to 13 years, whether natural or man-made restorations. For the entire upper Indus Valley, 2012 is an important time node. The cumulative disturbance area since 1990 is equal to the cumulative recovery area this year, reaching a point of equilibrium. After 2012, the cumulative recovery area was slightly greater than the cumulative disturbance area.

3.2. Temporal and Spatial Characteristics of Different Levels of Disturbance and Recovery

Figure 7 shows the spatiotemporal information distribution of 16 types of disturbance and recovery processes, including three different levels of combinations of disturbance and recovery. The results showed that 17.93% of the forests in the study area underwent two processes of disturbance and recovery from 1990 to 2020, especially in the areas bordering forests and farmland in the southern Himalayas and the sides of the Kashmir Valley.

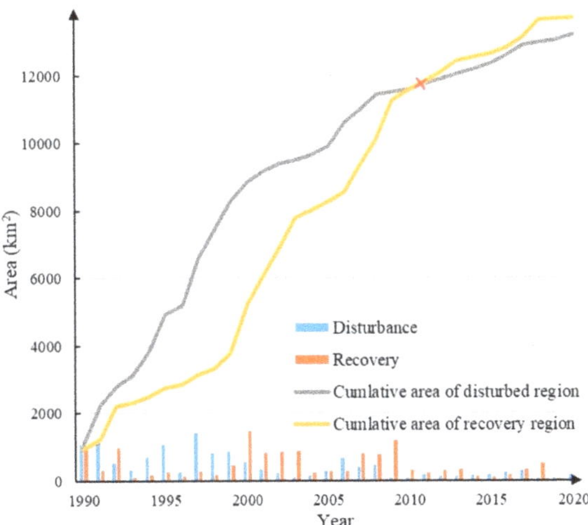

Figure 6. Interannual variation characteristics of disturbance and recovery. The red cross represents the cumulative area of the disturbed region equal to the cumulative area of the recovery region in the upper Indus Valley.

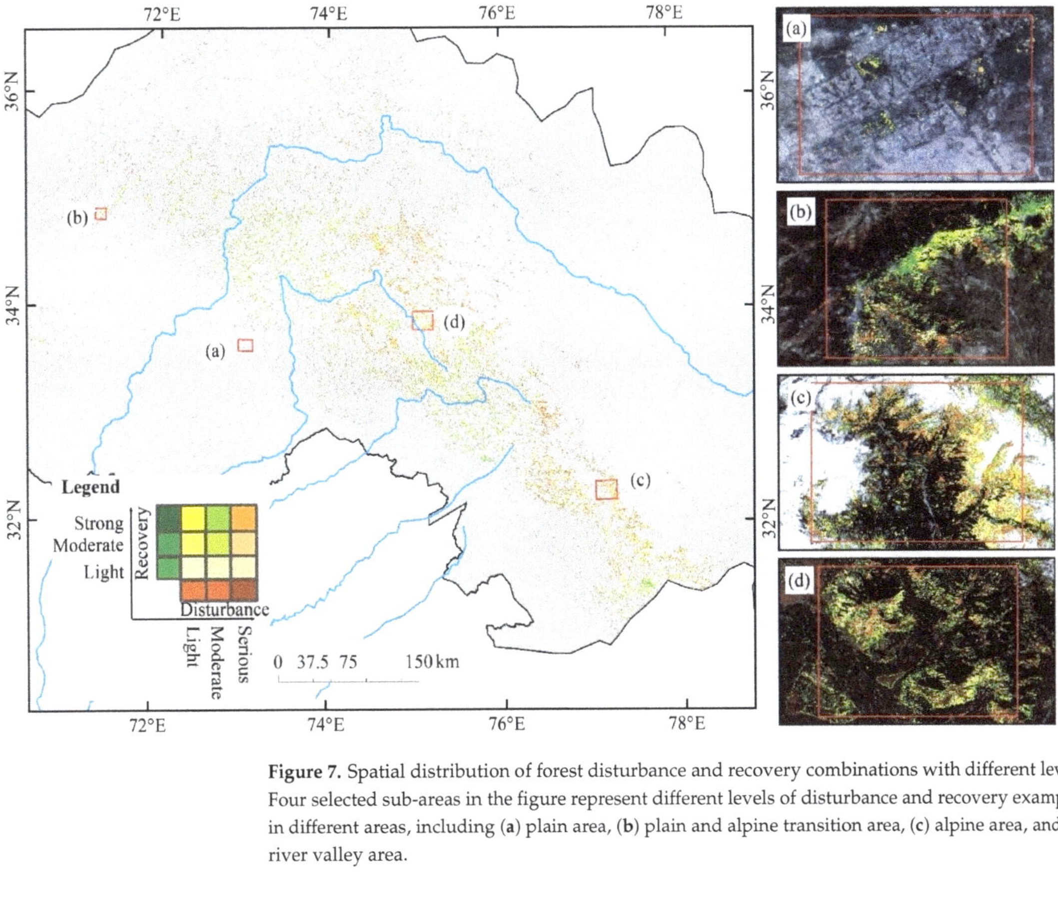

Figure 7. Spatial distribution of forest disturbance and recovery combinations with different levels. Four selected sub-areas in the figure represent different levels of disturbance and recovery examples in different areas, including (**a**) plain area, (**b**) plain and alpine transition area, (**c**) alpine area, and (**d**) river valley area.

Figure 8 shows the proportion of different combinations in the total forest area. A total of 59.71% of the forests remained stable in 31 years without disturbance or recovery. In the period from 1990 to 2020, forests that only underwent disturbances accounted for 10.72%, of which 86.51% were light, 11.62% were moderate, and only 1.86% were serious disturbances. The forests that only underwent recovery accounted for 11.64% of the total forest area, which was 0.92% more than the disturbance area. The results of different recovery level were as follows: light recovery (91.74%) > moderate recovery (7.63%) > strong recovery (0.61%).

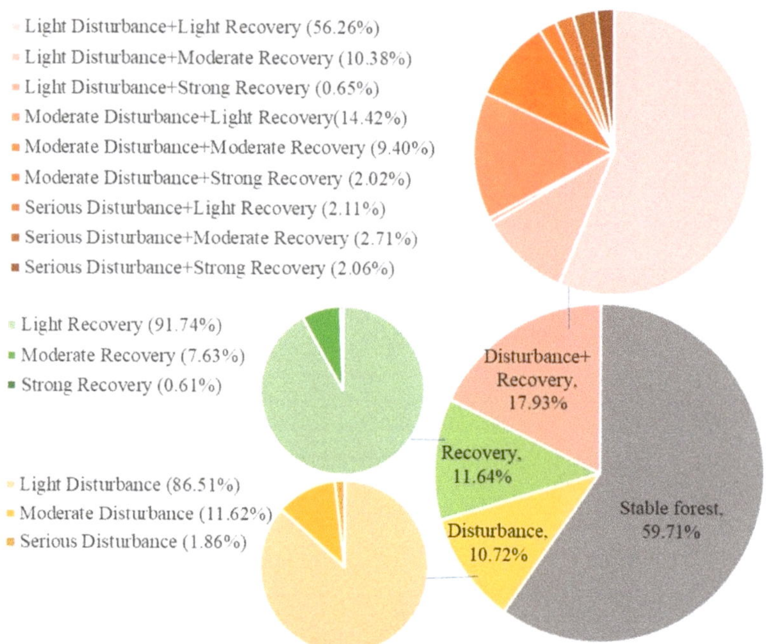

Figure 8. Proportion of disturbance and recovery of different levels in forest area.

The area where both disturbance and recovery occurred accounted for 17.93% of the total forest area, indicating that this part of the forest underwent a process of disturbance and recovery conversion. The areas of light disturbance and light recovery were the largest, accounting for 56.26%, followed by moderate disturbance + light recovery (14.42%), light disturbance + moderate recovery (10.38%), and moderate disturbance + moderate recovery (9.4%). The proportions of the other five combinations were all less than 3%. The results showed that forest disturbance and recovery in the study area were mainly light disturbances, and serious disturbances and strong recovery accounted for very few (<2%).

LandTrendr not only obtained spatiotemporal information of severely disturbed and obviously restored forest areas, but also captured the natural growth trend of forests or forest degradation caused by climate change and drought. Although there is no effective way to classify these areas in detail, obtaining varying degrees of disturbance and recovery is crucial for understanding the processes of forest change.

3.3. Accuracy Assessment of LandTrendr Results

The accuracy evaluation of the HGFC data and Google Earth images on LandTrendr segmentation results shows that the algorithm can effectively monitor forest disturbance and recover the footprint (Table 4). The different classes demonstrated high producer and user accuracies.

Table 4. The accuracy assessment result was based on LandTrendr segmentation results and HGFC datasets.

		Reference Data: HGFC Datasets (Pixels)			
		Disturbance	Recovery	Disturbance + Recovery	User Accuracy
LandTrendr results (pixels)	Disturbance	7265	593	65	91.69%
	Recovery	1014	4617	29	81.57%
	Disturbance + Recovery	250	83	623	65.16%
Producer accuracy		85.18%	87.22%	86.88%	
Overall accuracy				86.01%	
Kappa				0.73	

Analysis of the HGFC data showed that the highest user accuracy disturbance class was 91.69%, whereas the user accuracy for identifying both disturbance and recovery was 65.16%. There was a 26.15% confusion between the disturbance + recovery and disturbance classes, indicating that only one disturbance process was identified in the disturbance + recovery class. There was little difference among the three classes of producer accuracy: the highest accuracy of the recovery category was 87.22%, and the lowest accuracy of the disturbance category was 85.18%. In Google image-based evaluation, the highest user accuracy of 90.05% was achieved in the disturbance class (Table 5). In producer accuracy assessment, the highest accuracy was 91.84% for the disturbance class and the lowest was 61.86% for the recovery class.

Table 5. The accuracy assessment result was based on LandTrendr segmentation results and samples from Google Earth images.

		Reference Data: Google Earth Images (Pixels)			
		Disturbance	Recovery	Disturbance + Recovery	User Accuracy
LandTrendr results (pixels)	Disturbance	1520	107	61	90.05%
	Recovery	35	633	36	89.91%
	Disturbance + Recovery	100	102	837	80.56%
Producer accuracy		91.84%	61.86%	89.61%	
Overall accuracy				87.17%	
Kappa				0.79	

The high producer and user accuracies of the disturbance and recovery categories indicate that the LandTrendr method is robust in obtaining disturbance and recovery footprint information.

4. Discussion

4.1. Forest Disturbance and Recovery in Different Regions

The results showed that, although the overall trend of forest disturbance and recovery was balanced, there were significant differences in forest disturbance and recovery in different regions due to the geographical environment and management policies. The forests in the upper Indus Valley are managed by five countries. To explore the changes in forests in different countries, we created statistics and analyzed results from the perspective of forest management.

Table 6 shows the area of stability, disturbance, recovery, and both disturbance and recovery have occurred in India, Pakistan, Afghanistan, China, and Nepal. The disturbed and recovered forests of the five countries showed significantly different characteristics. From 1990 to 2020, the forest stability in Nepal was the highest (71.03%), Pakistan and China had little difference (66.97% and 64.59%, respectively), and India had the least stability

(52.98%). Figure 9 shows the proportion of disturbance and recovery areas in the forest area of the upper Indus Valley in different countries. The highest proportion of disturbance was 13.18% in India, and the least were 5.95% and 5.33% in China and Nepal, respectively. The area of forest recovery was highest in China (17.49%), followed by Pakistan (11.81%), Nepal (11.74%), India (11.41%), and Afghanistan (8.81%). In the region, where both disturbance and recovery have occurred, India accounted for the highest proportion at 22.42%, and Nepal for the lowest proportion at 11.89%.

Table 6. Forest disturbance and recovery statistics in the upper Indus Valley (unit: km^2).

Region	Forest Area	Stable	Disturbance	Recovery	Disturbance + Recovery
India	23,345.81	12,368.8 (52.98%)	3078.08 (13.18%)	2664.1 (11.41%)	5234.9 (22.42%)
Pakistan	20,366.96	13,640.44 (66.97%)	1646.3 (8.08%)	2405.44 (11.81%)	2678.15 (13.15%)
Afghanistan	1463.67	945.44 (64.59%)	167.42 (11.44%)	128.88 (8.81%)	221.41 (15.13%)
China	1009.24	626.36 (62.06%)	60.09 (5.95%)	176.51 (17.49%)	146.25 (14.49%)
Nepal	6.57	4.67 (71.03%)	0.35 (5.33%)	0.77 (11.74%)	0.78 (11.89%)

Disturbance + Recovery means both disturbance and recovery have occurred

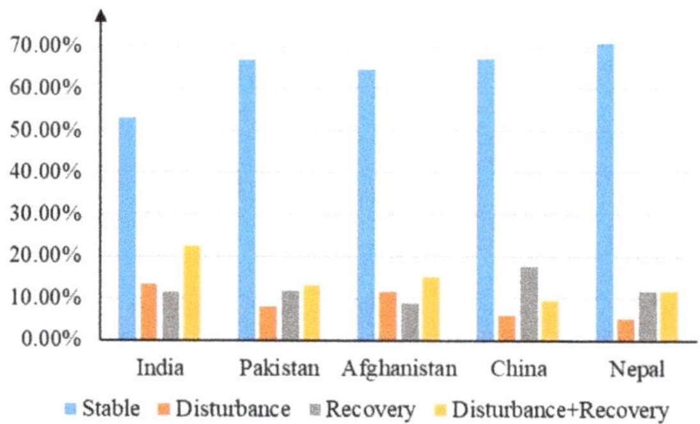

Figure 9. The proportion of forest disturbance and recovery area in the upper Indus Valley.

Figure 10 shows the interannual variation characteristics of forest disturbance and recovery in the upper Indus Valley in different forest management regions. In India, the disturbance forest area has been larger than the recovery forest area for 31 years since 1990. By 2020, the disturbance and recovery forest area showed a trend towards balance, but there was still a 387.86 km^2 gap. From 1990 to 1994, the accumulative recovery area of Afghanistan was larger than the accumulative recovery area. After 1994, the accumulative disturbance area continued to be larger than the accumulative recovery area. Although Afghanistan's forest area in the upper Indus Valley is relatively small, the disturbance and recovery of forests did not reach equilibrium until 2020. There were two inflection points in the forest change trend in Pakistan. In 1993, the cumulative disturbance area in Pakistan exceeded the cumulative recovery area and lasted for 16 years. After reaching equilibrium in 2009, the forest change trend in Pakistan tended to recover. The interannual variation of forest in China and Nepal was similar, and the cumulative recovery area was always larger than the cumulative disturbance area after 1993. Many studies have pointed to forest disturbances in Nepal, but this study only focused on a very small portion of Nepal's forests located in the upper Indus Valley, and the results represent only that portion of forests.

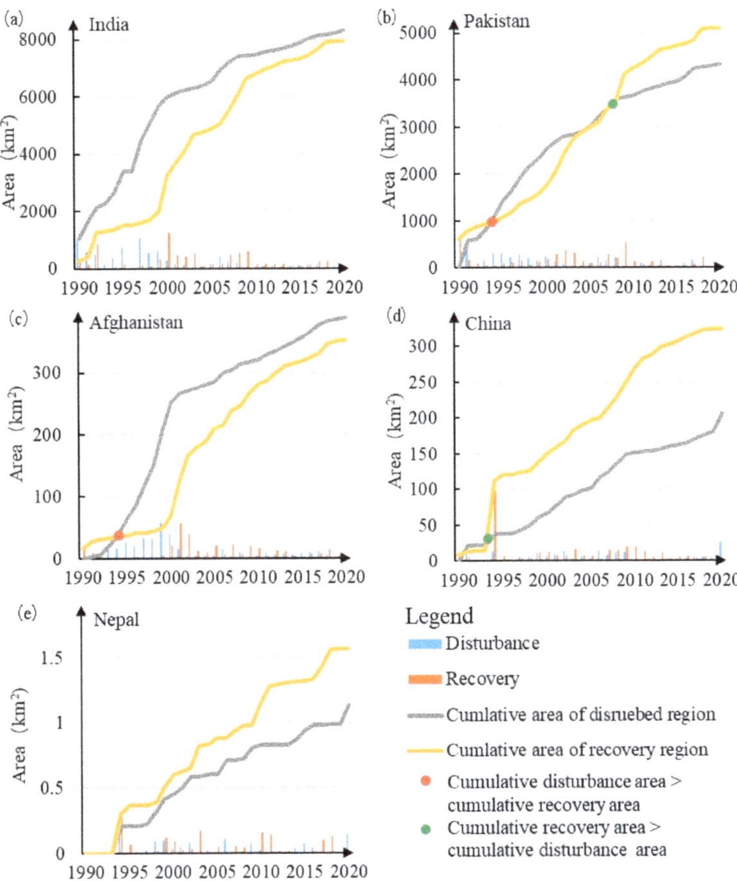

Figure 10. Interannual variation characteristics of disturbance and recovery in in the upper Indus Valley. (**a**) India (**b**) Pakistan (**c**) Afghanistan (**d**) China (**e**) Nepal.

4.2. Analysis of the Causes of Disturbance and Recovery

There is a close relationship between the climate and vegetation in the Indus Valley. In the upper Sindh and Punjab, overgrazing and deforestation have led to the destruction of many natural vegetation types. In addition, humans have long interfered with natural water systems, and in Shivalik, deforestation has led to a marked deterioration of groundwater and vegetation cover. From 1990 to 2020, 59.71% of the forests in the upper Indus Valley remained stable, and the restored area was 0.92% greater than the disturbed area. Although the overall trend of forest change was balanced, there was still an imbalance among the five countries, especially in different provinces.

Table 7 shows the data of forest change in different administrative regions. Himachal Pradesh and Jammu and Kashmir are the two regions with the most disturbance area exceeding the recovery area, covering 362.77 km^2 and 40.89 km^2, respectively. Pathania et al. (2012) noted that the forest area in this region decreased with the passage of time, and grazing caused an obvious loss of plantation forest [42]. Other studies have shown that the large-scale introduction of horticultural cash crops in the region between 1998 and 2010 led to some significant changes in forest composition, with commercial cultivation leading to the decline of some important plant species [43]. We also observed that some reports indicate that the forest coverage in this area increased from 1991 to 2015, which can be explained by commercial tree planting [44]. Forest fires, landslides, and other natural disasters are also significant causes of forest disturbance in this region [45], but it is difficult to quantify the specific area of disturbance currently caused by these reasons. These regions should be aware that changes in the composition of forest species and ecological imbalances caused by deforestation and forest degradation may cause irreversible damage to unstable and fragile mountainous areas.

Table 7. Forest disturbance and recovery area of 28 administrative regions in the upper Indus Valley (km^2).

Forest Management Country	Province	Forest Area	Stable	Disturbance	Recovery	Recovery − Disturbance
	Bamian	0.32	0.21	0.10	0.07	−0.02
	Badakhshan	10.10	7.43	1.53	2.21	0.68
	Baghlan	17.48	11.75	4.91	3.53	−1.38
	Ghazni	0.17	0.06	0.10	0.07	−0.02
	Kabol	4.89	2.31	1.19	2.21	1.02
	Kapisa	59.49	41.02	12.22	13.12	0.90
	Konarha	920.04	627.05	212.51	209.93	−2.58
Afghanistan	Laghman	132.28	100.70	21.86	20.38	−1.47
	Lowgar	8.22	4.53	1.76	3.20	1.44
	Nangarhar	102.24	56.92	34.70	27.47	−7.23
	Paktia	124.15	50.63	64.76	43.06	−21.70
	Paktika	22.67	6.49	15.08	7.61	−7.47
	Parvan	42.82	26.51	11.63	11.29	−0.34
	Takhar	0.01	0.01	0.00	0.00	0.00
	Vardak	18.79	10.08	6.69	6.32	−0.37
China	Xinjiang	9.18	6.10	2.60	1.76	−0.84
	Xizang	1000.05	620.26	203.74	321.00	117.26
	Himachal Pradesh	8259.54	3997.95	3423.46	3060.69	−362.77
	Haryana	0.10	0.03	0.06	0.06	0.00
India	Jammu and Kashmir	14,960.13	8318.08	4831.74	4784.82	−46.92
	Punjab	11.70	4.83	4.24	5.87	1.63
	Uttar Pradesh	114.34	47.71	53.43	47.49	−5.93
Nepal	Karnali	6.57	4.67	1.13	1.55	0.42
	Azad Kashmir	4517.45	3036.43	948.27	1117.16	168.89
Pakistan	Federally Administered Tribal Areas	1017.79	734.91	160.99	234.34	73.35
	Gilgit Baltistan	4248.70	2689.32	1126.95	1086.06	−40.89
	Khyber Pakhtunkhwa	10,073.75	6802.69	2002.93	2539.98	537.05
	Punjab	509.27	374.69	84.74	105.21	20.47
Total		46,192.26	27,583.36	4952.12	5375.30	423.18

In the Gilgit Baltistan of Pakistan, the forest disturbance area was 40.89 km^2 more than the recovery area, and the disturbance was mainly distributed in the Western border area with Afghanistan. Qamer et al. (2016) showed that the area was felled or severely degraded from 1990 to 2010 [18]. The results of this study and those of Qamer et al. (2016) show consistent temporal and spatial characteristics in the Gilgit Baltistan. The forests of Afghanistan are mainly distributed in the Eastern region, and the disturbance and recovery of the provinces are balanced, except in the Paktia, Paktika, and Nangarhar districts. A previous study on forest change in Northern Pakistan from 1990 to 2010 similarly showed that some assessment reports have grossly overstated deforestation rates, which is consistent with the findings of our study [18].

The main contribution of forest recovery came from the Khyber Pakhtunkhwa, Azad Kashmir, and Tibet, and the recovery areas were 537.05 km^2, 168.89 km^2, and 117.26 km^2 more than the disturbance area, respectively. Forest recovery in the Khyber Pakhtunkhwa Province was primarily driven by the Khyber Pakhtunkhwa Billion Tree Forestation Project, which aims to plan, design, initiate, and implement the Green Growth Initiative in the forestry sector [46]. Khan et al. (2020) studied the temporal changes in forest cover, carbon storage, and corresponding CO_2 emission/retention trends (1989–2018) in Azad Kashmir and showed that both forest cover and carbon storage increased significantly, and the research results were consistent with this study [47]. The Tibet Ali region, carried out from 2002 biological sand prevention engineering, implemented a five-phase afforestation project, which is consistent with the results of this study.

4.3. Method Limitations and Its Application

The LandTrendr spectral-temporal segmentation algorithm results may vary greatly according to different parameters and spectral indices [48]. The fixed parameters cannot obtain the optimal result, which needs to be further determined by combining the UI program provided by LandTrendr with real ground samples. For forest disturbance and recovery, the information extraction method used in this study can obtain spatiotemporal cognition of the forest change footprint in the upper Indus Valley. However, this approach does not distinguish between the possible causes of disturbance, such as deforestation, climate change, and geological hazards. The rate band in the results provided by LandTrendr provides a possible way to distinguish between these possible causes, which require further analysis.

5. Conclusions

To describe the disturbance and recovery of forests quantitatively and objectively in the upper Indus Valley, we used multi-source remote sensing data, the LandTrendr spectral-temporal segmentation algorithm, and a remote sensing big data computing platform to complete the monitoring of forest change footprint in the upper Indus Valley. The main conclusions are as follows:

(1) The LandTrendr algorithm combined with multi-source remote sensing data and the GEE big data platform completed forest change footprint tracking of forests in the upper Indus Valley for 31 years, and the algorithm showed stable robustness and portability.

(2) Forest disturbance and recovery were widespread in 1990–2020, most of the disturbance occurred in 1990–2000, the recovery was in 2000–2010, and equilibrium was widely attained in 2012.

(3) Forest disturbance and recovery in different forest management regions showed significant differences, and forest disturbance in India and Afghanistan did not reach equilibrium by 2020. Pakistan reached equilibrium in 2009, while Nepal and China showed relatively stable and continuous trends in forest recovery.

This study can further contribute to more effective forest management policy development by identifying the spatial and temporal patterns of disturbance and recovery for quantitative assessment.

Author Contributions: Conceptualization, methodology, and resources, X.Y. and J.W.; software, validation, data curation, and writing —original draft preparation, X.Y.; writing—review and editing, J.W. and X.Y. All authors have read and agreed to the published version of the manuscript.

Funding: Special acknowledgement should be expressed to the China–Pakistan Joint Research Center on Earth Sciences and the Construction Project of the China Knowledge Center for Engineering Sciences and Technology (Grant number CKCEST-2020-2-4), which supported the implementation of this study.

Institutional Review Board Statement: Not applicable.

Informed Consent Statement: Not applicable.

Data Availability Statement: The data presented in this study are available on request from the corresponding author.

Conflicts of Interest: The authors declare no conflict of interest.

References

1. Bala, G.; Caldeira, K.; Wickett, M.; Phillips, T.J.; Lobell, D.B.; Delire, C.; Mirin, A. Combined climate and carbon-cycle effects of large-scale deforestation. *Proc. Natl. Acad. Sci. USA* **2007**, *104*, 6550–6555. [CrossRef] [PubMed]
2. Bonan, G.B. Forests and climate change: Forcings, feedbacks, and the climate benefits of forests. *Science* **2008**, *320*, 1444–1449. [CrossRef] [PubMed]

3. Mondal, P.; McDermid, S.S.; Qadir, A. A reporting framework for Sustainable Development Goal 15: Multi-scale monitoring of forest degradation using MODIS, Landsat and Sentinel data. *Remote Sens. Environ.* **2020**, *237*, 111592. [CrossRef]
4. Attiwill, P.M. The disturbance of forest ecosystems—The ecological basis for conservative management. *For. Ecol. Manag.* **1994**, *63*, 247–300. [CrossRef]
5. Seidl, R.; Fernandes, P.M.; Fonseca, T.F.; Gillet, F.; Jonsson, A.M.; Merganicova, K.; Netherer, S.; Arpaci, A.; Bontemps, J.D.; Bugmann, H.; et al. Modelling natural disturbances in forest ecosystems: A review. *Ecol. Model.* **2011**, *222*, 903–924. [CrossRef]
6. Munoz, R.; Bongers, F.; Rozendaal, D.M.A.; Gonzalez, E.J.; Dupuy, J.M.; Meave, J.A. Autogenic regulation and resilience in tropical dry forest. *J. Ecol.* **2021**, *109*, 3295–3307. [CrossRef]
7. Vina, A.; McConnell, W.J.; Yang, H.B.; Xu, Z.C.; Liu, J.G. Effects of conservation policy on China's forest recovery. *Sci. Adv.* **2016**, *2*, 1500965. [CrossRef]
8. Chen, Y.; Luo, G.; Maisupova, B.; Chen, X.; Mukanov, B.M.; Wu, M.; Mambetov, B.T.; Huang, J.; Li, C. Carbon budget from forest land use and management in Central Asia during 1961–2010. *Agric. For. Meteorol.* **2016**, *221*, 131–141. [CrossRef]
9. Hansen, M.C.; Potapov, P.V.; Moore, R.; Hancher, M.; Turubanova, S.A.; Tyukavina, A.; Thau, D.; Stehman, S.V.; Goetz, S.J.; Loveland, T.R.; et al. High-resolution global maps of 21st-century forest cover change. *Science* **2013**, *342*, 850–853. [CrossRef]
10. Immerzeel, W.W.; Lutz, A.F.; Andrade, M.; Bahl, A.; Biemans, H.; Bolch, T.; Hyde, S.; Brumby, S.; Davies, B.J.; Elmore, A.C.; et al. Importance and vulnerability of the world's water towers. *Nature* **2020**, *577*, 364–369. [CrossRef]
11. FAO. *State of the World's Forests 2011*; Food and Agricultural Organization of the United Nations: Rome, Italy, 2011.
12. Rashid, I.; Bhat, M.A.; Romshoo, S.A. Assessing changes in the above ground biomass and carbon stocks of Lidder valley, Kashmir Himalaya, India. *Geocarto Int.* **2017**, *32*, 717–734. [CrossRef]
13. Zeb, A.; Hamann, A.; Armstrong, G.W.; Acuna-Castellanos, D. Identifying local actors of deforestation and forest degradation in the Kalasha valleys of Pakistan. *For. Policy Econ.* **2019**, *104*, 56–64. [CrossRef]
14. Masek, J.G.; Huang, C.Q.; Wolfe, R.; Cohen, W.; Hall, F.; Kutler, J.; Nelson, P. North American forest disturbance mapped from a decadal Landsat record. *Remote Sens. Environ.* **2008**, *112*, 2914–2926. [CrossRef]
15. Griffiths, P.; Kuemmerle, T.; Baumann, M.; Radeloff, V.C.; Abrudan, I.V.; Lieskovsky, J.; Munteanu, C.; Ostapowicz, K.; Hostert, P. Forest disturbances, forest recovery, and changes in forest types across the Carpathian ecoregion from 1985 to 2010 based on Landsat image composites. *Remote Sens. Environ.* **2014**, *151*, 72–88. [CrossRef]
16. Pflugmacher, D.; Cohen, W.B.; Kennedy, R.E.; Yang, Z.Q. Using Landsat-derived disturbance and recovery history and lidar to map forest biomass dynamics. *Remote Sens. Environ.* **2014**, *151*, 124–137. [CrossRef]
17. Joshi, A.K.; Joshi, P.K.; Chauhan, T.; Bairwa, B. Integrated approach for understanding spatio-temporal changes in forest resource distribution in the central Himalaya. *J. For. Res.* **2014**, *25*, 281–290. [CrossRef]
18. Qamer, F.M.; Shehzad, K.; Abbas, S.; Murthy, M.S.R.; Xi, C.; Gilani, H.; Bajracharya, B. Mapping Deforestation and Forest Degradation Patterns in Western Himalaya, Pakistan. *Remote Sens.* **2016**, *8*, 385. [CrossRef]
19. Gorelick, N.; Hancher, M.; Dixon, M.; Ilyushchenko, S.; Thau, D.; Moore, R. Google Earth Engine: Planetary-scale geospatial analysis for everyone. *Remote Sens. Environ.* **2017**, *202*, 18–27. [CrossRef]
20. Huang, C.; Coward, S.N.; Masek, J.G.; Thomas, N.; Zhu, Z.; Vogelmann, J.E. An automated approach for reconstructing recent forest disturbance history using dense Landsat time series stacks. *Remote Sens. Environ.* **2010**, *114*, 183–198. [CrossRef]
21. Verbesselt, J.; Hyndman, R.; Newnham, G.; Culvenor, D. Detecting trend and seasonal changes in satellite image time series. *Remote Sens. Environ.* **2010**, *114*, 106–115. [CrossRef]
22. Hughes, M.J.; Kaylor, S.D.; Hayes, D.J. Patch-Based Forest Change Detection from Landsat Time Series. *Forests* **2017**, *8*, 166. [CrossRef]
23. Deng, C.B.; Zhu, Z. Continuous subpixel monitoring of urban impervious surface using Landsat time series. *Remote Sens. Environ.* **2020**, *238*, 21. [CrossRef]
24. Zhu, Z.; Zhang, J.; Yang, Z.; Aljaddani, A.H.; Cohen, W.B.; Qiu, S.; Zhou, C. Continuous monitoring of land disturbance based on Landsat time series. *Remote Sens. Environ.* **2020**, *238*, 11116. [CrossRef]
25. Nature, W.W.F.f. As the World's Population Grows, Forests Are Coming under More Pressure Than Ever. Available online: https://www.wwfpak.org/our_work_/forests/ (accessed on 18 December 2021).
26. Roy, D.P.; Kovalskyy, V.; Zhang, H.K.; Vermote, E.F.; Yan, L.; Kumar, S.S.; Egorov, A. Characterization of Landsat-7 to Landsat-8 reflective wavelength and normalized difference vegetation index continuity. *Remote Sens. Environ.* **2016**, *185*, 57–70. [CrossRef]
27. Zhu, Z.; Wang, S.; Woodcock, C.E. Improvement and expansion of the Fmask algorithm: Cloud, cloud shadow, and snow detection for Landsats 4–7, 8, and Sentinel 2 images. *Remote Sens. Environ.* **2015**, *159*, 269–277. [CrossRef]
28. Zhu, Z.; Woodcock, C.E.; Olofsson, P. Continuous monitoring of forest disturbance using all available Landsat imagery. *Remote Sens. Environ.* **2012**, *122*, 75–91. [CrossRef]
29. Yan, X.; Wang, J. Dynamic monitoring of urban built-up object expansion trajectories in Karachi, Pakistan with time series images and the LandTrendr algorithm. *Sci. Rep.* **2021**, *11*, 1–15. [CrossRef]
30. Kennedy, R.E.; Yang, Z.G.; Cohen, W.B. Detecting trends in forest disturbance and recovery using yearly Landsat time series: 1. LandTrendr-Temporal segmentation algorithms. *Remote Sens. Environ.* **2010**, *114*, 2897–2910. [CrossRef]
31. Kennedy, R.E.; Yang, Z.; Gorelick, N.; Braaten, J.; Cavalcante, L.; Cohen, W.B.; Healey, S. Implementation of the LandTrendr Algorithm on Google Earth Engine. *Remote Sens.* **2018**, *10*, 691. [CrossRef]

32. Cohen, W.B.; Yang, Z.; Heale, S.P.; Kennedy, R.E.; Gorelic, N. A LandTrendr multispectral ensemble for forest disturbance detection. *Remote Sens. Environ.* **2018**, *205*, 131–140. [CrossRef]
33. Lopez-Garcia, M.; Caselles, V. Mapping burns and natural reforestation using Thematic Mapper data. *Geocarto Int.* **1991**, *6*, 31–37. [CrossRef]
34. Wilson, E.H.; Sader, S.A. Detection of forest harvest type using multiple dates of Landsat TM imagery. *Remote Sens. Environ.* **2002**, *80*, 385–396. [CrossRef]
35. Tucker, C.J. red and photographic infrared linear combinations for monitoring vegetation. *Remote Sens. Environ.* **1979**, *8*, 127–150. [CrossRef]
36. Salomonson, V.V.; Appel, I. Estimating fractional snow cover from MODIS using the normalized difference snow index. *Remote Sens. Environ.* **2004**, *89*, 351–360. [CrossRef]
37. Huete, A.; Didan, K.; Miura, T.; Rodriguez, E.P.; Gao, X.; Ferreira, L.G. Overview of the radiometric and biophysical performance of the MODIS vegetation indices. *Remote Sens. Environ.* **2002**, *83*, 195–213. [CrossRef]
38. Huang, C.; Wylie, B.; Yang, L.; Homer, C.; Zylstra, G. Derivation of a tasselled cap transformation based on Landsat 7 at-satellite reflectance. *Int. J. Remote Sens.* **2002**, *23*, 1741–1748. [CrossRef]
39. Crist, E.P. A Tm Tasseled Cap Equivalent Transformation for Reflectance Factor Data. *Remote Sens. Environ.* **1985**, *17*, 301–306. [CrossRef]
40. Baig, M.H.A.; Zhang, L.; Shuai, T.; Tong, Q. Derivation of a tasselled cap transformation based on Landsat 8 at-satellite reflectance. *Remote Sens. Lett.* **2014**, *5*, 423–431. [CrossRef]
41. Liu, S.; Wei, X.; Li, D.; Lu, D. Examining Forest Disturbance and Recovery in the Subtropical Forest Region of Zhejiang Province Using Landsat Time-Series Data. *Remote Sens.* **2017**, *9*, 479. [CrossRef]
42. Pathania, M.S.; Sharma, S.K. Impact of migratory animals on grazing lands: An evidence from Himachal Pradesh. *Indian J. Anim. Res.* **2012**, *46*, 40–45.
43. Shah, S.; Sharma, D.P. Land use change detection in Solan Forest Division, Himachal Pradesh, India. *Forest Ecosystems* **2015**, *2*, 327–338. [CrossRef]
44. Sharma, P.; Mahajan, A.; Lata, K.; Bharti, H.; Randhawa, S.S. *An Analysis of the Temporal Changes in the Forests of Himachal Pradesh—A Review*; State Centre on Climate Change State Council for Science: Shimla, India, 2015; Volume 10, pp. 63–76.
45. Nandy, S.; Singh, C.; Das, K.K.; Kingma, N.C.; Kushwaha, S.P.S. Environmental vulnerability assessment of eco-development zone of Great Himalayan National Park, Himachal Pradesh, India. *Ecol. Indic.* **2015**, *57*, 182–195. [CrossRef]
46. Khan, M.I.; Hussain, S.K.; Saad, H.; Rukh, G.; Ahmed, M.M.; Ahmad, I. *Third Party Monitoring of Billion Trees Afforestation Project in Khyber Pakhtunkhwa Phase-Ii*; World Wide Fund for Nature Pakistan (WWF-Pakistan): Lahore, Pakistan, 2017.
47. Khan, I.A.; Khan, M.R.; Baig, M.H.A.; Hussain, Z.; Hameed, N.; Khan, J.A. Assessment of forest cover and carbon stock changes in sub-tropical pine forest of Azad Jammu & Kashmir (AJK), Pakistan using multi-temporal Landsat satellite data and field inventory. *PLoS ONE* **2020**, *15*, e0226341. [CrossRef]
48. Hislop, S.; Jones, S.; Soto-Berelov, M.; Skidmore, A.; Haywood, A.; Nguyen, T.H. A fusion approach to forest disturbance mapping using time series ensemble techniques. *Remote Sens. Environ.* **2019**, *221*, 188–197. [CrossRef]

remote sensing

MDPI

Article

Spatiotemporal Dynamics of Land Surface Albedo and Its Influencing Factors in the Qilian Mountains, Northeastern Tibetan Plateau

Jichun Li [1,2,3], Guojin Pang [1,2,3,*], Xuejia Wang [4], Fei Liu [1,2,3] and Yuting Zhang [1,2,3]

[1] Faculty of Geomatics, Lanzhou Jiaotong University, Lanzhou 730070, China; 11200852@stu.lzjtu.edu.cn (J.L.); 12201855@stu.lzjtu.edu.cn (F.L.); 12211951@stu.lzjtu.edu.cn (Y.Z.)

[2] National-Local Joint Engineering Research Center of Technologies and Applications for National Geographic State Monitoring, Lanzhou 730070, China

[3] Gansu Provincial Engineering Laboratory for National Geographic State Monitoring, Lanzhou 730070, China

[4] Key Laboratory of Western China's Environmental Systems (Ministry of Education), College of Earth and Environmental Sciences, Lanzhou University, Lanzhou 730000, China; wangxuejia@lzu.edu.cn

* Correspondence: panggj@mail.lzjtu.cn

Abstract: Land surface albedo directly determines the distribution of radiant energy between the surface and the atmosphere, and it is a key parameter affecting the energy balance on the land surface. However, the spatiotemporal dynamics of land surface albedo and associated influencing factors in the Qilian Mountains (QM) have been rarely reported. By using the long-time series data products of MODIS shortwave albedo, normalized difference vegetation index (NDVI), and snow cover with a spatial resolution of 0.05° from 2001 to 2020, this paper analyzes the temporal and spatial variations of land surface albedo in the QM over the past 20 years and its influencing factors. The analysis results show that the multi-year average surface albedo in the QM has obvious differences in spatial distribution: it increases with the altitude, and it is high in the west (at the west of 98° E) and low in the east. Meanwhile, the surface albedo has different distribution characteristics in different seasons: the spatial distribution of surface albedo is similar in spring and autumn; the areas with a high surface albedo in summer are significantly fewer than those in other seasons. Besides, in the past 20 years, the annual average surface albedo has shown a weak growth trend in the QM, with a change rate of $5 \times 10^{-3}/10a$, and the minimum and maximum values were reached in 2001 and 2019, respectively. In addition, the annual variation of the surface albedo in the QM showed a "U" shape, with the largest variation in January and the smallest variation in August. The annual variation of surface albedo is significantly positively correlated with snow cover (r = 0.96) and significantly negatively correlated with NDVI (r = −0.91). Moreover, the interannual variation of the surface albedo in the QM is closely related to land surface cover and is greatly affected by snow cover. Spatially, the annual variation of surface albedo in most areas of the QM is dominated by the change of snow cover, and the increase of surface albedo in the middle area is consistent with the increase of snow cover, while the decrease of albedo in the edge area is related to the improvement of vegetation cover. The results of this study provide a scientific basis for studying the climate and environmental changes caused by changes in the surface of the QM and making ecological environment restoration strategies.

Keywords: land surface albedo; MODIS; Qilian Mountains; spatiotemporal variation; snow cover; NDVI

Citation: Li, J.; Pang, G.; Wang, X.; Liu, F.; Zhang, Y. Spatiotemporal Dynamics of Land Surface Albedo and Its Influencing Factors in the Qilian Mountains, Northeastern Tibetan Plateau. *Remote Sens.* **2022**, *14*, 1922. https://doi.org/10.3390/rs14081922

Academic Editor: Wenxin Zhang

Received: 31 March 2022
Accepted: 11 April 2022
Published: 15 April 2022

Publisher's Note: MDPI stays neutral with regard to jurisdictional claims in published maps and institutional affiliations.

1. Introduction

Since the Industrial Revolution, global warming has become increasingly serious, and multiple indicators of the Earth's climate system such as the atmosphere, ocean, land, and cryosphere have shown the impact of human activities (IPCC). The surface albedo is the ratio of the reflected solar radiation on the surface to the incident solar radiation, which regulates the radiation energy balance between the ground and the atmosphere [1,2]. It is a key parameter that affects the surface energy budget and the interaction between the ground and the atmosphere [3,4]. Therefore, surface albedo plays an extremely important role in the climate system, and it is influenced by solar elevation angle, topography, vegetation changes, ice/snow cover changes, soil moisture, soil properties, and human activities [5,6]. Meanwhile, the change of surface albedo will also react to the surface radiation balance and affect other climate variables, thus forming a complex loop feedback mechanism [7,8]. In particular, the feedback effect of surface albedo changes will further amplify its impact on climate, even though subtle changes are fed back into the climate system, thereby affecting local, regional, and even global climate change [9]. Generally, the improvement of vegetation coverage and the melting of ice/snow will lead to a decrease in the surface albedo, and the corresponding sensible heat flux and latent heat flux will also increase; the surface absorbs more solar radiation and increases the surface temperature, thus promoting the growth of vegetation and the melting of ice and snow [10,11]. Conversely, the increase of surface albedo will weaken atmospheric convergence, reduce cloud, precipitation, and soil moisture, thus exacerbating drought in arid regions [12]. From 1700 to 2005, the global albedo increased by about 0.00106 because of land cover change, leading to a radiative cooling at the top of the atmosphere of $-0.15\ \mathrm{Wm}^{-2}$ [13].

The strong reflection characteristic of snow has a remarkable impact on the surface albedo. The snow location (above or below the canopy), extent, and state (i.e., snow age, depth, water content and purity, etc.) can greatly change the spectral characteristics to modulate the surface albedo [8,14,15]. Changes in snow cover significantly affect the surface albedo, and there is a positive correlation between the two [5,6,16]. In addition, studies in the complex terrain area of the Tibetan Plateau (TP) found that the surface albedo under the coexistence of snow cover and vegetation is more sensitive to the response of snow cover [6]. Bond et al. (2013) found that incomplete combustion of fossil fuels and biofuels can reduce snow albedo, thereby accelerating Arctic snow melting [17,18]. Due to climate warming, the reduction of surface albedo caused by snow retreat and snow albedo feedback is considered to be an important reason for the amplification of Arctic climate warming [19,20]. Therefore, a deep understanding of the positive feedback of snow albedo becomes crucial.

Vegetation change is another important factor affecting surface albedo. Large parts of the Earth are experiencing a greening trend because of the changes in carbon dioxide fertilization, nitrogen deposition, climate change, and land cover [21]. Numerous studies have found a negative correlation between surface albedo and vegetation cover [5,6,16,22]. Under increased vegetation, surface albedo decreases, and net surface radiation increases, thereby heating the atmosphere and providing positive feedback for climate warming [11]. In the TP, the increase of vegetation cover has a negative effect on the surface albedo, and the surface albedo is more sensitive to the change of NDVI in the vegetation coverage area than in the coexistence area of snow cover and vegetation [6]. In high latitudes, climate warming increases the vegetation area, causing a significant reduction in surface albedo [5], but the radiative forcing warming effect caused by the decrease in surface albedo can compensate for the cooling effect of the carbon sink [23]. In temperate and boreal zones of the Northern Hemisphere the albedo effect is stronger and deforestation induces a cooling, which is related to the difference in albedo between trees and grasslands being amplified by the presence of snow; in tropical regions, deforestation has a significant effect on local warming effects by increasing surface albedo and reducing evaporation [24–26].

The QM are the ecological barrier of the TP and the throat of the Silk Road Economic Belt [27]. In the past 50 years, the temperature in the QM has increased significantly, with a rate of about 0.5 °C decade^{-1}, and the precipitation has also increased significantly, with a rate of about 6.95 mm decade^{-1} [28,29]. The climate of the whole region shows a warming and wetting trend, which promotes the climbing of the tree line and enhances the stability of the grassland [30]. On the whole, vegetation in the QM is greening. However, solid reservoirs (glaciers and permafrost) are in the process of aggravating degradation, and glaciers in mountainous areas below 4000 m have completely disappeared. Compared with the 1960s, the area of glaciers in the QM has decreased by 20.5% $\pm$ 6.04%, and the loss rate between 2007 and 2015 is as high as 5.82%/10a [29]. In addition, with the intensification of human activities, the areas of industrial and mining, housing, transportation, and other land uses have increased [27]. Furthermore, the extensive development of mineral resources, overgrazing of local grasslands, irregular operation of tourism facilities, and overloaded groundwater use has seriously damaged the ecological environment of the QM and posed serious challenges to the sustainable development of the society and economy [31].

Affected by various driving factors, surface albedo exhibits obvious temporal variability and spatial heterogeneity. To obtain credible and accurate albedo data, satellite remote sensing is indispensable. Compared with measured and simulated data, satellite remote sensing has a wider coverage, and fewer natural limitations and uncertainties. Thus, it has become an efficient method to acquire continuous albedo [32]. There is good consistency between the satellite-retrieved annual mean and field measurements, when the surface albedo varies seasonally, forest-covered areas are better matched than non-forested areas [33]. In satellite datasets, all latitude assessments show good agreement in summer, while albedo assessments have significant uncertainty in winter, especially at high latitudes [34]. The MODIS V006 albedo product has an improved temporal resolution and accuracy, and the daily MODIS V006 product can reproduce the dynamics of albedo well [32,35,36]. In recent years, some scholars have studied the albedo of glaciers in the QM. It was found that the dust on the surface of the glacier significantly reduced the albedo [18], and there was a significant negative correlation between the diurnal variation of glacial albedo and air temperature [37]. Another study found a significant positive correlation between glacier area change and annual mean albedo [38]. However, the spatiotemporal variation characteristics of surface albedo and its influencing factors in the QM are less studied. Therefore, this paper uses the MODIS MCD43C3 V006 product provided by NASA (National Aeronautics and Space Administration, NASA, Washington, DC, USA), combined with NDVI and snow cover data, to analyze the spatiotemporal distribution of surface albedo and its influencing factors in the QM. The results of the study will help to understand the feedback mechanism of surface albedo and reveal the relation between climate change and human activities, thus providing an important basis for ecological barrier construction and sustainable development in the QM.

2. Overview of the Study Area

The QM are located on the border between the northeastern Qinghai Province and the western Gansu Province in China, and it is also located at the intersection of three major plateaus, i.e., Qinghai-Tibet, Mengxin, and Loess. It consists of a number of parallel mountains and wide valleys from northwest to southeast (Figure 1a). Because of the high altitude of the overall mountain range, the peaks over 4000 m are covered with snow all year round. Meanwhile, there are 3306 modern glaciers [39], which are the source of the Shiyang River, Heihe River, Shule River, and other rivers [40]. As the altitude decreases from northwest to southeast, the land cover types are bare land, grassland, and cultivated land in sequence [41]. Bare land is mainly in the west of 98° E, and grassland is mainly in the east of 98° E (Figure 1b). The annual average temperature in the QM is below 4 °C, higher in barren/rocky areas, lower in snow/glacier areas [42], and the temperature gradually decreases with the increase of elevation. Different from air temperature, precipitation is not only affected by altitude, aspect, and slope but also by latitude, longitude, and atmospheric circulation; it shows a decreasing trend from east to west [28].

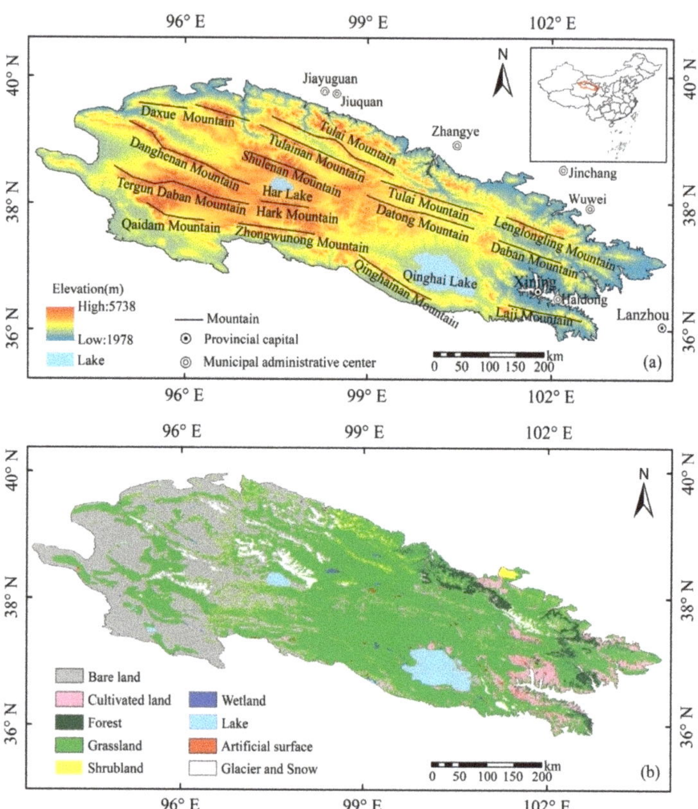

Figure 1. (**a**) Topography of the QM and (**b**) the types of vegetation cover.

3. Materials and Methods

3.1. Dataset and Preprocessing

This study used three sets of remote sensing products and reanalysis datasets (Table 1), including MODIS surface albedo, snow cover, normalized difference vegetation index (NDVI) products, and National Centers for Environmental Prediction (NCEP) downward solar radiation reanalysis data.

Table 1. The data sets used in this study.

Parameter	Dataset	Spatial Resolution	Temporal Resolution	References
Albedo	MCD43C3	0.05°	Daily	[43]
Snow cover	MOD10CM	0.05°	Monthly	[44]
NDVI	MOD13C2	0.05°	Monthly	[45]
Downward solar radiation	NCEP	T62 Gaussian grid 192 × 94	Daily	[46]

3.1.1. Remote Sensing Products

The albedo data and NDVI data used in this paper are obtained from the 2001–2020 MODIS MCD43C3 and MOD13C2 data provided by NASA (https://ladsweb.modaps.eosdis.nasa.gov (accessed on 1 November 2020)). The HEG tool was adopted to extract the black sky albedo (BSA) and the white sky albedo (WSA) in the albedo data. BSA represents the albedo under complete direct solar radiation, and WSA represents the albedo under complete diffusion of solar radiation. Then, MATLAB software was used to calculate the daily surface albedo, and the daily data were spatially aggregated into monthly, seasonal, and annual data. For NDVI data, considering that the areas with NDVI < 0.1 are bare soil and sparse vegetation areas, the values of NDVI > 0.1 were screened [47]. However, since this study took the MODIS MOD10CM product provided by the EOS/MODIS data center (http://nsidc.org/NASA/MODIS (accessed on 29 November 2020)) in the United States as snow cover data, a value between 0 and 100 was used to represent the snow cover rate of each pixel.

3.1.2. NCEP Reanalysis Products

To obtain the true surface albedo, the diffuse skylight ratio needs to be calculated from the downward solar flux reanalysis data provided by NCEP [46,48]. Meanwhile, the calculation of the regional average albedo requires surface downward radiation [34]. The data can be downloaded from the NCEP data website (https://psl.noaa.gov/data/gridded/data.ncep.reanalysis.html (accessed on 5 December 2020)). Because the spatial resolution of the NCEP reanalysis data is inconsistent with that of the albedo data, this paper used the nearest-neighbor interpolation to interpolate all NCEP reanalysis data to the same resolution of 0.05° as MODIS data.

3.1.3. Vegetation Coverage Data

The vegetation cover uses the global land cover data GlobeLand30 developed by China in 2020 with a spatial resolution of 30 m (http://www.webmap.cn/mapDataAction.do?method=globalLandCover (accessed on 3 November 2020)). The data covers the land range of 80 degrees north-south latitude, including 10 surface coverage types: cultivated land, forest, grassland, shrub land, wetland, water body, tundra, artificial surface, bare land and, glacier and snow.

3.2. Research Methods

3.2.1. Calculation of Surface Albedo

The true albedo (blue sky albedo) is approximately equal to the weighted combination of the black sky albedo and the white sky albedo [48], and the surface albedo can be used the following formula calculates:

$$\alpha = (1 - f_{dif})\mathrm{BSA} + f_{dif}\mathrm{WSA} \tag{1}$$

$$f_{dif} = \frac{\mathrm{DD}_v + \mathrm{DD}_n}{\mathrm{DD}_v + \mathrm{DD}_n + \mathrm{BD}_v + \mathrm{BD}_n} \tag{2}$$

where BSA and WSA are the black sky albedo and white sky albedo, f_{dif} is the diffuse skylight ratio, and DD_v, DD_n, BD_v, and BD_n are the visible diffuse downward solar flux, near IR diffuse downward solar flux, visible beam downward solar flux and near IR beam downward solar flux, respectively.

3.2.2. Calculation of Regional Average Albedo

The monthly average surface albedo of an area is calculated according to Equation (3) [34]. For Equation (3), we need to consider the downward radiation.

$$\bar{\alpha} = \frac{\sum A^i F_d^i \left[(1 - f_{dif}^i)\mathrm{BSA}^i + f_{dif}^i\mathrm{WSA}^i\right]}{\sum A^i F_d^i} \tag{3}$$

where $\bar{\alpha}$ is the spatially aggregated shortwave albedo. For pixel i, A^i is the area of the pixel, F_d^i is the surface downward radiation under all-sky condition, BSA^i and WSA^i are the BSA and WSA, respectively.

3.2.3. Trend Analysis

In this paper, the univariate linear regression method is used to estimate the interannual change rates of surface albedo, NDVI and snow cover in the QM in the past 20 years. The calculation formula of the change rate θ_{slope} is as follows:

$$\theta_{slope} = \frac{n \sum\limits_{i=1}^{n} (i\alpha_i) - \left(\sum\limits_{i=1}^{n} i \right) \sum\limits_{i=1}^{n} \alpha_i}{n \sum\limits_{i=1}^{n} i^2 - \left(\sum\limits_{i=1}^{n} i \right)^2} \tag{4}$$

where n is the total number of years in the study period, and α_i is the mean value of the variable in year i. $\theta_{slope} > 0$ means that the change of the variable in n years is an increasing trend; on the contrary, $\theta_{slope} < 0$ means that the variable is in a decreasing trend. Then, at the confidence level of 0.05, the F-test method is used to test the significance of the change trend of each pixel.

4. Results

4.1. Multi-Year Average Characteristics of Surface Albedo

First, the albedo data of the QM from 2001 to 2020 was used to calculate the monthly average data, and then the multi-year average and spring (March–May), summer (June–August), autumn (September–November), and winter (December–February of the following year) surface albedo was synthesized. The multi-year average surface albedo represents the overall albedo situation of the QM. According to the results shown in Figure 2a, the multi-year average surface albedo of the QM is about 0.25, showing obvious differences in the spatial distribution. The whole surface albedo increases with altitude, and the surface albedo at northwest is high and the southeast is low. The areas with a high-value of surface albedo are mainly distributed in the Daxue Mountain, Shulenan Mountain, Danghenan Mountain, Tergun Daban Mountain, Hark Mountain, and Lenglongling Mountain, etc., showing a northwest-southeast trend similar to the trend of the mountains. Meanwhile, the land cover is mainly desert, bare rock, glacier, and snow, and the surface albedo value is between 0.4 and 0.6. The areas with a low value of surface albedo are mainly distributed at Qinghai Lake, Har Lake, Laji Mountain, and Heihe River Basin, with values between 0.05 and 0.1. The land cover is dominated by lakes, rivers, and high vegetation coverage. Since the lake is similar to a black body and has a strong ability to absorb short waves, Qinghai Lake is the area with a minimum surface albedo.

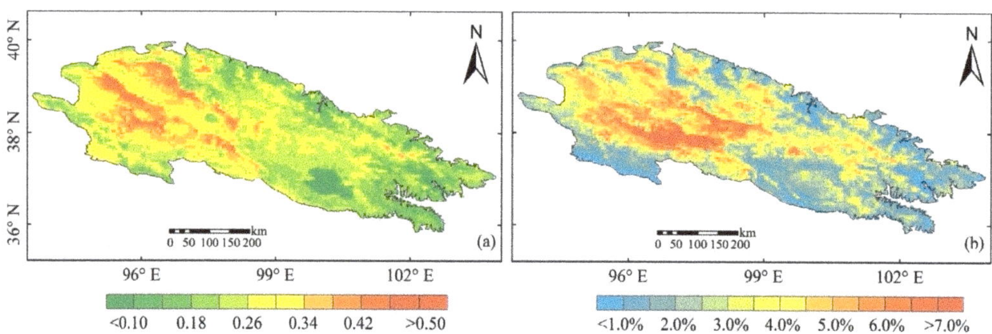

Figure 2. Spatial distributions of (**a**) multi-year averaged land surface albedo and (**b**) standard deviation of annual averaged land surface albedo in the QM.

Then, the standard deviation of the surface albedo was used to represent the degree of abnormality and deviation of the surface albedo from the average value. It can be seen from Figure 2b that the surface albedo sensitive areas in the QM are mainly distributed in the alpine belts of the Tergun Daban Mountain, Hark Mountain, Qaidam Mountain, and Zhongwunong Mountain. Also, the surface albedo has been relatively stable over the past 20 years in the Heihe River Basin, Shiyang River Basin, Laji Mountain, the vicinity of the Qinghai Lake, and the west of Qaidam Mountain with low elevations.

The spatial distribution of the multi-year average surface albedo in the QM in different seasons was further analyzed. As shown in Figure 3, the surface albedo in the QM is high in the west and low in the east in all four seasons, but there are differences in different seasons. The average values of surface albedo in spring, summer, autumn, and winter in the QM are 0.25, 0.18, 0.24, and 0.30, respectively, and the variation curve is in a single-valley "V" shape. It is the smallest in summer and the largest in winter. The spatial distribution of surface albedo in spring and autumn is similar, but the surface albedo is slightly higher in spring. As shown in Figure 3b, the spatial distribution of surface albedo in summer is different from other seasons. The number of areas with a high surface albedo is obviously smaller than that in other seasons. In the Shulenan Mountain and Tergun Daban Mountain, the overall spatial distribution of surface albedo is high in the west and low in the east, which is consistent with the spatial distribution in other seasons. The areas with a high surface albedo in winter are similar to those in spring and autumn. They are concentrated in Daxue Mountain, Shulenan Mountain, Danghenan Mountain, and Tergun Daban Mountain in the west and Lenglongling Mountain in the east. However, the areas with a high surface albedo in winter cover a wider range.

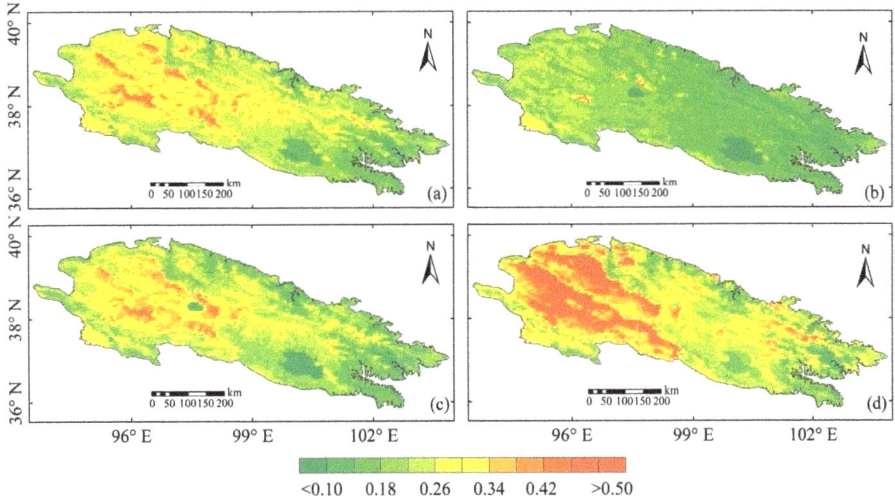

Figure 3. Spatial distributions of multi-year averaged land surface albedo in the QM for (**a**) spring (March–May), (**b**) summer (June–August), (**c**) autumn (September–November), and (**d**) winter (December–February of the following year).

4.2. Annual Variation Characteristics of Surface Albedo in the QM

As shown in Figure 4, the annual variation of the average surface albedo for many years exhibits a "U" shaped, with the "left valley slope" indicating the variation from January to July, the "valley bottom" indicating the variation from July and August, and the "right valley slope" indicating the variation from September to December. The surface albedo was the largest in January and the smallest in August, with a value of 0.31 and 0.17, respectively. It can be seen from the variation trend line that the change slopes from January to August and August to December are respectively −0.02 and 0.03. The surface albedo shows a slow downward trend from January to August and reaches the minimum value in August, and the albedo in July and August is not much different; then, it shows a strong upward trend from September to November and then slowly rises, reaching the maximum in January of the following year.

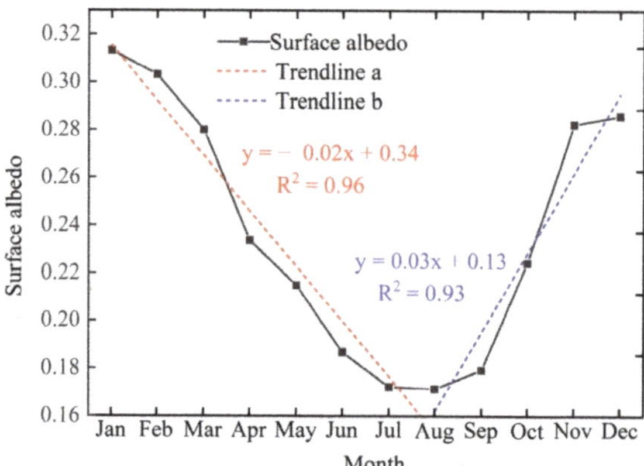

Figure 4. Annual variation characteristics of the land surface albedo in the QM.

The influence of the factors (i.e., NDVI and snow cover) on the annual change of surface albedo in the QM were further analyzed. Figure 5a shows that the annual change trend of snow cover and surface albedo is consistent, and there is a significant positive correlation between the two at the 0.01 level (both sides), with a correlation coefficient of 0.96. When the snow cover gradually decreases from January to August, the surface albedo also decreases slowly; when the snow cover increases rapidly from August to December, the surface albedo also increases rapidly. The annual change of NDVI shows a trend of increasing first and then decreasing, which is opposite to the variation trend of surface albedo (Figure 5b). Through Pearson correlation analysis, it was found that the annual change of surface albedo is significantly correlated with that of NDVI at the 0.01 level (both sides), with a correlation coefficient of −0.91. In autumn, the vegetation begins to turn yellow and new snowfall occurs, and the surface albedo gradually increases and peaks in January of the following year. The above analysis results show that the annual change of surface albedo is positively related to surface cover and negatively related to NDVI.

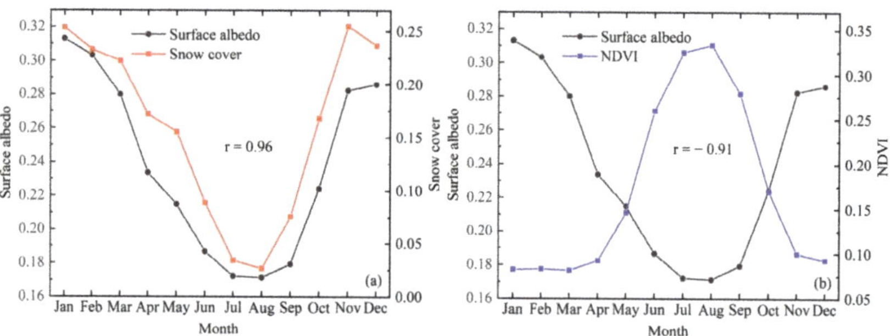

Figure 5. Annual variations of land surface albedo and snow cover (**a**) and NDVI (**b**) in the QM.

4.3. Characteristics of Interannual Variation of Surface Albedo in QM

The annual average value of surface albedo, snow cover, and NDVI in the QM from 2001 to 2020 was calculated to obtain the interannual variation trend (Figure 6). It can be seen that the annual average surface albedo in the QM shows a slight upward trend, with a change rate of $5.0 \times 10^{-3}/10a$. Since the surface albedo reached the minimum and maximum respectively in 2001 and 2019, the annual average surface albedo of the

QM from 2001 to 2020 showed a significant upward trend. However, the growth rate differs in different time periods. The change rate of from 2001 to 2010 and from 2010 to 2020 is $1.5 \times 10^{-2}/10a$ and $2.7 \times 10^{-2}/10a$, respectively. From 2001 to 2010, the surface albedo showed an "up-down-up-down" fluctuation; from 2010 to 2013, the surface albedo decreased substantially; from 2016 to 2019, the surface albedo showed a strong upward trend. Overall, the surface albedo fluctuated significantly in the last 10 years (Figure 6a).

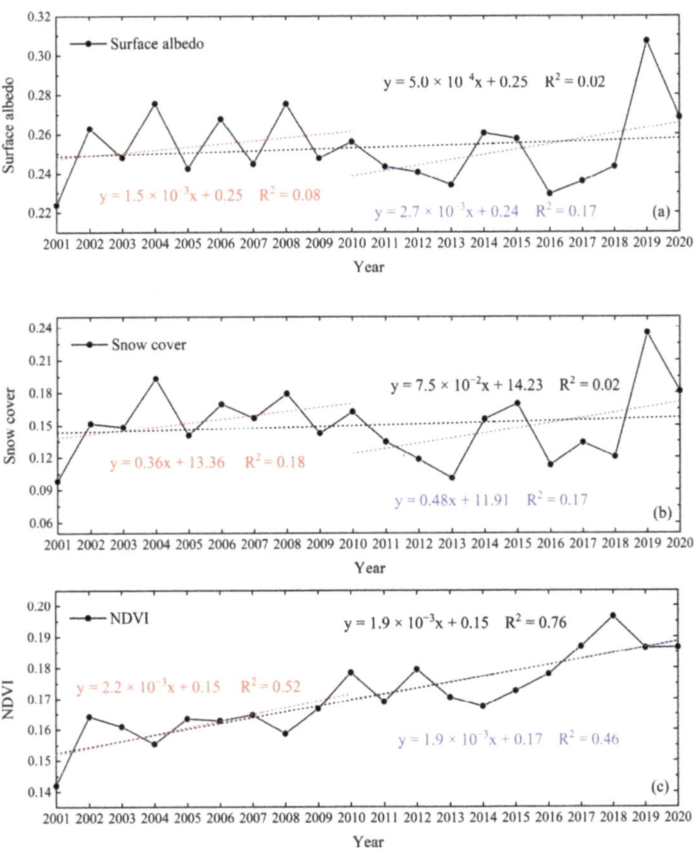

Figure 6. The inter-annual variations of regional averaged land surface albedo (**a**), snow cover (**b**), and NDVI (**c**) in the QM.

The snow cover rate in the QM showed a slight upward trend in general from 2001 to 2020, with a change rate of 0.75/10a. It reached the minimum and maximum in 2001 and 2019, respectively. The snow cover rate increased from 2001 to 2010 with a change rate of 3.6/10a, which is slightly smaller than that from 2010 to 2020 (i.e., 4.8/10a). From 2001 to 2004 and from 2013 to 2015, there were two rising periods of snow cover rate. From 2010 to 2013, the snow cover showed a significant downward trend (Figure 6b), which is consistent with the research results of Liang et al. (2019) on the temporal and spatial variation of snow cover in the QM [49]. The comparison of the interannual changes of surface albedo and snow cover indicates that the two change trends are highly consistent. Through Pearson correlation analysis, it was found that there is a significant positive correlation between the surface albedo and snow cover rate, with a correlation coefficient of 0.95. The indicates that the interannual variation of the surface albedo in the QM is largely affected by the snow cover changes.

The NDVI in the QM changed significantly from 2001 to 2020. It continued to increase and reached the maximum in 2018, with a 20-year change rate of 0.02/10a. There was a strong upward trend in NDVI from 2001 to 2005 and 2008 to 2010, and a slight downward trend from 2005 to 2008 and 2012 to 2016 (Figure 6c). This result is consistent with the observation of Zhang et al. (2021) in analyzing the variation trend of NDVI in the QM during the growing season. Through Pearson correlation analysis, it was found that there is a weak positive correlation between the surface albedo and NDVI, with a correlation coefficient of 0.12. When the linear trend of the two is eliminated, the surface albedo and NDVI are negatively correlated (r = −0.02). In addition, the spatial resolution of the surface albedo and NDVI data is still low, which may not accurately describe the interannual fluctuations of the surface conditions in the QM, resulting in a weak positive correlation between the two in this region [5]. Since the interannual variation cannot fully reflect the relationship between the surface albedo and NDVI, it is necessary to further explore their relationship by using high-precision remote sensing data.

4.4. Spatial Variation Trend of Surface Albedo in QM

The univariate linear regression method was used to calculate the interannual variation trends of the annual average surface albedo, NDVI, and snow cover in the QM. Meanwhile, the F-test method was employed to test the significance of the variation trend of each pixel at a confidence level of 0.05. As shown in Figure 7, the interannual variation of the average annual surface albedo in the QM shows obvious spatial heterogeneity, and the areas with a decreasing annual surface albedo are mainly distributed in the edge of the QM, such as the Qinghai Nanshan, the Shule River Basin, and the Heihe River Basin, accounting for about 39.6% of the total area. Most regions show an increasing trend, and the areas with increased annual surface albedo are mainly distributed in Hark Mountain, Shulenan Mountain, Tulainan Mountain, Tulai Mountain, Datong Mountain and Daban Mountain, accounting for about 60.4% of the total area. Besides, the insignificant areas account for 85.7% of the total area, while the significant areas only account for 14.3% of the total area. The areas with significantly increased annual surface albedo are located in the Tulai Mountain, Datong Mountain and Daban Mountain, accounting for 8.2% of the total area, and areas with significantly decreased annual surface albedo are located in the Qinghainan Mountain, Shule River Basin, and Heihe River Basin, accounting for 6.1% of the total area (Figure 7b).

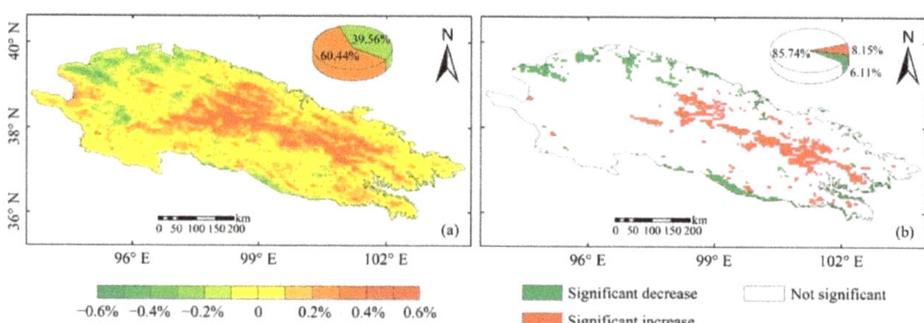

Figure 7. Spatial distributions of (**a**) interannual trend of land surface albedo and (**b**) the areas with the trend that passed the significance test at the 0.05 level in the QM.

4.5. Analysis of Influencing Factors of Surface Albedo at a Spatial Scale

Section 4.3 discusses the influencing factors for the regional average interannual variation of the surface albedo. This part focuses on the analysis of the spatially influencing factors of the surface albedo variation. The surface albedo of the QM is not only sensitive to changes in vegetation and snow cover, but also has a differentiation law with changes in terrain such as altitude. To analyze the distribution characteristics of surface albedo and the influencing factors in QM with different altitude gradients, the surface albedo and the influencing factors were divided according to the altitude of QM at an interval of 500 m (Figure 8). With the increase of altitude, the vertical distribution of surface albedo, snow cover, and vegetation cover in the QM changes significantly. A total of 81.8% of the QM are located at the altitude of 3000–4500 m. Overall, the surface albedo and snow cover increase with the increase of altitude; the NDVI shows a trend of first increase and then decrease with the increase of altitude; the snow cover dominates the spatial variation of surface albedo. Further research found that the vertical changes of the surface albedo and its influencing factors in different seasons are consistent with the average vertical changes in many years. There are significant positive correlations between albedo and snow cover and elevation. Therefore, altitude is another important influencing factor of the surface albedo variation in the QM. It is worth noting that when the altitude exceeds 4500 m, the increment rate of surface albedo in winter is not as obvious as that in other seasons. Compared with Figure 9b, it is found that the snow cover rate in winter also decreases in high altitude areas, and it is even smaller than that in other seasons. This may be the reason why the surface albedo in winter is lower than that in spring in the area above 5000 m above sea level.

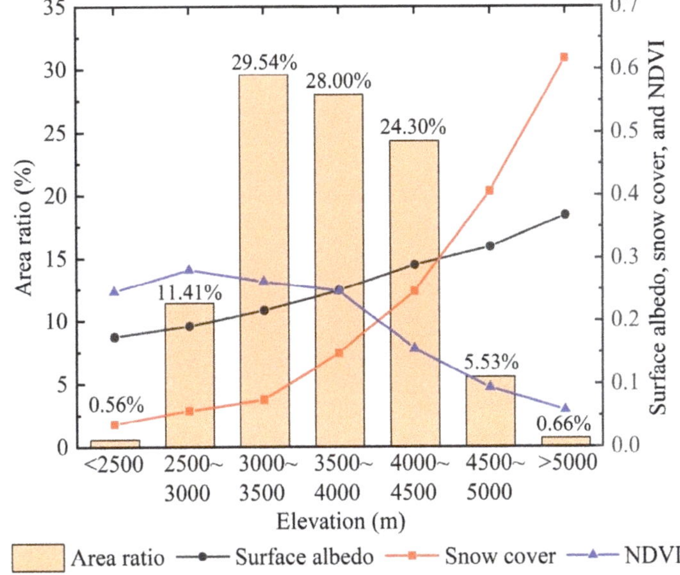

Figure 8. The vertical distributions of regionally annual average land surface albedo snow cover, and NDVI in the QM.

The spatial distribution of the interannual variation trend of snow cover has a good correspondence with the surface albedo. Specifically, the significant increase in snow cover is distributed in the south of Har Lake, the south of Qinghai Lake, Tulai Mountain, Datong Mountain, and Daban Mountain (as shown in Figure 10a), and the surface albedo also shows an increasing trend in these regions. This indicates that snow cover is an influencing factor of the significant increase in surface albedo in the central QM. Comparing the interannual variation trend of surface albedo and NDVI (Figure 10b), it is found that the spatial distribution of surface albedo and the interannual variation trend of NDVI also has a good correspondence: the areas with improved vegetation coverage is almost the same as those with reduced surface albedo. This suggests that the changes in the vegetation cover in these regions is also an important factor of surface albedo changes.

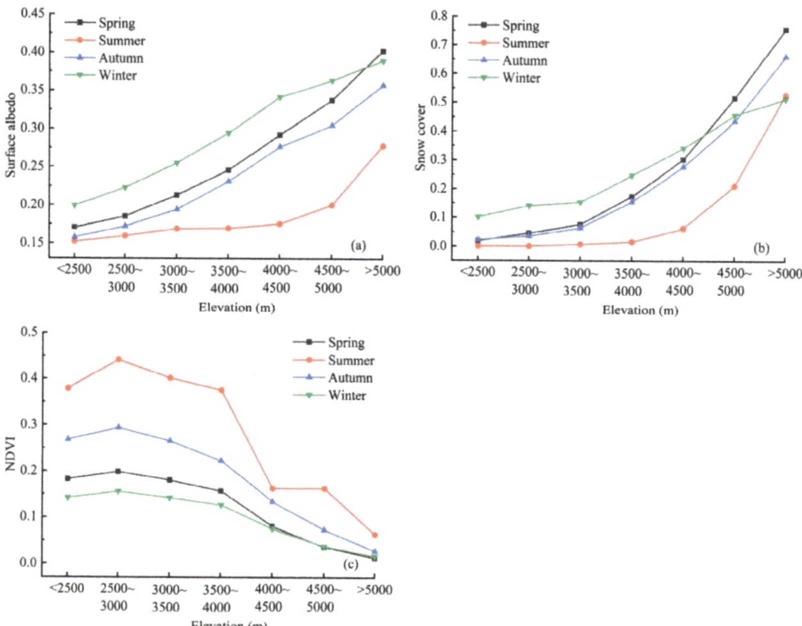

Figure 9. The vertical distributions of (**a**) land surface albedo and its influencing factors ((**b**) snow cover and (**c**) NDVI) in four seasons in the QM.

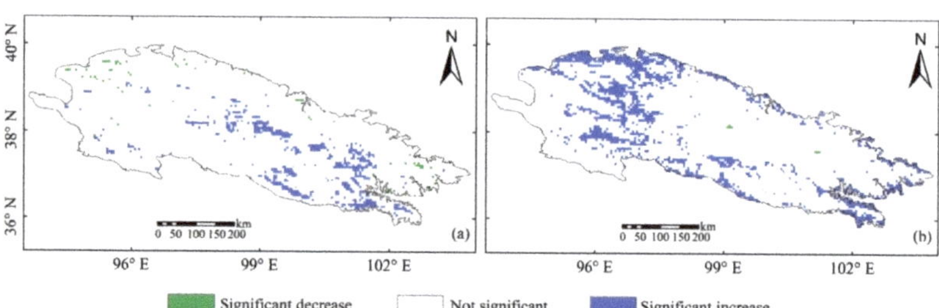

Figure 10. Spatial distributions of interannual trends of annual average (**a**) snow cover and (**b**) NDVI showing the interannual change rates that passed significance test at the 0.05 level in the QM.

The distribution of spatial correlation coefficients between surface albedo and influencing factors (Figure 11) shows that snow cover and vegetation have significant feedback on the changes of surface albedo. Surface albedo is significantly positively correlated with snow cover in most parts of the QM but negatively correlated with NDVI (Figure 11). Specifically, the surface albedo and snow cover rate are significantly positively correlated (with a correlation coefficient higher than 0.8) in the areas including Datong Mountain, Tulai Mountain, Hark Mountain, Zhongwunong Mountain, and the west of Tergun Daban Mountain (Figure 11a), accounting for about 72% of the total area. The areas showing a negative correlation between surface albedo and NDVI are distributed in the Shule Lake Basin, Heihe River Basin, Qinghainan Mountain, etc. (Figure 11b), which correspond to the areas with significantly improved vegetation coverage. Besides, there are sporadic positive correlations near the Qaidam Basin and Qinghai Lake, and these areas may have a combined effect of vegetation and snow cover on surface albedo changes [6]. Combining Figures 7b and 10b, it is found that an increase in NDVI in the low vegetation coverage area in the northern QM contributes to a decrease in albedo.

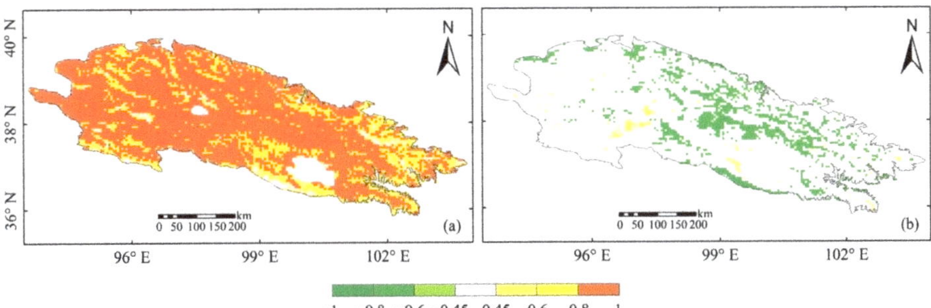

Figure 11. Spatial distributions of correlation coefficients between annual average land surface albedo and influencing factors ((**a**) snow cover and (**b**) NDVI) in the QM. The blank area in the study area indicates the area where the correlation coefficient fails the 0.05 significance level test.

5. Conclusions and Discussion

In this paper, the temporal and spatial dynamic distribution of surface albedo and its influencing factors in the QM from 2001 to 2020 are analyzed by using the MODIS MCD43C3 V006 product, combined with snow cover and NDVI data. The following conclusions are drawn.

(1) The multi-year average surface albedo in the QM is about 0.25, and there are obvious differences in the spatial distribution. Overall, the surface albedo increases with the altitude, and it is high in the west and low in the east. The areas with significant interannual changes include Daxue Mountain, Tulainan Mountain, Tergun Daban Mountain, Shulenan Mountain, and other high-value areas.

(2) The spatial distribution of surface albedo in the QM differs in different seasons. The order of surface albedo in the four seasons is winter > spring > autumn > summer, where the spatial distribution of surface albedo in spring and autumn is similar.

(3) From 2001 to 2020, the interannual variation of the annual average surface albedo in the QM showed a slight upward trend, with a change rate of $5.0 \times 10^{-3}/10a$. The fluctuation of the surface albedo in the study period was obviously more significant than that in the previous 10 years. Snow cover showed a slight increase during this period, and there was a significant positive correlation between surface albedo and snow cover rate. NDVI showed a significant upward trend, indicating that the overall vegetation was improving. Before removing the linear trend of the surface albedo and NDVI, there was a weak positive correlation between the two; when the linear trend was removed, a negative correlation was found between the two.

(4) The annual and interannual variations of the surface albedo in the QM are closely related to the surface cover. The annual variation of surface albedo is "U" shaped, with the largest variation in January and the smallest variation in August, which is positively correlated with snow cover and negatively correlated with NDVI. As for interannual variation, the increase of the regional average surface albedo is significantly related to the increase of snow cover. In terms of spatial distribution, the interannual variation of the surface albedo in most areas of the QM is mainly affected by the change of snow cover. The improvement of vegetation cover in marginal areas is the main factor of the significant decrease of surface albedo in these areas.

The MODIS surface albedo maintains a good consistency with the ground observations in most time periods, and the inversion accuracy is high [33,50]. Zhang et al. (2021) found that MODIS remote sensing data has good applicability in the QM, and it can be used to study the temporal and spatial changes of snow albedo. In the plateau area, the accuracy of MODIS surface albedo inversion results is not high enough, but after manual analysis and elimination of errors, the root mean square error of the surface observation results can be reduced to 0.00329, which can fully meet the accuracy requirements of climate and land surface process models [51]. In addition, long-time-series surface albedo datasets remove the uncertainty of trend analysis greatly, which is crucial for evaluating surface albedo trends [8].

The Earth's surface comprehensively manifests various ground features and is complex and changeable, which makes the surface albedo vary greatly in space and time. NDVI is one of the most commonly used vegetation indices, which can accurately reflect characteristics such as vegetation cover density and growth [41,47]. When NDVI gradually increases from a small value, the surface albedo decreases rapidly. Then, as the NDVI continues to increase, the surface albedo decreases slowly. When NDVI > 0.5, the surface albedo almost does not change with the change of NDVI [52]. In addition, ice and snow are highly positively correlated with surface albedo. Usually, the albedo of old snow is 0.7–0.8, and that of new snow is 0.8–0.9 [19]. As time passes, the new snow will be granulated, the grains will be deformed and continuously densified, the particle size will increase, and the pollution will deteriorate. The relationship between the albedo with ice and snow states is as follows: fresh snow > old snow > grain snow > glacier ice > contaminated glacier ice [5,52]. In recent years, under the warming and humidification in the plateau region, the significant changes in the surface cover have caused corresponding changes in the surface albedo, and the surface albedo will react to climate change, thus affecting the regional climate system substantially [2,5,8].

The change of surface albedo in different regions is closely related to the terrain and surface coverage [53]. This study mainly focuses on the relationship between surface albedo and its influencing factors (snow cover, vegetation, and altitude). However, different vegetation cover types make different contributions to surface albedo [54], and further investigation is needed to more accurately monitor vegetation dynamics and understand the impact of surface albedo on ecosystems. In recent years, the local ecological environment damage in the QM has attracted much attention. In response to the ecological problems in the QM, the state has successively adopted a number of remediation measures to strengthen ecological environment protection and ecological restoration in the QM. The rectification and supervision of human activities such as mineral resources development, tourism activities, grazing, and water and soil resources in the QM have achieved initial results. From 2001 to 2020, the vegetation in the QM showed significant improvement, mainly distributed in the Shule River Basin, the Heihe River Basin, and the relatively low-altitude mountains in the Qinghai Lake Basin. Meanwhile, the snow cover increased significantly in the relatively high-altitude Qinghainan Mountain, Datong Mountain, Daban Mountain, and other areas. The improvement of vegetation coverage will cause the surface albedo to decrease, and through the negative feedback of the surface albedo, it will further promote the vegetation, so that the albedo will continue to decrease. The increase of ice and snow coverage will cause the surface albedo to increase, and positive feedback will be formed between ice and snow, albedo and temperature. Based on this, the area of ice and snow will be further expanded, and the albedo will continue to increase. Besides, other factors, such as soil moisture and slope aspect, also affect the surface albedo, and future studies will investigate these factors with high-resolution data. Through the understanding of the surface albedo feedback mechanism, the relationship between climate change and human activities can be deeply understood, thereby providing a scientific basis for the win-win green development of ecological livelihoods in the QM.

Author Contributions: Conceptualization, G.P. and J.L.; methodology, G.P. and X.W.; software, J.L. and F.L.; validation, G.P., J.L. and Y.Z.; formal analysis, J.L.; data curation, J.L.; writing—original draft preparation, J.L.; writing—review and editing, G.P. and X.W.; visualization, J.L.; supervision, G.P. and X.W.; funding acquisition, G.P. and X.W. All authors have read and agreed to the published version of the manuscript.

Funding: This research was funded by the National Natural Science Foundation of China (Grants Nos. 42161025), the National Natural Science Foundation of China (U21A2006), the National Key Research and Development Program of China (2019YFC0507401), the Strategic Priority Research Program of Chinese Academy of Sciences (XDA20100102), the Natural Science Foundation of Gansu

Province in China (20JR5RA405), and the Second Tibetan Plateau Scientific Expedition and Research Program (STEP) (2019QZKK0208).

Institutional Review Board Statement: Not applicable.

Informed Consent Statement: Not applicable.

Data Availability Statement: The data (the MCD43C3, MOD10CM, MOD13C2 products, and downward solar radiation reanalysis data) presented in this study are openly available in (NASA EOSDIS Land Processes DAAC), (NASA National Snow and Ice Data Center Distributed Active Archive Center), and (National Centers for Environmental Prediction) at (https://doi.org/10.5067/MODIS/MCD43C3.006), (https://doi.org/10.5067/MODIS/MOD13C2.006.), (https://doi.org/10.5067/MODIS/MOD10CM.061), and (https://psl.noaa.gov/data/gridded/data.ncep.reanalysis.html (accessed on 5 December 2020)), respectively.

Acknowledgments: The authors would like to acknowledge the National Aeronautics and Space Administration (NASA) for providing the MODIS MCD43C3, MOD10CM, and MOD13C2 products, and the National Centers for Environmental Prediction (NCEP) for the downward solar radiation reanalysis data.

Conflicts of Interest: The authors declare no conflict of interest.

References

1. Liang, S.; Wang, K.; Zhang, X.; Wild, M. Review on estimation of land surface radiation and energy budgets from ground measurement, remote sensing and model simulations. *IEEE J. Sel. Top. Appl. Earth Obs. Remote Sens.* **2010**, *3*, 225–240. [CrossRef]
2. Bonan, G.B. Forests and climate change: Forcings, feedbacks, and the climate benefits of forests. *Science* **2008**, *320*, 1444–1449. [CrossRef] [PubMed]
3. Trenberth, K.E.; Fasullo, J.T.; Kiehl, J. Earth's global energy budget. *Bull. Am. Meteorol. Soc.* **2009**, *90*, 311–324. [CrossRef]
4. Wang, J.; Gao, F. Discussion on the problems on land surface albedo retrieval by remote sensing Data. *Remote Sens. Technol. Appl.* **2004**, *19*, 295–300.
5. Li, Q.; Ma, M.; Wu, X.; Yang, H. Snow cover and vegetation-induced decrease in global albedo from 2002 to 2016. *J. Geophys. Res.* **2018**, *123*, 124–138. [CrossRef]
6. Pang, G.; Chen, D.; Wang, X.; Lai, H. Spatiotemporal variations of land surface albedo and associated influencing factors on the Tibetan Plateau. *Sci. Total Environ.* **2022**, *804*, 150100. [CrossRef]
7. Meng, X.; Lyu, S.; Zhang, T.; Zhao, L.; Li, Z.; Han, B.; Li, S.; Ma, D.; Chen, H.; Ao, Y. Simulated cold bias being improved by using MODIS time-varying albedo in the Tibetan Plateau in WRF model. *Environ. Res. Lett.* **2018**, *13*, 044028. [CrossRef]
8. Li, X.; Zhang, H.; Qu, Y. Land surface albedo variations in SanJiang plain from 1982 to 2015: Assessing with glass data. *Chin. Geogr. Sci.* **2020**, *30*, 876–888. [CrossRef]
9. Charney, J.; Stone, P.H.; Quirk, W.J. Drought in the Sahara: A biogeophysical feedback mechanism. *Science* **1975**, *187*, 434–435. [CrossRef]
10. Hall, A. The role of surface albedo feedback in climate. *J. Clim.* **2004**, *17*, 1550–1568. [CrossRef]
11. Chapin III, F.S.; Sturm, M.; Serreze, M.C.; McFadden, J.P.; Key, J.R.; Lloyd, A.H.; McGuire, A.D.; Rupp, T.S.; Lynch, A.H.; Schimel, J.P.; et al. Role of land-surface changes in Arctic summer warming. *Science* **2005**, *310*, 657–660. [CrossRef] [PubMed]
12. Courel, M.F.; Kandel, R.S.; Rasool, S.I. Surface albedo and the Sahel drought. *Nature* **1984**, *307*, 528–531. [CrossRef]
13. Ghimire, B.; Williams, C.A.; Masek, J.; Gao, F.; Wang, Z.; Schaaf, C.; He, T. Global albedo change and radiative cooling from anthropogenic land cover change, 1700 to 2005 based on MODIS, land use harmonization, radiative kernels, and reanalysis. *Geophys. Res. Lett.* **2014**, *41*, 9087–9096. [CrossRef]
14. Moody, E.G.; King, M.D.; Schaaf, C.B.; Hall, D.K.; Platnick, S. Northern Hemisphere five-year average (2000–2004) spectral albedos of surfaces in the presence of snow: Statistics computed from Terra MODIS land products. *Remote Sens. Environ.* **2007**, *111*, 337–345. [CrossRef]
15. Warren, S.G. Optical properties of snow. *Rev. Geophys.* **1982**, *20*, 67–89. [CrossRef]
16. Lin, X.; Wen, J.; Liu, Q.; You, D.; Wu, S.; Hao, D.; Xiao, Q.; Zhang, Z.; Zhang, Z. Spatiotemporal variability of land surface albedo over the Tibet Plateau from 2001 to 2019. *Remote Sens.* **2020**, *12*, 1188. [CrossRef]
17. Bond, T.C.; Doherty, S.J.; Fahey, D.W.; Forster, P.M.; Berntsen, T.; DeAngelo, B.J.; Flanner, M.G.; Ghan, S.; Kärcher, B.; Koch, D.; et al. Bounding the role of black carbon in the climate system: A scientific assessment. *J. Geophys. Res.* **2013**, *118*, 5380–5552. [CrossRef]
18. Takeuchi, N.; Matsuda, Y.; Sakai, A.; Fujita, K. A large amount of biogenic surface dust (cryoconite) on a glacier in the Qilian Mountains, China. *Bull. Glaciol. Res.* **2005**, *22*, 1–8.
19. Zhang, R.; Wang, H.; Fu, Q.; Rasch, P.J.; Wang, X. Unraveling driving forces explaining significant reduction in satellite-inferred Arctic surface albedo since the 1980s. *Proc. Natl. Acad. Sci. USA* **2019**, *116*, 23947–23953. [CrossRef]

20. Pithan, F.; Mauritsen, T. Arctic amplification dominated by temperature feedbacks in contemporary climate models. *Nat. Geosci.* **2014**, *7*, 181–184. [CrossRef]

21. Zhu, Z.; Piao, S.; Myneni, R.B.; Huang, M.; Zeng, Z.; Canadell, J.G.; Ciais, P.; Sitch, S.; Friedlingstein, P.; Zeng, N.; et al. Greening of the Earth and its drivers. *Nat. Clim. Chang.* **2016**, *6*, 791–795. [CrossRef]

22. Zheng, L.; Zhao, G.; Dong, J.; Ge, Q.; Tao, J.; Zhang, X.; Qi, Y.; Doughty, R.B.; Xiao, X. Spatial, temporal, and spectral variations in albedo due to vegetation changes in China's grasslands. *ISPRS J. Photogramm. Remote Sens.* **2019**, *152*, 1–12. [CrossRef]

23. Betts, R.A. Offset of the potential carbon sink from boreal forestation by decreases in surface albedo. *Nature* **2000**, *408*, 187–190. [CrossRef] [PubMed]

24. Davin, E.L.; de Noblet-Ducoudré, N. Climatic impact of global-scale deforestation: Radiative versus nonradiative processes. *J. Clim.* **2010**, *23*, 97–112. [CrossRef]

25. Jiao, T.; Williams, C.A.; Ghimire, B.; Masek, J.; Gao, F.; Schaaf, C. Global climate forcing from albedo change caused by large-scale deforestation and reforestation: Quantification and attribution of geographic variation. *Clim. Chang.* **2017**, *142*, 463–476. [CrossRef]

26. Zeng, Z.; Wang, D.; Yang, L.; Wu, J.; Ziegler, A.D.; Liu, M.; Ciais, P.; Searchinger, T.D.; Yang, Z.; Wood, E.F.; et al. Deforestation-induced warming over tropical mountain regions regulated by elevation. *Nat. Geosci.* **2021**, *14*, 23–29. [CrossRef]

27. Li, X.; Gou, X.; Wang, N.; Sheng, Y.; Jin, H.; Qi, Y.; Song, X.; Hou, F.; Li, Y.; Niu, X.; et al. Tightening ecological management facilitates green development in the Qilian Mountains. *Chin. Sci. Bull.* **2019**, *64*, 2928–2937. (In Chinese)

28. Wang, X.; Pang, G.; Yang, M.; Wan, G.; Liu, Z. Precipitation changes in the Qilian Mountains associated with the shifts of regional atmospheric water vapour during 1960–2014. *Int. J. Climatol.* **2018**, *38*, 4355–4368. [CrossRef]

29. Sun, M.; Liu, S.; Yao, X.; Guo, W.; Xu, J. Glacier changes in the Qilian Mountains in the past half-century: Based on the revised First and Second Chinese Glacier Inventory. *J. Geogr. Sci.* **2018**, *28*, 206–220. [CrossRef]

30. Wang, T.; Gao, F.; Wang, B.; Wang, P.; Wang, Q.; Song, H.; Yin, C. Status and suggestions on ecological protection and restoration of Qilian Mountains. *J. Glaciol. Geocryol.* **2017**, *39*, 229–234.

31. Li, Z.; Feng, Q.; Li, Z.; Wang, X.; Gui, J.; Zhang, B.; Li, Y.; Deng, X.; Xue, J.; Liang, P.; et al. Reversing conflict between humans and the environment-The experience in the Qilian Mountains. *Renew. Sustain. Energy Rev.* **2021**, *148*, 111333.

32. Feng, Z.M.; Wen, J.G.; Xiao, Q. Comparison of global albedo products of MODIS V006 and V005 based on FLUXNET. *J. Remote Sens.* **2018**, *22*, 97–109. (In Chinese)

33. Cescatti, A.; Marcolla, B.; Vannan, S.K.S.; Pan, J.Y.; Román, M.O.; Yang, X.; Ciais, P.; Cook, R.B.; Law, B.E.; Matteucci, E.; et al. Intercomparison of MODIS albedo retrievals and in situ measurements across the global FLUXNET network. *Remote Sens. Environ.* **2012**, *121*, 323–334. [CrossRef]

34. He, T.; Liang, S.; Song, D.X. Analysis of global land surface albedo climatology and spatial-temporal variation during 1981–2010 from multiple satellite products. *J. Geophys. Res.* **2014**, *119*, 10281–10298. [CrossRef]

35. Mira, M.; Weiss, M.; Baret, F.; Courault, D.; Hagolle, O.; Gallego-Elvira, B.; Olioso, A. The MODIS (collection V006) BRDF/albedo product MCD43D: Temporal course evaluated over agricultural landscape. *Remote Sens. Environ.* **2015**, *170*, 216–228. [CrossRef]

36. Wang, Z.; Schaaf, C.B.; Sun, Q.; Shuai, Y.; Román, M.O. Capturing rapid land surface dynamics with Collection V006 MODIS BRDF/NBAR/Albedo (MCD43) products. *Remote Sens. Environ.* **2018**, *207*, 50–64. [CrossRef]

37. Jiang, X.; Wang, N.; Pu, J.; He, J.; Chen, L. Variations of albedo and spectral reflectance on Qiyi Glacier in Qilian Mountains during the ablation season. *Sci. Cold Arid. Reg.* **2009**, *1*, 59–70.

38. Zhang, T.; Gao, T.; Diao, W.; Zhang, Y. Snow/ice albedo variation and its impact on glacier mass balance in the Qilian Mountains. *J. Glaciol. Geocryol.* **2021**, *43*, 145–157.

39. Tian, H.; Yang, T.; Liu, Q. Climate change and glacier area shrinkage in the Qilian mountains, China, from 1956 to 2010. *Ann. Glaciol.* **2014**, *55*, 187–197. [CrossRef]

40. Qian, D.; Du, Y.; Li, Q.; Guo, X.; Cao, G. Alpine grassland management based on ecosystem service relationships on the southern slopes of the Qilian Mountains, China. *J. Environ. Manag.* **2021**, *288*, 112447. [CrossRef]

41. Zhang, L.; Yan, H.; Qiu, L.; Cao, S.; He, Y.; Pang, G. Spatial and temporal analyses of vegetation changes at multiple time scales in the Qilian Mountains. *Remote Sens.* **2021**, *13*, 5046. [CrossRef]

42. Filonchyk, M.; Hurynovich, V. Validation of MODIS aerosol products with AERONET measurements of different land cover types in areas over Eastern Europe and China. *J. Geovisualization Spat. Anal.* **2020**, *4*, 10. [CrossRef]

43. Schaaf, C.; Wang, Z. MCD43C3 MODIS/Terra+Aqua BRDF/Albedo Albedo Daily L3 Global 0.05Deg CMG V006. *NASA EOSDIS Land Processes DAAC* **2015**. [CrossRef]

44. Hall, D.K.; Riggs, G.A. *MODIS/Terra Snow Cover Monthly L3 Global 0.05Deg CMG*; Version 6.1; NASA National Snow and Ice Data Center Distributed Active Archive Center: Boulder, CO, USA, 2021.

45. Didan, K. MOD13C2 MODIS/Terra Vegetation Indices Monthly L3 Global 0.05Deg CMG V006. *NASA EOSDIS Land Processes DAAC* **2015**. [CrossRef]

46. Kalnay, E.; Kanamitsu, M.; Kistler, R.; Collins, W.; Deaven, D.; Gandin, L.; Iredell, M.; Saha, S.; White, G.; Woollen, J.; et al. The NCEP/NCAR 40-year reanalysis project. *Bull. Am. Meteorol. Soc.* **1996**, *77*, 437–472. [CrossRef]

47. Pang, G.; Wang, X.; Yang, M. Using the NDVI to identify variations in, and responses of, vegetation to climate change on the Tibetan Plateau from 1982 to 2012. *Quat. Int.* **2017**, *444*, 87–96. [CrossRef]

48. Yang, W.; Wang, Y.; Liu, X.; Zhao, H.; Shao, R.; Wang, G. Evaluation of the rescaled complementary principle in the estimation of evaporation on the Tibetan Plateau. *Sci. Total Environ.* **2020**, *699*, 134367. [CrossRef]
49. Liang, P.; Li, Z.; Zhang, H. Temporal-spatial variation characteristics of snow cover in Qilian Mountains from 2001 to 2017. *Arid Land Geogr.* **2019**, *42*, 56–66.
50. Rongali, G.; Keshari, A.K.; Gosain, A.K.; Khosa, R. Split-window algorithm for retrieval of land surface temperature using Landsat 8 thermal infrared data. *J. Geovisualiz. Spat. Anal.* **2018**, *2*, 14. [CrossRef]
51. Chen, A.; Meng, W.; Hu, S.; Bian, L. Comparative analysis on land surface albedo from MODIS and GLASS over the Tibetan Plateau. *Trans. Atmos Sci* **2020**, *43*, 932–942.
52. Xiao, D.; Tao, F.; Moiwo, J.P. Research progress on surface albedo under global change. *Adv. Earth Sci.* **2011**, *26*, 1217.
53. Hu, Y.; Hou, M.; Zhao, C.; Zhen, X.; Yao, L.; Xu, Y. Human-induced changes of surface albedo in Northern China from 1992-2012. *Int. J. Appl. Earth Obs. Geoinf.* **2019**, *79*, 184–191. [CrossRef]
54. Wang, S.; Hu, Y.; Wang, S.; Shang, K.; Yan, D. Influence of the difference of plant functional types on surface albedo variation. *Remote Sens. Technol. Appl.* **2015**, *30*, 932–938.

remote sensing

Article

Direct and Legacy Effects of Spring Temperature Anomalies on Seasonal Productivity in Northern Ecosystems

Hanna Marsh and Wenxin Zhang *

Department of Physical Geography and Ecosystem Science, Lund University, Sölvegatan 12, SE-223 62 Lund, Sweden; hanna.marsh@nateko.lu.se
* Correspondence: wenxin.zhang@nateko.lu.se; Tel.: +46-46-222-16-87

Abstract: Warmer or cooler spring in northern high latitudes will, for the most part, directly impact gross primary productivity (GPP) of ecosystems, but also carry consequences for the upcoming seasonal GPP. Spatiotemporal patterns of these legacy effects are still largely unknown but important for improving our understanding of how plant phenology is associated with vegetation dynamics. In this study, impacts of spring temperature anomalies on spring, summer and autumn GPP were investigated, and the dominant drivers of summer and autumn GPP including air temperature, vapor pressure deficit and soil moisture have been explored for northern ecosystems (>30°N). Three remote sensing products of seasonal GPP (GOSIF-GPP, NIRv-GPP and FluxSat-GPP) over 2001–2018, all based on a spatial resolution of 0.05°, were employed. Our results indicate that legacy effects from spring temperature are most pronounced in summer, where they have stimulating effects on the Arctic ecosystem productivity. Spring warming likely lessens the harsh climatic constraints that govern the Arctic tundra and extends the growing season length. Further south, legacy effects are mainly negative. This strengthens the hypothesis that enhanced vegetation growth in spring will increase plant water demand and stress in summer and autumn. Soil moisture is the dominant control of summer GPP in temperate regions. However, the dominant meteorological variables controlling vegetation growth may differ depending on the GPP products, highlighting the need to address uncertainties among different methods of estimating GPP.

Keywords: carry-over effects; gross primary productivity; phenology; GOSIF; NIRv; FluxSat

Citation: Marsh, H.; Zhang, W. Direct and Legacy Effects of Spring Temperature Anomalies on Seasonal Productivity in Northern Ecosystems. *Remote Sens.* **2022**, *14*, 2007. https://doi.org/10.3390/rs14092007

Academic Editor: Michael Sprintsin

Received: 7 March 2022
Accepted: 18 April 2022
Published: 21 April 2022

Publisher's Note: MDPI stays neutral with regard to jurisdictional claims in published maps and institutional affiliations.

1. Introduction

The last decades have seen a rapid increase in global temperatures, particularly more pronounced in the northern high latitudes [1]. The rapid warming may lead to changes in the timing of thermal growing seasons and phenological cycles of plants [2]. Earlier onsets of spring, increased frequency of droughts as well as changing precipitation patterns have been observed [1]. These changes may induce direct and legacy effects on ecosystem gross primary production (GPP—the gross uptake of carbon dioxide (CO_2) by plant photosynthesis) as well as various types of climate feedback associated with ecosystem responses to climatic warming [3,4].

Direct impacts on growing-season vegetation productivity due to seasonal warming or drought have been studied previously [5,6]. Spring warming directly affects the onset of leaf growth by advancing bud-burst, leading to increases in spring GPP (GPP_{spring}) and a generally earlier greening of vegetation [7,8]. Summer warming may have positive effects on Arctic and alpine biomes. Based on annual shrub ring-width measurements and long-term meteorological records, growth of both evergreen and deciduous dwarf shrubs were found to be stimulated due to summer warmth [9]. However, severe and long-lasting warming events, i.e., heat waves, may dampen vegetation productivity due to soil moisture deficits, or reach the thermal tolerance threshold of photosynthetic decline in forest growth [10,11]. Similarly, autumn warming can enhance photosynthesis rates via the delayed senescence of plants, however, soil decomposition can also increase, leading to net carbon losses of the ecosystems during this season [12].

Nevertheless, the legacy effects of spring temperature (T_{spring}) anomalies on seasonal GPP is known to a lesser degree. Currently, even the state-of-the-art terrestrial biosphere models struggle to implement them [3]. As noted by [5], most carbon-cycle models seem to overestimate the positive legacy effects on GPP caused by spring warming and underestimate the potential influence of built-up water stress. Accurately assessing direct and legacy effects of seasonal warming on vegetation productivity is important, as vegetation is a key component of the global carbon, water and energy cycles, and its responses may trigger biogeochemical and biophysical climate feedback [13]. A proper understanding of these highly dynamic systems is crucial, particularly for our general ability to accurately predict and mitigate climate impacts, as well as adapting our society to a changing climate [14].

Recent works [5,6,15] indicate that annual GPP can be suppressed or stimulated by both cooler and warmer springs. For example, boreal forests have seen a dampening effect on peak summer productivity related to early springs [16,17]. The suppressed vegetation productivity in summer can be caused by soil moisture deficits resulting from winter precipitation and early onset of vegetation. Other northern ecosystems, mainly dominated by graminoids, lichens and deciduous shrubs, have been shown to benefit from a fairly early date of snowmelt, as many such species are unable to accelerate or adapt their rate of phenological development [18,19]. The delayed outcome, or carry-over effect, thus depends on a combination of several factors, including meteorological and environmental conditions, vegetation composition and phenological responses of vegetation, but as of now, a comprehensive understanding of how these factors interplay remains lacking. Earlier research in this area has attempted to address these questions utilizing data sets with relatively coarse resolution (50×50 km^2) based on the Normalized Difference Vegetation Index (NDVI) (e.g., [10,16]). As vegetation response to warming is spatially non-uniform, such a coarse resolution may fail to represent explicit details of greening and browning trends of different ecosystems. Moreover, NDVI is often claimed to misrepresent the nonlinear nature of vegetation responses, and is thus regarded as less efficient to be a proxy of GPP [20,21].

Recently, a few novel products have been tested and regarded as more robust proxies for canopy structure, leaf pigment content, and, subsequently, plant photosynthetic potential. Among these products, GOSIF (Global Orbiting Carbon Observatory-2 Solar-Induced Chlorophyll Fluorescence data set)-GPP, NIRv (Near-Infrared Reflectance of terrestrial vegetation)-GPP and FluxSat (Fluxnet data fused with satellite images)-GPP, based on Solar-Induced chlorophyll Fluorescence (SIF), canopy near-infrared reflectance, eddy flux and satellite data fusion, respectively, are shown to better reflect photosynthetic activities [22–25]. In this study, we have utilized these three products, and the approach of each product is distinctly different from the others. Based on a higher spatial resolution (0.05°), we aim to explore the direct and legacy effects of spring temperature anomalies on seasonal GPP with greater detail and accuracy.

We have hypothesized that direct and legacy effects of spring warming or cooling, will either stimulate or suppress GPP in summer, autumn or across both seasons. This, in turn, depends on amplified or dampened water stress, the sensitivity of plant species to summer warmth and biological conditions of spring. The summer response of GPP can further affect the autumn phenology by advancing or delaying the senescence of plants. Therefore, the main aims are to answer the following questions: (1) What are the geographical distribution and patterns of direct and lagged responses of GPP to spring temperature anomalies? (2) How does each biome respond to these direct and lagged effects? (3) How do the dominant drivers (environmental variables, current and pre-seasonal GPP) explain summer and autumnal GPP? For environmental variables, this study only considers warming and drought-related variables, that is, temperature, soil moisture and vapor pressure deficit (VPD).

2. Materials and Methods

2.1. Study Area and Biome Classification

The study area covers land areas ($\sim$41 million km^2) located north of 30°N, encompassing several land cover types from the Arctic/subarctic tundra in the high latitudes to temperate forests and grasslands in the south. Regions located north of 30°N have manifested a strong seasonality for climatic thermal conditions and vegetation productivity in the growing season. To make our results comparable to many other studies, the seasons were defined as: spring (March, April, and May), summer (June, July, and August) and autumn (September, October, and November). The biome classification used in this study was based on the global 0.05° land cover data set from the Terra and Aqua Combined Moderate Resolution Imaging Spectroradiometer (MODIS) Land Cover Climate Modeling Grid (MCD12C1) Version 6 Yearly L3 Product [26] and the Köppen–Geiger climate map (1×1 km^2) of the world [27,28] for the year 2011. The MODIS land cover data set uses the biome classification scheme of the International Geosphere-Biosphere Programme (IGBP), which was to generate 17 land classes to meet the needs of the IGBP core science projects relevant to climate, carbon cycle, and others [29]. To separate alpine and high arctic tundra from low-laying temperate grasslands or shrublands, the MODIS grasslands were divided into temperate and arctic grasslands and the MODIS shrublands were divided into temperate and arctic and boreal shrublands. Overall, there are 12 terrestrial biomes that were included in the study area: evergreen needleleaf forests (ENF), evergreen broadleaf forests (EBF), deciduous needleleaf forests (DNF), deciduous broadleaf forests (DBF), mixed forests (MF), arctic and boreal shrublands (ABS), temperate shrublands (TS), savanna (SV), arctic grasslands (AG), temperate grasslands (TG), permanent wetlands (PW), and croplands (CRO) (Figure A1). A more thorough description of each biome type is given in Table 1.

Table 1. Definitions of the 12 terrestrial biomes were developed by [26]. Grass- and shrublands were further divided in Arctic/boreal and temperate groups according to [27,28].

Name	Short Name	Description
Evergreen needleleaf forests	ENF	Dominated by Evergreen conifer trees, (canopy > 2 m). Tree cover > 60%
Evergreen broadleaf forests	EBF	Dominated by evergreen broadleaf and palmate trees (canopy > 2 m). Tree cover > 60%
Deciduous needleleaf forests	DNF	Dominated by deciduous needleleaf (larch) trees (canopy > 2 m). Tree cover > 60%
Deciduous broadleaf forests	DBF	Dominated by deciduous broadleaf trees (canopy > 2 m). Tree cover > 60%
Mixed forests	MF	Mixed between deciduous and evergreen (40–60% of each tree type) (canopy > 2 m). Tree cover > 60%
Arctic and boreal shrublands	ABS	Dominated by woody perennials (1–2 m height) including both closed and open shrublands
Temperate shrublands	TS	Dominated by woody perennials (1–2 m height) including both closed and open shrublands
Savanna	SA	Tree cover 10–30% (canopy > 2 m).
Arctic grasslands	AG	Dominated by herbaceous annuals (<2 m).
Temperate grasslands	TG	Dominated by herbaceous annuals (<2 m).
Permanent Wetlands	PW	Permanently inundated lands with 30–60% water cover and >10% vegetated cover.
Croplands	CRO	At least 60% of area is cultivated cropland.

2.2. Data Sets of Gross Primary Productivity

All three GPP data sets used in this study can be retrieved freely from the links listed in the Data Availability Statement, and also viewed in Table 2. The time-span of the analysis (2001–2018) was confined by the availability of the NIRv-GPP product, which has only been processed to the year 2018.

2.2.1. GOSIF-GPP

The monthly 0.05° GOSIF-GPP product (2001–2018) used in this study is generated based on a global SIF data set from the Orbiting Carbon Observatory-2 (OCO-2) (i.e., GOSIF) and its biome-specific linear relationship with the measured GPP [30]. The GOSIF-GPP data set was estimated using a data-driven model, in which variables reflecting vegetation conditions, meteorological conditions, and land cover information were used as model input [31]. It is based on SIF observations made by NASA's sun-synchronous and polar Orbiting Carbon Observatory-2 (OCO-2) (launched 2014). SIF retrieved by OCO-2 has a higher resolution (1.3×2.25 km^2) and data acquisition rate compared to, e.g., GOSAT or Global Ozone Monitoring Experiment (GOME). These measurements are still too spatially sparse to provide a data set with 0.05° (spatial) resolution over a long-term period. The authors of [30] used eight biome-specific linear relationships between SIF and GPP, in combination with machine-learning techniques to create a global coverage (0.05°) of SIF-based GPP from the year 2001 to 2018. Climate data, such as air temperature, photosynthetic active radiation (PAR), and VPD, was used in combination with the Enhanced Vegetation Index (EVI) retrieved from MODIS MCD43C4v006 to help with calibration of SIF. Compared to GPP retrieved from Eddy Covariance (EC) measurements of 91 sites (https://fluxnet.org/data/fluxnet2015-dataset/, accessed on 6 March 2022, FLUXNET tier1) an overall correlation coefficient (R^2) between EC measurements and GOSIF-GPP is 0.71 ($p < 0.001$).

2.2.2. NIRv-GPP

The monthly global NIRv-GPP data set was derived based on the Advanced Very High Resolution Radiometer (AVHRR) reflectance from LTDR (Land Long Term Data Record v4) product [24]. NIRv is the product of total scene near-infrared reflectance (NIR) and NDVI, commonly used as a proxy to represent vegetation productivity [23]. Based on the established linear relationship between NIRv and EC flux-derived GPP from 104 flux towers, the global monthly 0.05° of GPP data set was estimated. This approach yielded a mean R^2, around 0.70, for the validation sites. The NIRv-GPP data set uses no climate data as input.

2.2.3. FluxSat-GPP

The third data set of monthly global GPP (0.05°) was derived based on the MODIS MCD43C4v006 Nadir Bidirectional Reflectance Distribution Function-Adjusted Reflectance (NBAR) product, SIF, PAR, plant and soil classification data, FLUXNET GPP, meteorological and hydrological fields. These data sets were used as input to a neural-network model to estimate GPP on a global scale [25].

Table 2. Main data sets used in the study. Bilinear interpolation was used to convert the ERA5land climate data from 0.1° to 0.05°. The time-span of the analysis (2001–2018) was confined by the availability of the GPP-data products.

Name	Spatial	Temporal	Time Span	Data Source
GOSIF-GPP	0.05°	monthly	2001–2018	[30]
NIRv-GPP	0.05°	monthly	2001–2018	[24]
FluxSat-GPP	0.05°	monthly	2001–2018	[25]
MODIS land cover (MCD12C1 v006)	0.05°	yearly	2011	[26]
the Köppen–Geiger climate map	1 km	static	2007	[27,28,32]
ERA5-land (air temperature, soil moisture and VPD)	0.1°	monthly	2001–2018	[33]

2.3. ERA5-Land Air Temperature, Soil Moisture and Vapor Pressure Deficit

ERA5-Land is the high-resolution land component product based on the fifth generation of the European ReAnalysis (ERA5) data set, which is used as forcing to drive the European Centre for Medium-Range Weather Forecasts (ECMWF) land surface model [33]. It provides a consistent view of the evolution of land variables over several decades at an enhanced resolution (9 km) compared to the reanalysis products such as ERA5 (31 km) and ERA-Interim (80 km). The added values of ERA5-land consist of improved representation of hydrological cycle, including soil moisture, river discharge and lake description [33]. We used 2 m air temperature, 2 m dew-point temperature, soil moisture (0–7 cm), and surface pressure from the ERA5-land product. VPD was calculated based on 2 m air and dew-point temperature and surface pressure [34].

2.4. Pre-Treatment of the Data

The spatial resolution of all the data sets in this study needs to be aligned and compatible to be able to produce viable correlations. The spatial scale chosen for this study is $0.05°$, and data with a lower resolution thus needs to be converted. This conversion was performed using bilinear interpolation. The spatial resolution ($0.05°$), that each pixel covers varies across the study-area, and mainly depends on the latitude. Hence, the pixels in the further north cover a smaller area than in the south. However, this is believed to have a marginal effect on the results, since the GPP products are measured as densities (meaning they are area-independent), rather than quantities.

The correlations between climate variables were calculated between anomalies (z-scores) relative to the mean of the period 2001–2018. z-scores were calculated according to Equation (1).

$$z = \frac{x - \mu}{\sigma} \tag{1}$$

where μ is the seasonal mean of the whole time period per grid-cell and σ is the standard deviation. The raw number x can thus be more easily interpreted in terms of standard deviations and whether the anomaly is positive or negative compared to the seasonal mean.

2.5. Pearson's Correlations between T_{spring} Anomalies to GPP_{summer} and GPP_{autumn} Anomalies

The main approach for assessing legacy affects of spring temperature on plant productivity and their associated drivers was to use Pearson's correlation and partial correlation. This method has often been adopted by previous studies (e.g., [35,36]) which attempt to identify the dominant drivers for GPP or plant phenology. GPP_{summer} and GPP_{autumn} anomalies were the predicted variables, and temperature, soil moisture, VPD, and pre-season GPP anomalies were predictive variables (Table 3). Direct correlations between T_{spring} anomalies and T_{summer} anomalies were calculated to gain an overview of the transition patterns for the legacy effects. Possible co-varying effects between environmental variables were accounted for when performing partial correlations. This yields a clear picture of the degree of correlation between any individual environmental variable and summer/autumn GPP. The correlations were based on 18 years of data (2001–2018) per individual grid-cell. The statistical significance was tested based on p-values, and only correlations with ($p < 0.05$) were kept. The set of variables was chosen to see how lagged effects from pre-season GPP anomalies correlate to summer/autumn GPP anomalies and how they compare to direct climate effects from soil moisture, air temperature, and VPD anomalies. The dominating variable per grid-cell was determined by selecting the maximum partial correlation value (absolute value) to create a spatial representation of which out of pre-season GPP, soil moisture, air temperature, and VPD that dominates in regulating seasonal plant productivity.

Table 3. Direct correlations between spring temperature and GPP anomalies, as well as two sets of variables (pertaining to summer and autumn GPP anomalies) were used in correlation analysis. All variables were converted into anomalies based on Equation (1) before Pearson's correlations were performed.

X: Predictive	Y	Type
Spring: T	Spring: GPP	Pearson, Direct
Spring: T	Summer: GPP	Pearson, Direct
Spring: T	Autumn: GPP	Pearson, Direct
Spring: T, GPP, Summer: T, SM, VPD	Summer: GPP	Pearson, Partial
Spring: T, GPP, Summer + Autumn: T, SM, VPD	Autumn: GPP	Pearson, Partial

The T_{spring} anomalies correlation to GPP_{summer} and GPP_{autumn} was further investigated by looking at the transition patterns between summer and autumn. Eight transition patterns were identified: $+-$, $-+$, $++$, $--$, $+/$, $-/$, $/+$ and $/-$. The plus, minus, and slash signs denote positive, negative, or non-significant impact, respectively, and the position reveals whether it affects GPP_{summer} or GPP_{autumn}. Further, the statistical spread of T_{spring} anomalies correlations was also determined for each biome to assess whether the aggregated overall impact could be considered neutral, positive or negative. The mean value of the correlation coefficients grouped per biome-type was determined as the defining factor of this.

2.6. General Overview of Methodology

An overview of the workflow in this study is given in Figure 1, which illustrates the order of the main procedures. The methodology follows a similar scheme of comparable studies [5,6]. The data analysis was performed using Mathworks MATLAB, version 2020a.

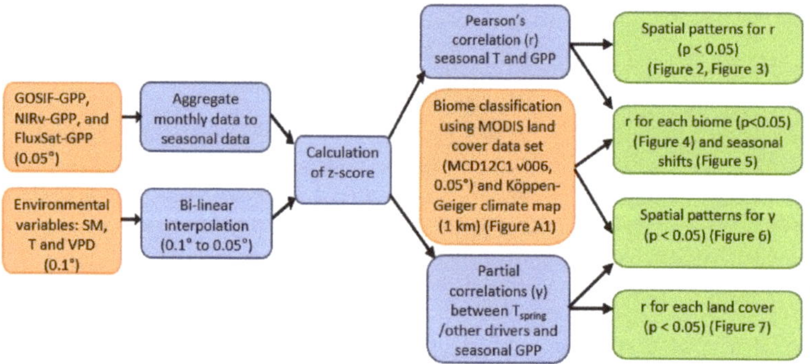

Figure 1. The diagram of the study workflow. The input data (depicted in orange color) underwent pre-processing and statistics calculations (depicted in blue color), eventually leading to the main results (depicted in green color).

3. Results

3.1. Effects of T_{spring} Anomalies on Current and Post-season GPP

Pearson's correlations between T_{spring} and GPP_{spring} showed similar spatial patterns in all three GPP products (Figure 2a–c). A high positive correlation dominated the mid- and high latitudes. Only the southwestern part of North America and South Asia showed a negative correlation. The same pattern was seen in all three products, however, NIRv-GPP had fewer significant correlations, resulting in a sparser pattern.

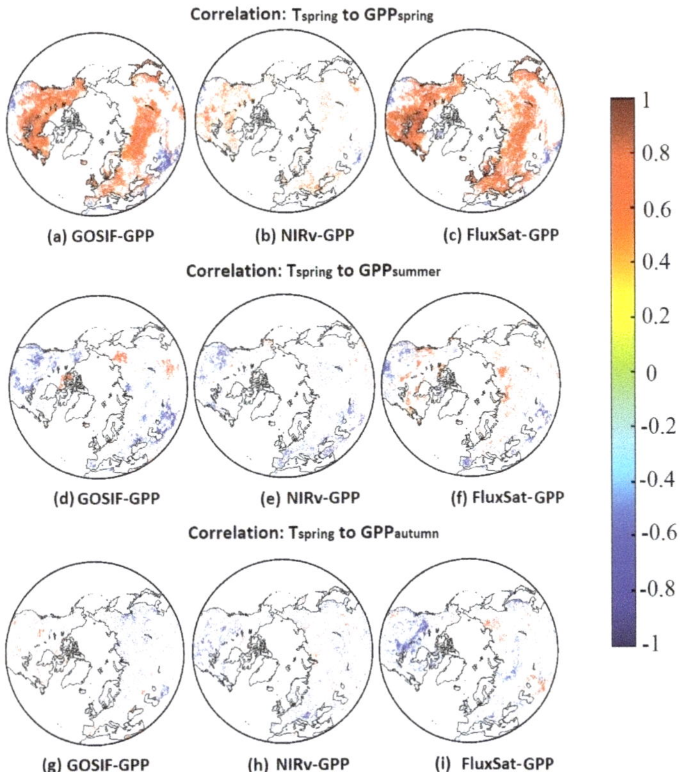

Figure 2. Pearson's correlations ($p < 0.05$) between T_{spring} anomalies and GPP_{spring} (**a–c**)/GPP_{summer} (**d–f**)/GPP_{autumn} (**g–i**).

Lagged effects of T_{spring} anomalies on GPP_{summer} showed fairly similar effects on certain land areas, and a larger difference among different GPP products was noticeable in the Arctic tundra and North Eurasia (Figure 2d–f). All three GPP products showed negative correlations in North America as well as Southeast Eurasia. The Arctic displayed clear positive effects in GOSIF-GPP and FluxSat-GPP, whereas negative effects dominated the response in NIRv-GPP.

Legacy effects from T_{spring} anomalies on GPP_{autumn} were consistent to legacy effects from summer in the large southern areas (Figure 2g–i). However, GOSIF-GPP showed positive effects in North America, where negative effects were found in NIRv-GPP and FluxSat-GPP. Particularly, PW (Hudson Bay and West Siberia) showed strong negative effects in FluxSat-GPP. Some sporadic patches of positive correlations in Eurasia were also found in NIRv-GPP and FluxSat-GPP.

3.2. Latitudinal Distributions for Legacy Effects of T_{spring} Anomalies on GPP_{summer} and GPP_{autumn}

Latitudinal distributions for legacy effects of T_{spring} anomalies on GPP_{summer} and GPP_{autumn} were quantified based on their Pearson's correlation coefficients (mean and one standard deviation) calculated across all the longitudes at each latitude interval of $0.05°$ (Figure 3). As the legacy effects largely oscillated in the high latitudes, the results were shown based on a moving average of every five pixels along the latitude. In general, all three GPP products showed positive legacy effects in most areas at north of $70°$ and negative legacy effects in lower latitudes. GOSIF-GPP and NIRv-GPP had a fairly similar distribution of the legacy effects on GPP_{summer}, however, NIRv-GPP and FluxSat-GPP had a similar distribution of the legacy effects on GPP_{autumn}. FluxSat-GPP differed from the other two in regards to GPP_{summer}, where positive legacy effects began to dominate further south, at $\sim$43°N compared to >60°N for GOSIF-GPP and NIRv-GPP. FluxSat-GPP also displayed a strong negative peak at 55°N–60°N for GPP_{autumn}, which almost has an inverse pattern of the legacy effects seen in summer.

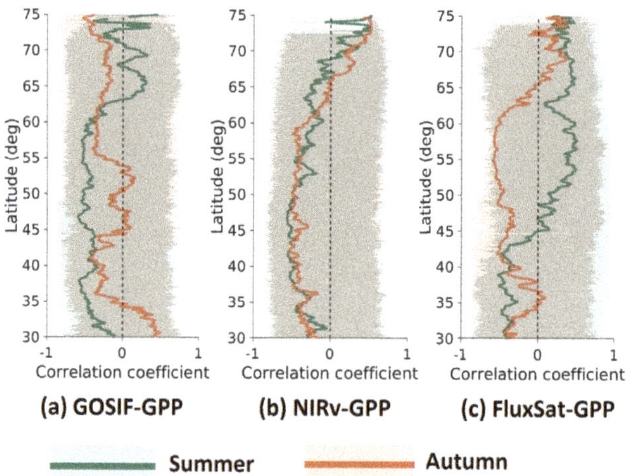

Figure 3. Latitudinal distribution of Pearson's correlations ($p < 0.05$) between T_{spring} anomaly to summer (green) and autumn (orange) GPP anomaly, for: (**a**) GOSIF-GPP, (**b**) NIRv-GPP, and (**c**) FluxSat-GPP. The shaded area represents one standard deviation for the pixels with each latitude intervals (0.05°). The solid line represents the mean values for the pixels with each latitude intervals. All the values have been processed using the moving average of every five consecutive latitudes.

3.3. Effects of T_{spring} Anomalies on GPP_{spring}, GPP_{summer}, and GPP_{autumn} for Each Biome

Effects of T_{spring} anomalies on GPP_{spring}, GPP_{summer} and GPP_{autumn} were aggregated for each biome based on all three GPP products (Figure 4). Results between GOSIF-GPP, NIRv-GPP and FluxSat-GPP were similar, but differed on certain biomes. In general, the forest biomes had a smaller statistical spread around negative correlations, meaning that these negative effects were specific to trees rather than herbaceous vegetation. The other biomes had a wider statistical spread, stretching across both positive and negative impacts, meaning that the overall impact is more neutral. FluxSat-GPP, on average, showed more positive correlations, particularly in summer. ABS and AG stood out from the rest with overall positive impacts, implying that drought propagation from increased T_{spring} is not promoted to the same extent here, as seen in biomes located further south.

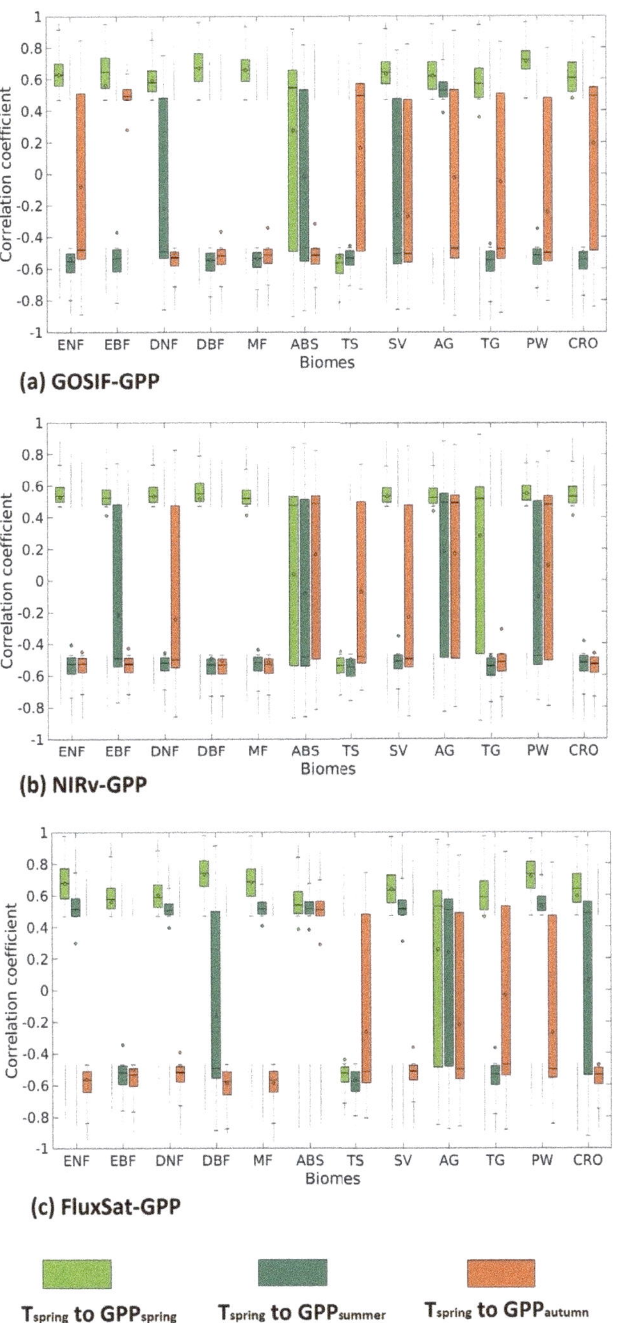

Figure 4. Boxplot of T_{spring} correlation coefficients to GPP_{spring}, GPP_{summer}, and GPP_{autumn} (spring: light green, summer: green and autumn: orange), for: (**a**) GOSIF-GPP, (**b**) NIRv-GPP, and (**c**) FluxSat-GPP. The mean value is represented by a small circle, and the median by a line.

3.4. Transition Patterns for Legacy Effects of T_{spring} on GPP_{summer} and GPP_{autumn}

How legacy effects of T_{spring} propagated through summer and autumn was investigated based on spatial distribution of transition patterns and their aggregation for different biomes (Figure 5). Insignificant correlations are denoted by /, positive by +, and negative by −. The majority of legacy effects was found either in summer or autumn, meaning that the legacy effects only propagated one season. GOSIF-GPP and NIRv-GPP were dominated by a negative impact in summer (−/), whereas FluxSat-GPP revealed more pixels of a positive impact in summer (+/) and a negative impact in autumn (/−). This means that the timing of water-stress induced by the pre-season stimulated vegetation growth may vary between the GPP data sets.

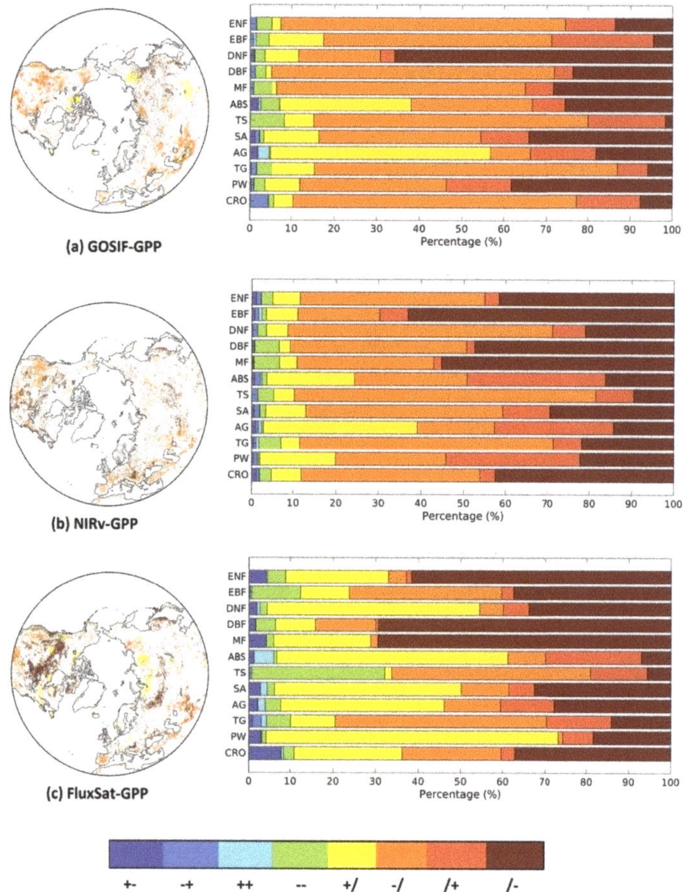

Figure 5. Transition patterns of legacy effects from T_{spring} anomalies on GPP_{summer} and GPP_{autumn} for each biome, for: (**a**) GOSIF-GPP, (**b**) NIRv-GPP, and (**c**) FluxSat-GPP. The labels should be interpreted as follows: +− means there was a positive significant impact seen in summer and a negative in autumn, /− means there was an insignificant correlation seen in summer and a significant negative correlation in autumn). There were a total of eight different transition patterns excluding //, which means no significant legacy for both summer and autumn.

Separating the transition patterns based on biomes further reveals the similarity between GOSIF-GPP and NIRv-GPP, that is, the northern biomes (ABS and AG) showed a clear positive impact during summer. However, FluxSat-GPP showed positive impact in wetlands (PW), savannas (SV) and DNF during summer.

3.5. Dominant Drivers for GPP_{summer} and GPP_{autumn}

The dominant drivers for GPP_{summer} showed similar spatial patterns in all three GPP products (Figure 6a–c). In summer, the arctic plant productivity was strongly correlated with temperature, whereas temperate regions showed soil moisture (SM) as the dominant driver. These temperature-dominant patterns also agreed with the patterns for positive legacy effects of T_{spring} on GPP_{summer} (Figure 3d–f). Legacy effects from GPP_{spring} were the dominant driver in many regions, but legacy effects from T_{spring} were not strong enough to surpass the importance of other drivers.

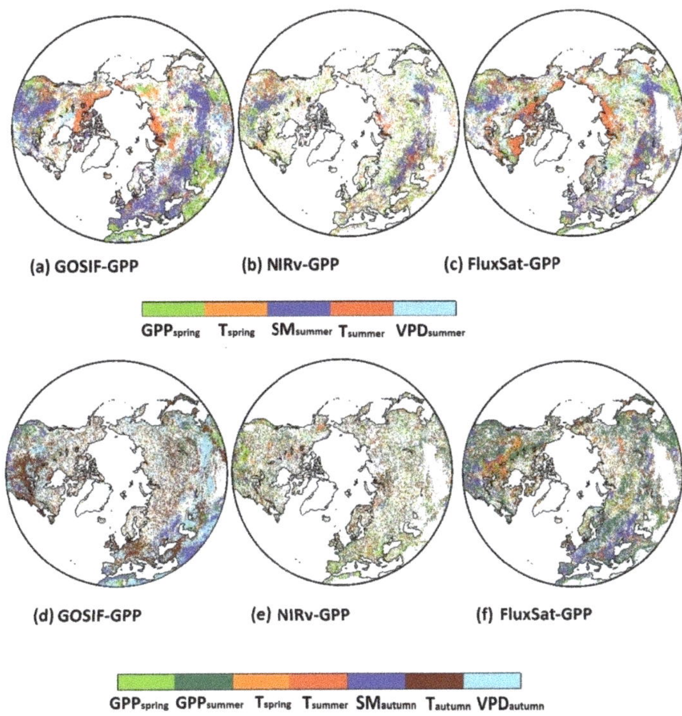

Figure 6. Dominant variables that explain variance of GPP_{summer} (**a–c**) and GPP_{autumn} (**d–f**) for the three GPP products (i.e., GOSIF-GPP, NIRv-GPP, and FluxSat-GPP). The dominant driver was identified based on the highest correlation coefficient (absolute value).

In contrast to summer, the dominant drivers of GPP_{autumn} were more diverse among the three GPP products (Figure 6d–f). GOSIF-GPP showed strong dependence on current seasonal drivers (e.g., T_{autumn}, VPD_{autumn}, and SM_{autumn}). For NIRv-GPP, the legacy effects were mostly dominated by GPP_{spring}. For FluxSat-GPP, GPP_{summer}, T_{summer}, and SM_{summer} were found to be the most important drivers.

3.6. The Importance of Drivers on GPP_{summer} and GPP_{autumn} Aggregated on the Biome Level

The important drivers for GPP_{summer} and GPP_{autumn} were analyzed based on the biome level (Figure 7). Both environmental variables and biological variables (i.e., pre-season GPP) have been accounted for. Generally, the largest number of pixels that showed the significant correlation was found in FluxSat-GPP, followed by GOSIF-GPP and NIRv-GPP (Table 4). For the spatial patterns of the dominant driver, only the pixels with significant correlations were considered.

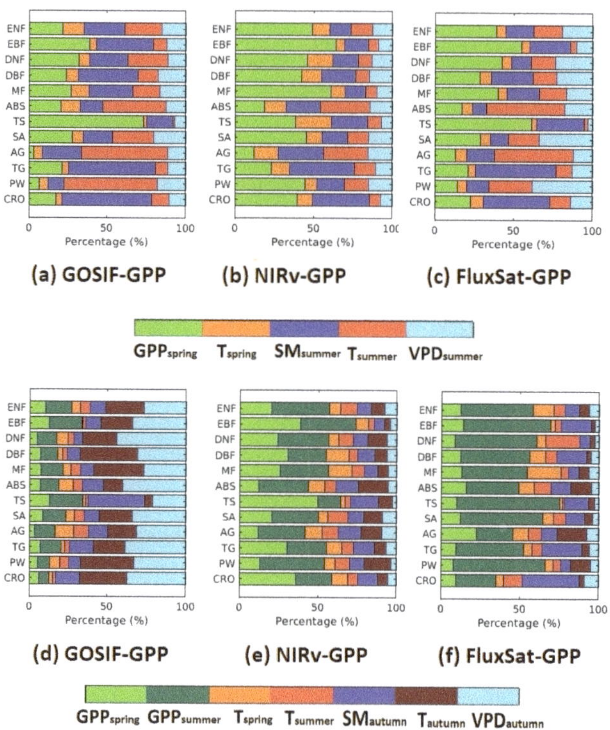

Figure 7. Relative importance (%) of environmental and biological drivers on GPP_{summer} (**a–c**) and GPP_{autumn} (**d–f**) based on different biomes.

Table 4. Percentage of pixels with significant correlations, as seen in Figure 7 divided by biome-type and season. The first three columns correspond to summer, and the following to autumn.

	Summer			Autumn		
	GOSIF-GPP	NIRv-GPP	FluxSat-GPP	GOSIF-GPP	NIRv-GPP	FluxSat-GPP
ENF	46.18%	36.48%	56.66%	48.55%	47.24%	69.47%
EBF	59.94%	45.12%	68.81%	78.18%	63.39%	79.33%
DNF	39.11%	32.87%	53.77%	57.42%	36.52%	49.64%
DBF	58.45%	47.54%	44.92%	70.68%	47.96%	73.17%
MF	42.02%	41.90%	53.44%	52.64%	38.39%	63.91%
ABS	57.10%	25.67%	73.06%	48.50%	33.46%	46.82%
TS	85.36%	40.04%	81.64%	89.61%	54.85%	81.69%
SA	51.80%	34.46%	64.13%	51.20%	43.88%	60.05%
AG	62.32%	18.18%	59.88%	36.20%	19.67%	40.11%
TG	81.17%	56.69%	76.83%	62.53%	40.58%	63.32%
PW	51.81%	18.57%	57.05%	48.99%	24.44%	62.68%
CRO	76.06%	55.04%	57.67%	78.99%	50.22%	73.97%

The dominant driver varied depending on biome types, but overall, all three GPP data sets showed that among all the biomes, the smallest legacy effects from pre-season GPP were found in ABS and AG. These effects persisted in both summer and autumn. For these two biomes, temperature was the dominant driver for GPP_{summer}. However, for the GPP_{autumn}, VPD was the dominant driver in GOSIF-GPP and GPP_{summer} was the dominant driver in NIRv-GPP and FluxSat-GPP. Generally, the forest biomes showed larger legacy effects from pre-season GPP than other biomes in summer and autumn. For the current season drivers (i.e., soil moisture, temperature, and VPD), the contributions to GPP_{autumn} were larger than GPP_{summer}. The contributions of these drivers were larger in GOSIF-GPP than in NIRv-GPP and FluxSat-GPP. Croplands showed the least amount of effects from the pre-season drivers.

4. Discussion

T_{spring} is reported to cause both negative and positive direct impacts on ecosystem productivity [6,16], but our results make a case for the latter being the dominant influence, at least in the mid- to high latitudes. Damage from, e.g., late spring frost events or acute frost desiccation does not seem to override the growth caused by warming, at least on a global scale. The only biome-type deviation from this trend is temperate shrublands, mainly located in north Africa and Turkey, in other words, close to the southernmost borders of our study area. These regions are known for their warm and dry climate. Warm springs in these regions likely bring adverse effects on water-stressed plants, causing a negative correlation.

The cohesive correlation trend between spring temperatures and spring productivity does not persist in the same vein for the legacy effects. A warm spring will not necessarily lead to the same legacy effects to summer and autumn productivity, as an unusually lush and productive spring would. The fact that these two usually go hand in hand makes this result somewhat surprising, but also highlights the importance of keeping these qualities separate. The legacy effects from warmer springs with high ecosystem productivity also differ strongly between GPP products. There are several uncertainties in the data sets that could contribute to this. One possible explanation, pointed out in [37], is the difficulty to separate snow cover decrease from leaf area increase. The spring months, particularly in the northernmost regions, are often characterized by partially covered snow. Increasing temperatures might lower the fraction of snow cover in spring, leading to a false signal of greening.

Our results also indicate that forests overall had the most prevalent negative responses from the legacy effects, whereas grasslands and shrublands showed an ambiguous response (Figure 4). The patterns (i.e., narrow spread of correlation coefficients in forests and wide spread of correlation coefficients in other types) are seen in all three GPP products. This result is much in line with previous studies (e.g., [16–19]). The authors of [16] used the Global Inventory Modeling and Mapping Studies (GIMMS) NDVI data set [38] and actual evapotranspiration derived from the FLUXNET latent heat flux measurements [39] to investigate the legacy effects of earlier springs on boreal forests in North America. They found that earlier spring and associated drying in summer can cause a decline of vegetation productivity due to increased tree mortality and fire activity. Possibly, the long seasonal time frames that the legacy effects operate on better match the slow growth seen in forests. In general, grasslands and shrublands are rapidly growing and changing across the seasons, and respond more directly to current-season precipitation and soil moisture conditions. The importance from previous seasons was then less evident. The legacy effects that were visible for grasslands and shrublands mostly occurred during summer. Moreover, the soil moisture data sets used in this study only reflected water contents of the topsoil horizon (0–7 cm), which may explain that grasslands or shrublands with the low rooting depth are more sensitive to the topsoil moisture condition than forests. This agrees with a study [40] which is based on eddy covariance measurements of carbon flux for Swiss forests and grasslands. They found that grasslands are more sensitive to spring drought because forests can reduce their evapotranspiration to increase the water use efficiency reflecting a better adaptive strategy.

Significant difference between GOSIF-GPP, NIRv-GPP, and FluxSat-GPP is a recurring theme for our results. Most likely, this is explained by fundamental differences in the models used to up-scale the data sets, as well as the measurement gaps that occur between different sensors. Satellite data sets of even higher spatial resolutions and quality might be necessary to bridge this gap and create a cohesive signal. On average, the least amount of significant correlations was seen using the NIRv-GPP product. Possibly, it has a weaker connection to surrounding climate variables than the other two products, as this data set relies more heavily on spectral radiation measurements. Further, outward disturbances on GPP, such as forest fires, anthropogenic land-cover change, or insect attacks, are not accounted for. Ideally, these factors should give rise to low significant correlations and thus be excluded from the analysis. However, the extent to which this has occurred is unclear. The modeling used to fill in the gaps for GOSIF-GPP and FluxSat-GPP could also, to some degree, explain the differences seen in dominating drivers between the three products. Climate variables such as soil moisture and temperature were generally considered more important for GOSIF-GPP and FluxSat-GPP, possibly an echo from utilizing climate variables as input to fill the gaps. Still, for the northern high latitudes, temperature increase has been shown to be the main factor facilitating increased plant productivity in summer [37,41]. However, our results also indicate a relatively cohesive north-south gradient trend of climate vegetation controls during summer, meaning ambiguity between different data-sets and types of vegetation indices can be bridged.

Further, the changes in the dominant climate drivers, particularly during the autumn season, could be the result of a relatively marginal difference between a cluster of controls, where the legacy effects from previous seasons also play an important part [42]. This ambiguity between controls of the late season is to some degree also reflected in previous research, e.g., work by [43] indicates that light limitation is also an important factor for autumn productivity in northern ecosystems. To include this driver in our analysis might therefore lead to a more cohesive result between products. In addition, [35,44] find that there is a complex coupling between soil moisture and temperature, both of which seem to be important controls towards the end of the growing season (a proxy for autumn productivity in this context). In an increasingly warmer world, the limitations imposed by temperature during autumn may shift to an expansion of mainly water-limited ecosystems, where the legacy effects from previous seasons may be amplified even further.

5. Conclusions

Propagating impacts from spring growth and temperature anomalies have been shown to affect summer and autumn ecosystem productivity in the Northern Hemisphere. These legacy effects are mostly negative and set in either in summer or autumn. The forest biomes (ENF, DBF, MF) show conclusive signals of negative impacts for all three GPP products in summer and autumn, whereas shrublands, croplands, and wetlands have a wider statistical spread between positive and negative impacts, leading to a more neutral overall impact. Only the northernmost biomes, AG and ABS, seem to conclusively show lagged positive impacts from spring temperature anomalies, which is important to account for in carbon-cycle models.

Results also seem to depend on the type of method used to quantify GPP, which somewhat may diminish the credibility of our findings. This mainly applies to the main drivers that affect browning and greening trends seen in the seasonal GPP. The effect on GPP from temperature change is most likely not linear and many factors are involved. Continuous development of higher-resolution GPP data sets is needed to further assess vegetation response to a warming climate.

Author Contributions: Conceptualization, W.Z. and H.M.; methodology, W.Z. and H.M.; software, H.M.; validation, W.Z. and H.M.; formal analysis, H.M.; investigation, H.M. and W.Z.; resources, W.Z.; data curation, W.Z.; writing—original draft preparation, H.M.; writing—review and editing, H.M. and W.Z.; visualization, H.M.; supervision, W.Z.; project administration, W.Z.; funding acquisition, W.Z. All authors have read and agreed to the published version of the manuscript.

Funding: This research was funded by a grant from the Swedish National Space Agency (209/19).

Data Availability Statement: The ERA5-land products can be downloaded via (https://cds.climate.copernicus.eu/cdsapp#!/dataset/reanalysis-era5-land?tab=overview). The GOSIF-GPP data sets can be downloaded via (https://globalecology.unh.edu/data/GOSIF-GPP.html). The NIRv-GPP data sets can be downloaded via (https://data.tpdc.ac.cn/en/data/d6dff40f-5dbd-4f2d-ac96-55827ab93cc5/). The FluxSat-GPP can be downloaded via (https://avdc.gsfc.nasa.gov/pub/tmp/FluxSat_GPP/, all accessed on 6 March 2022).

Acknowledgments: The author acknowledges the support by the Center for Scientific and Technical Computing at Lund University for data storage and computation via the project (SNIC 2021/6-341 and LU 2021/2-115).

Conflicts of Interest: The authors declare no conflict of interest.

Abbreviations

The following abbreviations are used in this manuscript:

ABS	Arctic and Boreal Shrublands
AG	Arctic Grasslands
AVHRR	Advanced Very High Resolution Radiometer
BRDF	Bidirectional Reflectance Distribution Function
CRO	Croplands
DBF	Deciduous Broadleaf Forests
DNF	Deciduous Needleleaf Forests
EBF	Evergreen Broadleaf Forests
EC	Eddy Co-variance
ECMWF	European Centre for Medium-Range Weather Forecasts
ENF	Evergreen Needleleaf Forests
EVI	Enhanced Vegetation Index
FluxSat	Fluxnet Data Fused with Satellite Images
GOME	Global Ozone Monitoring Experiment
GOSIF	the Global, OCO-2-based SIF Product
GPP	Gross Primary Productivity
IGBP	International Geosphere-Biosphere Programme
LTDR	Land Long-Term Data Record
MF	Mixed Forests
MODIS	Moderate Resolution Imaging Spectrometer
NDVI	Normalized Difference Vegetation Index
NIRv	Near-infrared Reflectance of Vegetation
OCO-2	Orbiting Carbon Observatory-2
PAR	Photosynthetic Active Radiation
PW	Permanent Wetlands
SIF	Solar Induced Fluorescence
SM	Soil Moisture
SV	Savannas
T	Temperature
TG	Temperate Grasslands
TS	Temperate Shrublands
VPD	Vapor Pressure Deficit

Appendix A

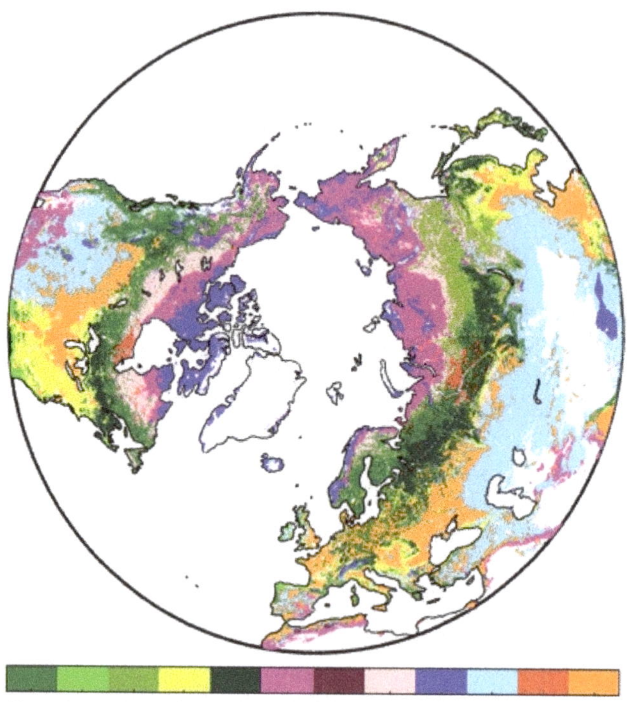

Figure A1. In total, 12 biomes comprise the study area—Evergreen Needleleaf Forests (ENF), Evergreen Broadleaf Forests (EBF), Deciduous Needleleaf Forests (DNF), Decidous Broadleaf Forests (DBF), Mixed forests (MF), Arctic and Boreal Shrublands (ABS), Temperate Shrublands (TS), Savannas (SV), Arctic Grasslands (AG), Temperate Grasslands (TG), Permanent Wetlands (PW), and Croplands (CRO).

References

1. IPCC. *Climate Change 2014: Synthesis Report. Contribution of Working Groups I, II and III to the Fifth Assessment Report of the Intergovernmental Panel on Climate Change*; Pachauri, R.K., Meyer, L.A., Eds.; Core Writing Team: Geneva, Switzerland, 2014.
2. Piao, S.; Liu, Q.; Chen, A.; Janssens, I.A.; Fu, Y.; Dai, J.; Liu, L.; Lian, X.; Shen, M.; Zhu, X. Plant phenology and global climate change: Current progresses and challenges. *Glob. Chang. Biol.* **2019**, *25*, 1922–1940. [CrossRef] [PubMed]
3. Lian, X.; Piao, S.; Chen, A.; Wang, K.; Li, X.; Buermann, W.; Huntingford, C.; Peñuelas, J.; Xu, H.; Myneni, R.B. Seasonal biological carryover dominates northern vegetation growth. *Nat. Commun.* **2021**, *12*, 983. [CrossRef] [PubMed]
4. Zhang, W.; Döscher, R.; Koenigk, T.; Miller, P.; Jansson, C.; Samuelsson, P.; Wu, M.; Smith, B. The interplay of recent vegetation and sea ice dynamics—results from a regional Earth system model over the Arctic. *Geophys. Res. Lett.* **2020**, *47*, e2019GL085982. [CrossRef]
5. Buermann, W.; Forkel, M.; O'sullivan, M.; Sitch, S.; Friedlingstein, P.; Haverd, V.; Jain, A.K.; Kato, E.; Kautz, M.; Lienert, S.; et al. Widespread seasonal compensation effects of spring warming on northern plant productivity. *Nature* **2018**, *562*, 110–114. [CrossRef]
6. Bastos, A.; Ciais, P.; Friedlingstein, P.; Sitch, S.; Pongratz, J.; Fan, L.; Wigneron, J.; Weber, U.; Reichstein, M.; Fu, Z.; et al. Direct and seasonal legacy effects of the 2018 heat wave and drought on European ecosystem productivity. *Sci. Adv.* **2020**, *6*, eaba2724. [CrossRef] [PubMed]
7. Piao, S.; Tan, J.; Chen, A.; Fu, Y.H.; Ciais, P.; Liu, Q.; Janssens, I.A.; Vicca, S.; Zeng, Z.; Jeong, S.J.; et al. Leaf onset in the northern hemisphere triggered by daytime temperature. *Nat. Commun.* **2015**, *6*, 6911. [CrossRef] [PubMed]
8. Jin, H.; Jönsson, A.M.; Olsson, C.; Lindström, J.; Jönsson, P.; Eklundh, L.; Yohe, G. New satellite-based estimates show significant trends in spring phenology and complex sensitivities to temperature and precipitation at northern European latitudes. *Int. J. Biometeorol.* **2019**, *63*, 763–775. [CrossRef] [PubMed]

9.   Weijers, S.; Myers-Smith, I.; Löffler, J. A warmer and greener cold world: Summer warming increases shrub growth in the alpine and high arctic tundra. *Erdkunde* **2018**, *72*, 63–85. [CrossRef]

10. Berner, L.T.; Massey, R.; Jantz, P.; Forbes, B.C.; Macias-Fauria, M.; Myers-Smith, I.; Kumpula, T.; Gauthier, G.; Andreu-Hayles, L.; Gaglioti, B.V.; et al. Summer warming explains widespread but not uniform greening in the Arctic tundra biome. *Nat. Commun.* **2020**, *11*, 4621. [CrossRef]

11. Kunert, N.; Hajek, P.; Hietz, P.; Morris, H.; Rosner, S.; Tholen, D. Summer temperatures reach the thermal tolerance threshold of photosynthetic decline in temperate conifers. *Plant Biol.* **2021**. [CrossRef] [PubMed]

12. Piao, S.; Ciais, P.; Friedlingstein, P.; Peylin, P.; Reichstein, M.; Luyssaert, S.; Margolis, H.; Fang, J.; Barr, A.; Chen, A.; et al. Net carbon dioxide losses of northern ecosystems in response to autumn warming. *Nature* **2008**, *451*, 49–52. [CrossRef] [PubMed]

13. Pongratz, J.; Reick, C.H.; Raddatz, T.; Claussen, M. Biogeophysical versus biogeochemical climate response to historical anthropogenic land cover change. *Geophys. Res. Lett.* **2010**, *37*, L08702. [CrossRef]

14. Zhang, Y.; Song, C.; Band, L.E.; Sun, G.; Li, J. Reanalysis of global terrestrial vegetation trends from MODIS products: Browning or greening? *Remote Sens. Environ.* **2017**, *191*, 145–155. [CrossRef]

15. Lian, X.; Piao, S.; Li, L.Z.; Li, Y.; Huntingford, C.; Ciais, P.; Cescatti, A.; Janssens, I.A.; Peñuelas, J.; Buermann, W.; et al. Summer soil drying exacerbated by earlier spring greening of northern vegetation. *Sci. Adv.* **2020**, *6*, eaax0255. [CrossRef] [PubMed]

16. Buermann, W.; Bikash, P.R.; Jung, M.; Burn, D.H.; Reichstein, M. Earlier springs decrease peak summer productivity in North American boreal forests. *Environ. Res. Lett.* **2013**, *8*, 024027. [CrossRef]

17. Parida, B.R.; Buermann, W. Increasing summer drying in North American ecosystems in response to longer nonfrozen periods. *Geophys. Res. Lett.* **2014**, *41*, 5476–5483. [CrossRef]

18. Kelsey, K.C.; Pedersen, S.H.; Leffler, A.J.; Sexton, J.O.; Feng, M.; Welker, J.M. Winter snow and spring temperature have differential effects on vegetation phenology and productivity across Arctic plant communities. *Glob. Chang. Biol.* **2021**, *27*, 1572–1586. [CrossRef]

19. Wipf, S.; Rixen, C. A review of snow manipulation experiments in Arctic and alpine tundra ecosystems. *Polar Res.* **2010**, *29*, 95–109. [CrossRef]

20. Camps-Valls, G.; Campos-Taberner, M.; Moreno-Martínez, Á.; Walther, S.; Duveiller, G.; Cescatti, A.; Mahecha, M.D.; Muñoz-Marí, J.; García-Haro, F.J.; Guanter, L.; et al. A unified vegetation index for quantifying the terrestrial biosphere. *Sci. Adv.* **2021**, *7*, eabc7447. [CrossRef] [PubMed]

21. Wu, X.; Guo, W.; Liu, H.; Li, X.; Peng, C.; Allen, C.D.; Zhang, C.; Wang, P.; Pei, T.; Ma, Y.; et al. Exposures to temperature beyond threshold disproportionately reduce vegetation growth in the northern hemisphere. *Natl. Sci. Rev.* **2018**, *6*, 786–795. [CrossRef]

22. Li, X.; Xiao, J. A Global, 0.05-Degree Product of Solar-Induced Chlorophyll Fluorescence Derived from OCO-2, MODIS, and Reanalysis Data. *Remote Sens.* **2019**, *11*, 517. [CrossRef]

23. Badgley, G.; Field, C.B.; Berry, J.A. Canopy near-infrared reflectance and terrestrial photosynthesis. *Sci. Adv.* **2017**, *3*, e1602244. [CrossRef] [PubMed]

24. Wang, S.; Zhang, Y.; Ju, W.; Qiu, B.; Zhang, Z. Tracking the seasonal and inter-annual variations of global gross primary production during last four decades using satellite near-infrared reflectance data. *Sci. Total Environ.* **2021**, *755*, 142569. [CrossRef] [PubMed]

25. Joiner, J.; Yoshida, Y.; Zhang, Y.; Duveiller, G.; Jung, M.; Lyapustin, A.; Wang, Y.; Tucker, C.J. Estimation of terrestrial global gross primary production (GPP) with satellite data-driven models and eddy covariance flux data. *Remote Sens.* **2018**, *10*, 1346. [CrossRef]

26. Friedl, M.; Sulla-Menashe, D. *MCD12C1 MODIS/Terra+Aqua Land Cover Type Yearly L3 Global 0.05Deg CMG V006*; NASA EOSDIS Land Processes DAAC: Sioux Falls, SD, USA, 2015. [CrossRef]

27. Kottek, M.; Grieser, J.; Beck, C.; Rudolf, B.; Rubel, F. World Map of the Köppen-Geiger Climate Classification Updated. *Meteorol. Zeitschrif* **2006**, *15*, 259–263. [CrossRef]

28. Peel, M.C.; Finlayson, B.L.; McMahon, T.A. Updated world map of the Köppen-Geiger climate classification. *Hydrol. Earth Syst. Sci.* **2007**, *11*, 1633–1644. [CrossRef]

29. Belward, A.S. The IGBP-DIS global 1-km land-cover data set DIS-Cover: A project overview. *Photogramm. Eng. Remote Sens.* **1999**, *65*, 1013–1020.

30. Li, X.; Xiao, J. Mapping photosynthesis solely from solar-induced chlorophyll fluorescence: A global, fine-resolution dataset of gross primary production derived from OCO-2. *Remote Sens.* **2019**, *11*, 2563. [CrossRef]

31. Li, J.; Tam, C.Y.; Tai, A.P.; Lau, N.C. Vegetation-heatwave correlations and contrasting energy exchange responses of different vegetation types to summer heatwaves in the Northern Hemisphere during the 1982–2011 period. *Agric. For. Meteorol.* **2021**, *296*, 108208. [CrossRef]

32. Friedl, M.A.; Gray, J.M.; Melaas, E.K.; Richardson, A.D.; Hufkens, K.; Keenan, T.F.; Bailey, A.; O'Keefe, J. A tale of two springs: Using recent climate anomalies to characterize the sensitivity of temperate forest phenology to climate change. *Environ. Res. Lett.* **2014**, *9*, 054006. [CrossRef]

33. Muñoz-Sabater, J.; Dutra, E.; Agustí-Panareda, A.; Albergel, C.; Arduini, G.; Balsamo, G.; Boussetta, S.; Choulga, M.; Harrigan, S.; Hersbach, H.; et al. ERA5-Land: A state-of-the-art global reanalysis dataset for land applications. *Earth Syst. Sci. Data* **2021**, *13*, 4349–4383. [CrossRef]

34. Yuan, W.; Zheng, Y.; Piao, S.; Ciais, P.; Lombardozzi, D.; Wang, Y.; Ryu, Y.; Chen, G.; Dong, W.; Hu, Z.; et al. Increased atmospheric vapor pressure deficit reduces global vegetation growth. *Sci. Adv.* **2019**, *5*, eaax1396. [CrossRef]
35. Liu, Q.; Fu, Y.H.; Zeng, Z.; Huang, M.; Li, X.; Piao, S. Temperature, precipitation, and insolation effects on autumn vegetation phenology in temperate China. *Glob. Chang. Biol.* **2016**, *22*, 644–655. [CrossRef] [PubMed]
36. Zhou, S.; Zhang, Y.; Ciais, P.; Xiao, X.; Luo, Y.; Caylor, K.; Huang, Y.; Wang, S. Dominant role of plant physiology in trend and variability of gross primary productivity in North America. *Sci. Rep.* **2017**, *7*, 41366. [CrossRef] [PubMed]
37. Piao, S.; Wang, X.; Park, T.; Chen, C.; Lian, X.; He, Y.; Bjerke, J.W.; Chen, A.; Ciais, P.; Tømmervik, H.; et al. Characteristics, drivers and feedbacks of global greening. *Nat. Rev. Earth Environ.* **2020**, *1*, 14–27. [CrossRef]
38. Tucker, C.J.; Pinzon, J.E.; Brown, M.E.; Slayback, D.A.; Pak, E.W.; Mahoney, R.; Vermote, E.F.; Saleous, N.E. An extended AVHRR 8-km NDVI dataset compatible with MODIS and SPOT vegetation NDVI data. *Int. J. Remote Sens.* **2005**, *26*, 4485–4498. [CrossRef]
39. Jung, M.; Reichstein, M.; Margolis, H.A.; Cescatti, A.; Richardson, A.D.; Arain, M.A.; Arneth, A.; Bernhofer, C.; Bonal, D.; Chen, J.; et al. Global patterns of land-atmosphere fluxes of carbon dioxide, latent heat, and sensible heat derived from eddy covariance, satellite, and meteorological observations. *J. Geophys. Res. Biogeosciences* **2011**, *116*, G00J07. [CrossRef]
40. Wolf, S.; Eugster, W.; Ammann, C.; Häni, M.; Zielis, S.; Hiller, R.; Stieger, J.; Imer, D.; Merbold, L.; Buchmann, N. Contrasting response of grassland versus forest carbon and water fluxes to spring drought in Switzerland. *Environ. Res. Lett.* **2013**, *8*, 035007. [CrossRef]
41. Xu, L.; Myneni, R.; Chapin Iii, F.; Callaghan, T.V.; Pinzon, J.; Tucker, C.J.; Zhu, Z.; Bi, J.; Ciais, P.; Tømmervik, H.; et al. Temperature and vegetation seasonality diminishment over northern lands. *Nat. Clim. Chang.* **2013**, *3*, 581–586. [CrossRef]
42. Liu, Q.; Fu, Y.H.; Zhu, Z.; Liu, Y.; Liu, Z.; Huang, M.; Janssens, I.A.; Piao, S. Delayed autumn phenology in the Northern Hemisphere is related to change in both climate and spring phenology. *Glob. Chang. Biol.* **2016**, *22*, 3702–3711. [CrossRef] [PubMed]
43. Zhang, Y.; Commane, R.; Zhou, S.; Williams, A.P.; Gentine, P. Light limitation regulates the response of autumn terrestrial carbon uptake to warming. *Nat. Clim. Chang.* **2020**, *10*, 739–743. [CrossRef]
44. Zhang, Y.; Parazoo, N.C.; Williams, A.P.; Zhou, S.; Gentine, P. Large and projected strengthening moisture limitation on end-of-season photosynthesis. *Proc. Natl. Acad. Sci. USA* **2020**, *117*, 9216–9222. [CrossRef] [PubMed]

remote sensing

MDPI

Article

Changes in Vegetation Dynamics and Relations with Extreme Climate on Multiple Time Scales in Guangxi, China

Leidi Wang [1,*,†], Fei Hu [1,†], Yuchen Miao [2], Caiyue Zhang [1], Lei Zhang [1] and Mingzhu Luo [1]

1   College of Agriculture, South China Agricultural University, Guangzhou 510642, China; hufei@scau.edu.cn (F.H.); caiyue@stu.scau.edu.cn (C.Z.); zhanglei@scau.edu.cn (L.Z.); lmzhd2701@scau.edu.cn (M.L.)
2   Faculty of Veterinary and Agricultural Sciences, The University of Melbourne, Melbourne, VIC 3010, Australia; yucmiao@student.unimelb.edu.au
*   Correspondence: wangld@mail.iap.ac.cn
†   These authors contributed equally to this work.

Abstract: Understanding the responses of vegetation to climate extremes is important for revealing vegetation growth and guiding environmental management. Guangxi was selected as a case region in this study. This study investigated the spatial-temporal variations of the Normalized Difference Vegetation Index (NDVI), and quantitatively explored effects of climate extremes on vegetation on multiple time scales during 1982–2015 by applying the Pearson correlation and time-lag analyses. The annual NDVI significantly increased in most areas with a regional average rate of 0.00144 year^{-1}, and the highest greening rate appeared in spring. On an annual scale, the strengthened vegetation activity was positively correlated with the increased temperature indices, whereas on a seasonal or monthly scale, this was the case only in spring and summer. The influence of precipitation extremes mainly occurred on a monthly scale. The vegetation was negatively correlated with both the decreased precipitation in February and the increased precipitation in summer months. Generally, the vegetation significantly responded to temperature extremes with a time lag of at least one month, whereas it responded to precipitation extremes with a time lag of two months. This study highlights the importance of accounting for vegetation-climate interactions.

Keywords: vegetation dynamics; multiple time scales; extreme climate; NDVI; correlation; Guangxi

Citation: Wang, L.; Hu, F.; Miao, Y.; Zhang, C.; Zhang, L.; Luo, M. Changes in Vegetation Dynamics and Relations with Extreme Climate on Multiple Time Scales in Guangxi, China. *Remote Sens.* **2022**, *14*, 2013. https://doi.org/10.3390/rs14092013

Academic Editors: Wenxin Zhang, Xuejia Wang, Tinghai Ou and Youhua Ran

Received: 6 April 2022
Accepted: 19 April 2022
Published: 22 April 2022

1. Introduction

As an irreplaceable component of terrestrial ecosystems, vegetation is a pivotal link between the atmosphere and the land's surface [1]. Vegetation shows a significant influence on the carbon cycle and the water balance, and changes in vegetation can alter the ecology balance and the water cycle [2,3]. Climate change, especially increased climate extremes [4–6], can have profound impacts on vegetation and ecosystems [7–14]. Therefore, analyzing the vegetation variation and the influence of climate extremes on vegetation dynamics can help evaluate ecological responses and guide environmental management.

The Normalized Difference Vegetation Index (NDVI) has been widely used to detect vegetation dynamics and the response of vegetation to climate extremes [13–19]. In the past decades, the regional average NDVI has showed an increasing trend in China on the national scale [20,21], but there was a significant spatial heterogeneity of NDVI trends [13,20–22]. Meanwhile, due to the uneven variation in the NDVI in each different growing period, the analyses focusing on the annual or longer scales were not sufficient to represent the variation details. Therefore, it is necessary to investigate the variation characteristics of vegetation dynamics by considering different temporal scales in each sensitive area.

The vegetation was greatly influenced by climate extremes, which varies by region, season and scale [7,15,18,23]. The degree of vegetation responses to climate extremes showed great spatial heterogeneity [13,14,18,24–27], and the impact of extreme climate indices varied among local, regional and national scales [18]. Extreme precipitation generally promotes vegetation growth in most dry areas but has a negative influence on vegetation in humid areas [18,20]. Nevertheless, extreme temperature is recognized to have a more extensive and complex effect on vegetation in China [12,18]. In addition, the influences of climate factors on vegetation often exhibit a time lag [28–30], and the time lag of vegetation responses is generally shorter than a quarter of a year on a monthly scale [30].

Regarding the vegetation responses to climate variables, many studies have respectively identified the temporal differences in the correlations between vegetation and extreme climate indices, and the different time-lag effects of climate extremes on vegetation growth in different spaces [12,20,21,28,31,32]. Exploring this relationship on a monthly scale can help better understand the main limiting factors for vegetation growth compared with those on a longer time scale [28]. An analysis of this correlation concentrating on only a single time scale may underestimate the effect of climate change and cannot accurately reflect the response mechanism of vegetation to climate variables [7,14,33]. Hence, it is of great significance to study the relationship between vegetation dynamics and climate extremes on different temporal scales. However, relatively little attention has been paid to the implications of climate extremes for vegetation under multiple temporal scales.

Guangxi, located in the subtropical humid monsoon climate zone, has complex and diverse landforms and mountains. It has widely distributed Karst landforms. Karst areas usually feature a thin soil layer, low soil fertility, and serious soil erosion with many exposed bedrocks highly vulnerable to climate extremes. Moreover, karst vegetation can provide a great carbon sink function [34]. Rocky desertification was identified as the most severe ecological problem threatening the productivity of agriculture, forestry, grassland and livestock husbandry in the karst areas of southwest China [35]. Karst ecosystems are characterized by low environmental capacity, high sensitivity, weak anti-interference ability and ecological vulnerability. Thus, the fragile ecological environment makes it extremely important to understand the responses of vegetation dynamics to extreme climate in this region. Multiple time-scale analysis is necessary for the vegetation variation and the assessment of climate extremes on vegetation over Guangxi. This kind of study can provide knowledge for vegetation conservation and restoration of fragile ecosystems in both Guangxi and other similar regions.

By taking Guangxi as a representative example of fragile regions, this study comprehensively investigated the spatial-temporal trends of NDVI and relations with climatic extremes on multiple time scales (annual, seasonal and monthly). With a focus on the differences in the effects of climate extremes on vegetation at different time scales, this study tried to explore whether the effects of extreme climate on vegetation can be disentangled from the baseline effect of climate on a time series. The layout of this paper is presented as follows: Section 2 introduces the materials and methods. Section 3 elaborates the spatial-temporal variations of NDVI and the correlations between NDVI and extreme climate indices on different time scales. Section 4 discusses the results and Section 5 summarizes the presented work.

2. Materials and Methods

2.1. Study Area and Data

Guangxi is located in the contiguous karst area of southwest China, with a Peak Cluster Depression, a Peak Forest Plain and non-karst landforms. The variety of vegetation is very comprehensive, such as coniferous forest, broadleaf forest, thicket, grass, and cultivated vegetation. Guangxi is a national key forest region and one of the world's top sugar-producing areas. Crops mainly include double cropping rice, sugar cane, and corn. The vegetation ecosystem in Guangxi is highly fragile because of the widely distributed karst landform. Guangxi has a warm climate and abundant rainfall with annual total precipitation above 1500 mm. Figure 1 shows the geographic location of the study area.

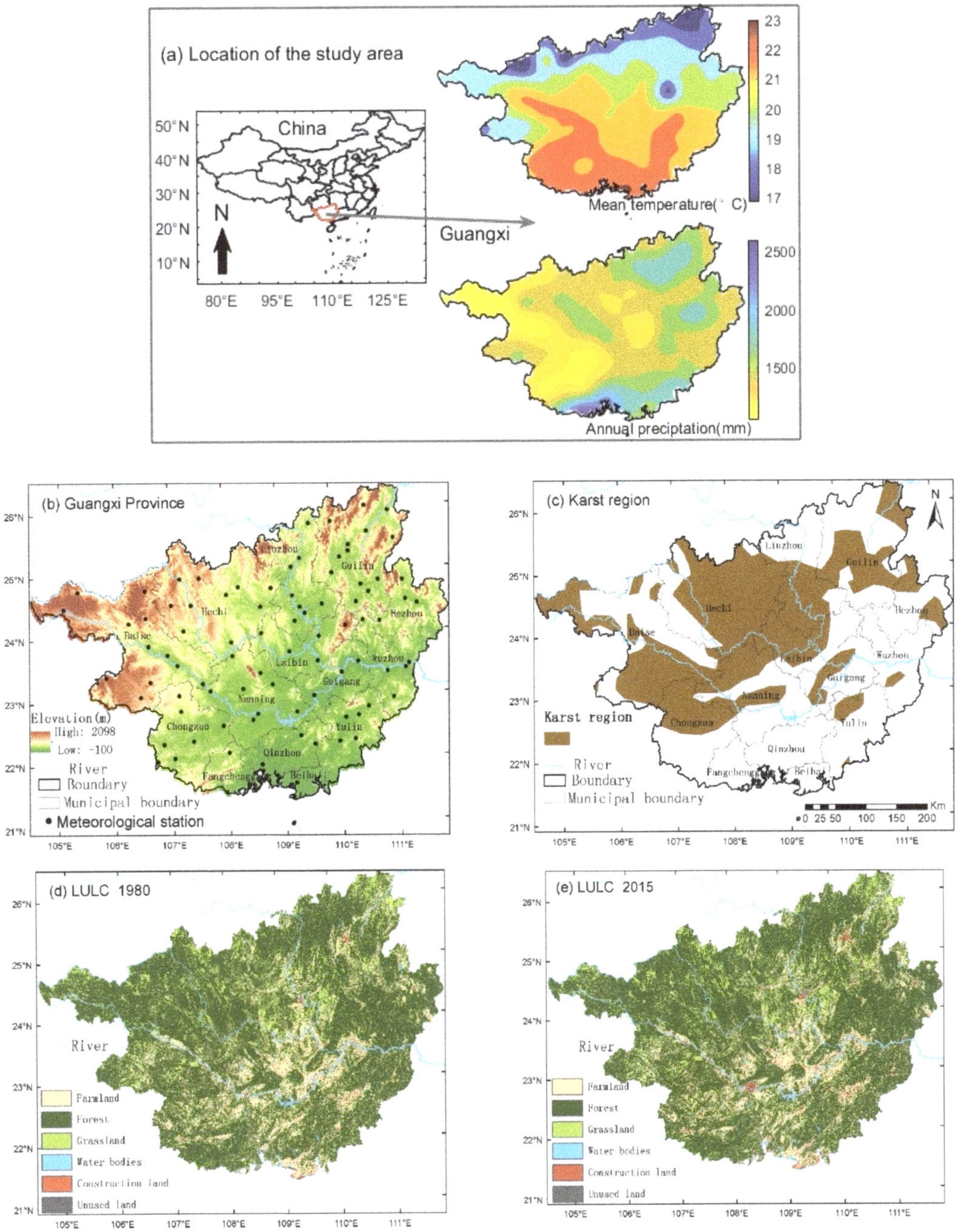

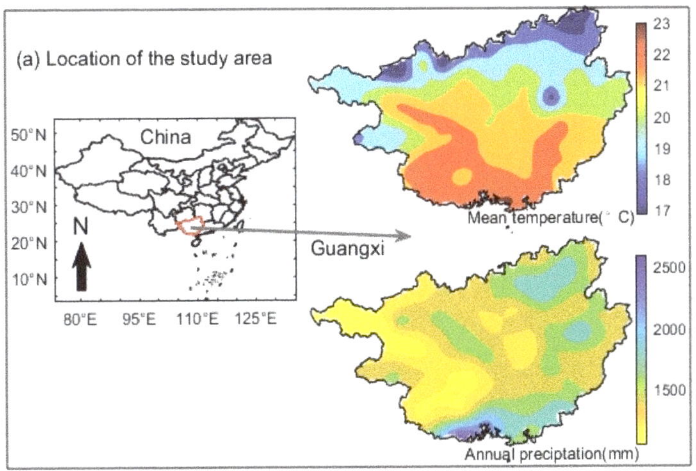

Figure 1. Geographic location of the study area, and spatial distribution of meteorological stations and land use land cover classes (LULC) (for the year 1980 and 2015, respectively). Note: The LULC data and the digital elevation data were provided by the Resource and Environment Science and Data Center, Chinese Academy of Sciences (https://www.resdc.cn/, accessed on 10 October 2021); the karst boundary data were obtained from the World Map of Carbonate Rock Outcrops (http://web.env.auckland.ac.nz/our-research/karst/, accessed on 14 June 2017).

All data used in this study are described in Table 1. The daily data obtained from the China Meteorological Administration include daily mean temperature, daily maximum temperature, daily minimum temperature, and daily precipitation (http://data.cma.cn/, accessed on 15 October 2019). This study selected 88 meteorological stations, each with a 58-year good and qualified data record (from 1961 to 2018). The karst boundary data were obtained from the World Map of Carbonate Rock Outcrops (http://web.env.auckland.ac.nz/our-research/karst/, accessed on 14 June 2017). The land use and land cover (LULC) data with 1 km spatial resolution were provided by the Resource and Environment Science and Data Center, Chinese Academy of Sciences (https://www.resdc.cn/, accessed on 10 October 2021). The sub-regions of the three vegetation types (farmlands, forests and grasslands) were extracted from the LULC data. According to LULC maps of 1980 and 2015 (Figure 1d,e), the spatial patterns of farmlands, forests and grasslands did not change very much in the past decades. In addition, urban, water and barren areas show little vegetation cover, and therefore these areas were generally excluded from the spatial analysis to diminish some effects of anthropological activities.

Table 1. Datasets used in this study.

Name	Data Source	Spatial Scale
Daily weather data	China Meteorological Administration (http://data.cma.cn/, accessed on 15 October 2019)	-
Karst boundary data	World Map of Carbonate Rock Outcrops (http://web.env.auckland.ac.nz/our-research/karst/, accessed on 14 June 2017)	-
LULC data	Resource and Environment Science and Data Center, Chinese Academy of Sciences (https://www.resdc.cn/, accessed on 10 October 2021)	1 km
Digital Elevation data	Resource and Environment Science and Data Center, Chinese Academy of Sciences (https://www.resdc.cn/, accessed on 10 October 2021)	250 m
GIMMS NDVI3g data	National Oceanic and Atmospheric Administration (https://www.nasa.gov/nex, accessed on 7 December 2020)	1/12° (about 8 km)

At present, although a series of vegetation indices, such as the NDVI, the Soil adjusted Vegetation Index (SAVI), and the Enhanced Vegetation Index (EVI), have been developed to reflect changes in vegetation activities [36,37], the NDVI is still a good indicator when dealing with large-scale vegetation coverage and greenness [9,14,18,27,38,39]. Thus, the NDVI was employed as an indicator to monitor changes in vegetation activities in this study. A higher NDVI value implies a higher density of green vegetation, and vice versa.

This study used NDVI data from the Global Inventory Monitoring and Modeling Studies (GIMMS) NDVI3g dataset [40], which originated from the Advanced Very High Resolution Radiometer (AVHRR) sensors of the National Oceanic and Atmospheric Administration (NOAA) (https://www.nasa.gov/nex, accessed on 7 December 2020). The GIMMS NDVI3g dataset has been corrected for radiometric calibration, atmospheric attenuation, cloud cover, sensor degradation, inter-sensor differences, view and illumination geometry, orbital drift, volcanic aerosols, and other non-vegetation effects [41]. It has a spatial resolution of 1/12° (about 8 km) and a temporal interval of 15 days. Although the coarse spatial resolution of the dataset cannot be helpful to detect small scale changes, the GIMMS NDVI dataset has a sufficient quality and the longest time series in the period from 1982 to 2015.

The GIMMS NDVI dataset has been evaluated through comparisons with other NDVI products or ground-based validations [42–44]. For instance, Fensholt and Proud (2012) [43] compared the performances of a time series of GIMMS NDVI with MODIS NDVI data and found that the trends of the two datasets are basically consistent; the GIMMS NDVI dataset has a significant correlation with ground-based observations in Eastern China [42]. Thus, the dataset performs well in exploring the long-term trend in vegetation greenness and its relationship with climate factors, and its accuracy is relatively reliable [17,40,45,46].

This study could use the GIMMS NDVI data directly. The monthly NDVI values were obtained through the maximum-value composites (MVCs) method, which is the same as that employed in Cui et al. (2019) [17]. The seasonal or yearly NDVI data were calculated by averaging the monthly NDVI over the corresponding periods.

2.2. Methods

Figure 2 provides the flowchart of the approach of this study. The Expert Team on Climate Change Detection and Indices (ETCCDI) has developed a suite of extreme climate indices, including 16 temperature indices and 11 precipitation indices [47]. Guangxi has a typical subtropical and Asian monsoon climate with generally sufficient rainfall (Figure 1a), and the average temperature in the coldest mouth is above 10 °C. Based on these climate characteristics of Guangxi, this study selected the 16 most relevant indices to reflect the intensity and duration in climate extremes (Table 2). This study also analyzed the mean value of the daily mean temperature (Tm) to supplement the extremes. All climate indices were divided into three categories: annual indices, seasonal indices and monthly indices. All these indices can be calculated on an annual basis and used as annual indices. In addition, the 12 indices (Tm, TXm, TNm, DTR, TXx, TNx, TXn, TNn, Rx1day, Rx5day, SDII and PRCPTOT) can be calculated not only on an annual basis, but also on a seasonal or monthly basis. Thus, the 12 indices can also be used as seasonal and monthly indices by being calculated for corresponding time scales. The kriging method was used to resample the climate data of the meteorological stations to a spatial resolution of 8 km to spatially match the NDVI dataset.

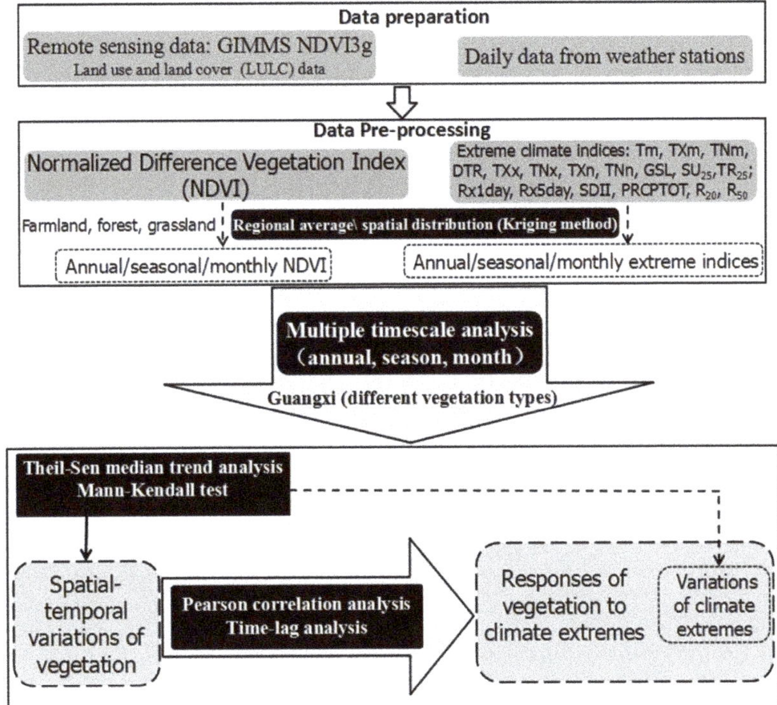

Figure 2. Flowchart of monitoring changes in vegetation dynamics and relations with extreme climate.

Table 2. Definitions of extreme temperature indices in this study.

Indices	Indicator Name	Definition	Unit
Tm	Mean temperature	Mean value of daily mean temperature	°C
TXm	Maximum temperature	Mean value of daily maximum temperature (TX)	°C
TNm	Minimum temperature	Mean value of daily minimum temperature (TN)	°C
DTR	Diurnal temperature range	Mean difference between TX and TN	°C
TXx	Max TX	Maximum value of daily maximum temperature	°C
TNx	Max TN	Maximum value of daily minimum temperature	°C
TXn	Min TX	Minimum value of daily maximum temperature	°C
TNn	Min TN	Minimum value of daily minimum temperature	°C
GSL	Growing season length	Annually count between first span of at least 6 consecutive days with Tm > 10 °C and first span after July 1 of 6 days with Tm < 10 °C	days
SU_{25}	Summer days	Number of days with daily maximum temperature > 25 °C	days
TR_{25}	Tropical nights	Number of days with daily minimum temperature > 25 °C	days
Rx1day	Max 1-day precipitation amount	Maximum 1-day precipitation	mm
Rx5day	Max 5-day precipitation amount	Maximum consecutive 5-day precipitation	mm
SDII	Simple daily intensity index	Total precipitation divided by the number of wet days (defined as PRCP ≥ 1.0 mm) in the year	mm/day
PRCPTOT	Total wet-day precipitation	Total precipitation in wet days (PRCP ≥ 1 mm)	mm
R20	Number of very heavy precipitation days	Annual count of days when PRCP ≥ 20 mm	days
R50	Rainstorm	Annual count of days when PRCP ≥ 50 mm	days

Note: PRCP represents the daily precipitation.

The Theil–Sen (TS) median trend analysis method is less sensitive to observations of missing time series and outliers in the time series [48]. Thus, this work used the TS slope estimator β to analyze the variation trends in NDVI and climate indices over large temporal scales. The calculation formula is as follows:

$$\beta = \text{Median}\left[\frac{x_i - x_j}{i - j}\right] \text{ for all } j < i \qquad (1)$$

where $1 < j < i < n$. β, the median of the slope of data combinations, indicates the variation rate within the time series; i and j are the ordinal numbers of years (one is held constant while the other is varied); x_i and x_j are the mean values within a certain time in the ith and jth year; and n is the number of monitoring years (equal to 34 in this study). A positive β-value indicates an increasing trend; a negative β-value indicates a decreasing trend; and $\beta + 0$ indicates no change. The unit of the NDVI values is 1, and the unit of the NDVI rate β is year^{-1} in this study.

The Mann–Kendall (MK) test [49] does not require the data to be distributed normally or linearly. Thus, this study used the MK trend test to determine whether the trends were significant. In addition, this study utilized the MK abrupt test to detect abrupt changes and potential turning points. The MK abrupt test calculates two statistical measures, which are the sequential values of a reduced or standardized variable. A forward sequential statistic is estimated using the original time series, and a backward sequential statistic is estimated using the reversed time series. If the progressive series and the retrograde series intersect at a certain point, the intersection point is within the significance level, and if the sequence curve of the progressive or retrograde series exceeds the 0.05 significance level after the intersection point, then the intersection points of the two curves represent the potential turning points in the times series.

Pearson correlation analysis was applied to measure the strength of the correlations between vegetation dynamics and extreme indices at both regional average and pixel scales. As the whole time series covered 34 years, the sample size was 34 for the correlation analysis for each time scale. Significance levels of $p = 0.05$ and 0.01 were taken as the thresholds to distinguish significance. As for the analysis of time-lag effects, the time lag for a climate index was defined as the number of months after which the NDVI showed the highest significant correlation coefficient with this climate index [30,32,50,51]. The time lag is generally shorter than a quarter, and 6 months at most [10,30,51]. Therefore, this study only considered the time lags of 0–6 months. Spring includes the months from March to May, summer from June to August, autumn from September to November, and winter from December to February of the following year.

3. Results

3.1. Spatiotemporal Variability of Vegetation Dynamics

3.1.1. Trends of NDVI on an Annual Scale

Based on the GIMMS NDVI data during 1982–2015, this study summarized the temporal variations and spatial distributions of the annual NDVI in Guangxi (Figure 3). The multiyear mean of the annual NDVI was 0.6775 during the past 34 years. The maximum was found in 2015 with a value of 0.7193, whereas the minimum was found in 1984 with a value of 0.6418. The annual NDVI showed a significant increasing trend with a rate of 0.00144 year^{-1}, and the trend exhibited several fluctuations (Figure 3a). According to the result of the MK abrupt test (Figure 3b), the annual NDVI had two significant transition points in 2004 and 2005. The annual NDVI was very low with a value of 0.659 in 2005. In order to explore the decadal difference of the NDVI, this study divided the time series (1982–2015) into two periods (1982–2004; 2005–2015) based on the turning point 2004. The multiyear average NDVI was 0.6688 for the years before 2004 and 0.6955 for the years after 2004. However, the regional averaged annual NDVI did not show any significant trend during each separate period.

The spatial distributions of the NDVI were extremely heterogeneous (Figure 3c). Inevitably, they were partly affected by the remote sensing data sources and the calibration. The NDVI increased with altitude in some studied regions. The NDVI mostly ranged from 0.40 to 0.80, and the NDVI at around 0.60 occurred in most regions. The higher NDVI at 0.60 and above was basically distributed on the areas covered by forests and grasses. To further illustrate the change in spatial distributions, the spatial distributions of the NDVI were analyzed during each separate period. The spatial distributions of the NDVI during 1982–2004 (Figure 3d) and 2005–2015 (Figure 3e) were similar to those during the whole period. However, during the period 2005–2015, the NDVI in most regions was higher than that during the former period. The highest NDVI can reach 0.80 or above in some forest and grassland areas.

The regional averaged annual NDVI for the three vegetation types (farmlands, forests and grasslands) was demonstrated to further explore differences in the greenness in different vegetation areas (Figure 3f). Generally, the forest had the highest NDVI, whereas the farmland had the lowest in each year. The NDVI trends in the three vegetation areas showed several fluctuations, and they were similar to those of the whole Guangxi region. The annual NDVI of farmlands showed the highest increasing rate, whereas that of forests was the lowest. The MK abrupt test indicated that the annual NDVI also had two weak turning points (in 2004 and 2005) in forests and grasslands (Figure not shown). In addition, the annual NDVI in forests and grasslands also did not show any significant trend during each separate period.

Furthermore, the variation trend in NDVI values was extracted respectively in karst and non-karst areas (figure not shown). It was 0.00153 year^{-1} in karst areas, and 0.00150 year^{-1} in non-karst areas. It seems that there was a weak greening difference between karst and non-karst areas. However, the altitudes (including topographic configuration and atmospheric environment) can affect the NDVI values, which may weaken the influences on the NDVI trends originating from the karst landscape itself. The variation rate of the NDVI exhibited significant spatial differences (Figure 4). Overall, all vegetation mainly had greening trends during 1982–2015, and vegetation afforestation occurred in most areas of Guangxi (Figure 4a). In forests and grasslands, the greening rates were mostly around 0.0005–0.0015 year^{-1}. The farmland showed higher greening rates, which were above 0.002 year^{-1} in some areas. About

83% of the experimental areas with vegetation afforestation passed the 0.05 significance test during 1982–2015 (Figure 4b).

However, only about 40% of the total pixels passed the significance test during the first period of 1982–2004 (Figure 4d), and only about 50% passed the significance test during the second period of 2005–2015 (Figure 4f). In these two separate periods, the significant increasing rate of the NDVI only occurred in the central and southeast of Guangxi (Figure 4c,e). During the second period, the annual NDVI experienced a great increase in rate, as high as 0.003 year^{-1} in many of the central and eastern regions. The greening rate of the second period was higher than that in the first period, which indicated increased green vegetation in Guangxi after 2004. Since the NDVI variation rarely passed the significance test during each separate period, hereafter, this study focused on the analysis of the whole period of 1982–2015.

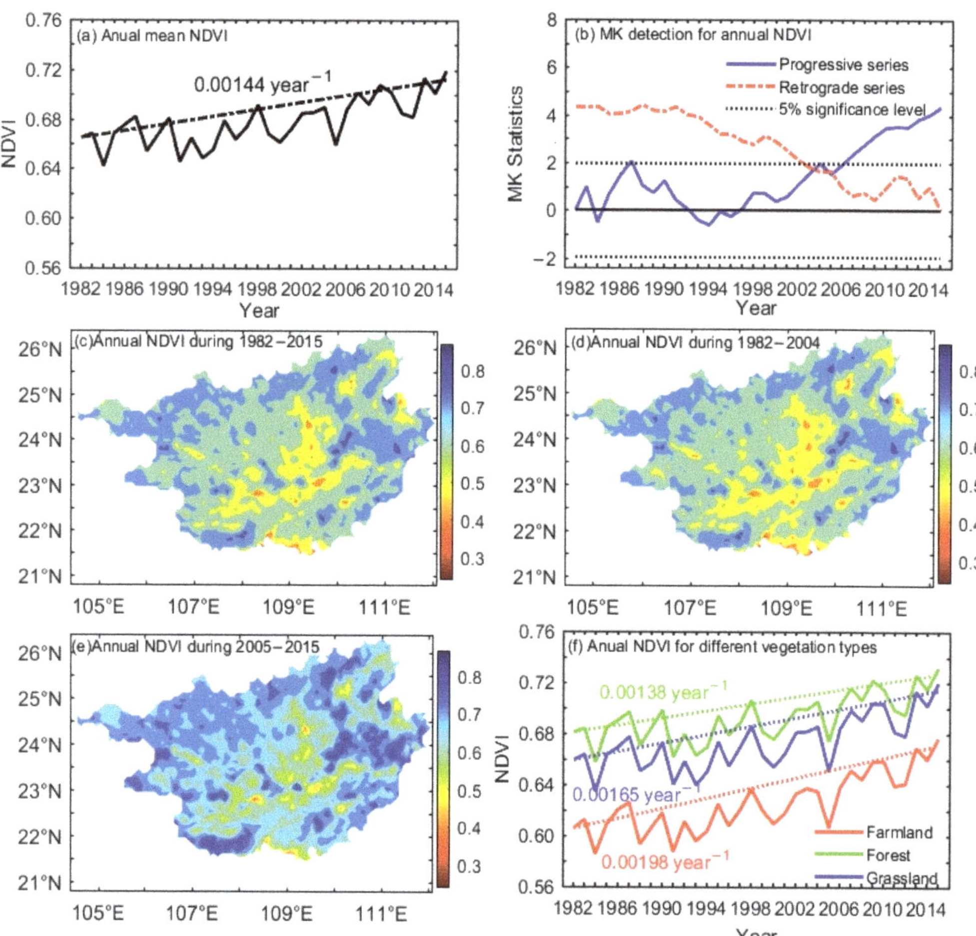

Figure 3. Annual mean NDVI in Guangxi during 1982–2015 (**a**); the Mann–Kendall (MK) abrupt change detection for the annual NDVI during 1982–2015 (**b**); spatial distributions of the annual NDVI during 1982–2015, 1982–2004, and 2005–2015 (**c–e**); and annual mean NDVI for different vegetation types during 1982–2015 (**f**). Note: in (**a,f**), the solid line represents the observed NDVI, and the dotted line represents the fitted NDVI.

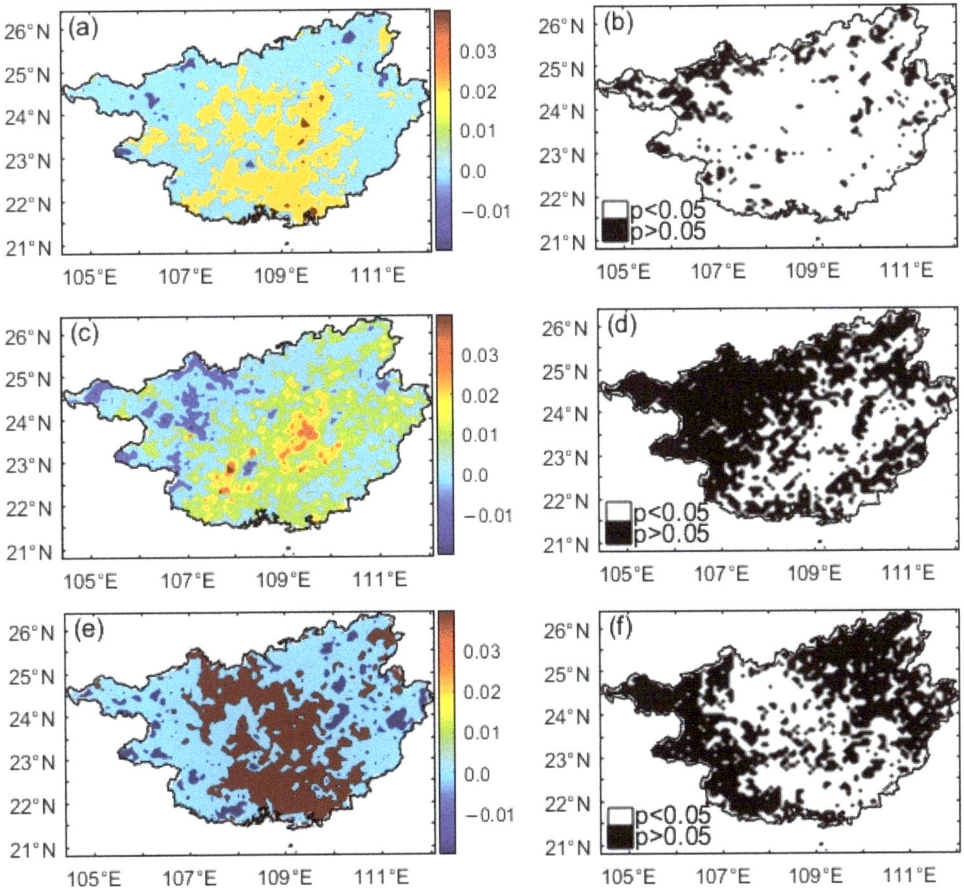

Figure 4. Spatial distributions of the trend rates of the annual NDVI (unit: $10 \times \text{year}^{-1}$) (**a,c,e**) and their significant tests (**b,d,f**) during 1982–2015, 1982–2004, and 2005–2015.

3.1.2. Trends of NDVI on a Seasonal Scale

The spatial distributions of the NDVI in each season were similar to those of the annual NDVI, with the NDVI mostly varying from 0.40 to 0.80 (Figure not shown). The winter NDVI was lower than that in other seasons over the whole study area. For the regional average, the highest NDVI appeared in autumn with a value of 0.738, whereas the lowest NDVI appeared in winter with a value of 0.618. The seasonal NDVI significantly increased in most regions during the past decades (Figure 5), and about 77% of the whole research area showed significant greening trends in spring, in contrast with around 50% of the area in other seasons. In spring, the trend rate of the NDVI showed the highest value and strongest spatial heterogeneity. It was as high as 0.0015 year^{-1} in most regions, especially farmland areas. In summer and winter, the trends showed relatively less inhomogeneity, with a greening rate of around 0.0005 year^{-1} in most areas. However, a remarkable greening rate occurred in some regions of central or eastern Guangxi, which was partly similar to that in spring. In contrast, the NDVI trends in autumn were uniformly distributed with a rate of mostly around 0.0005 year^{-1}.

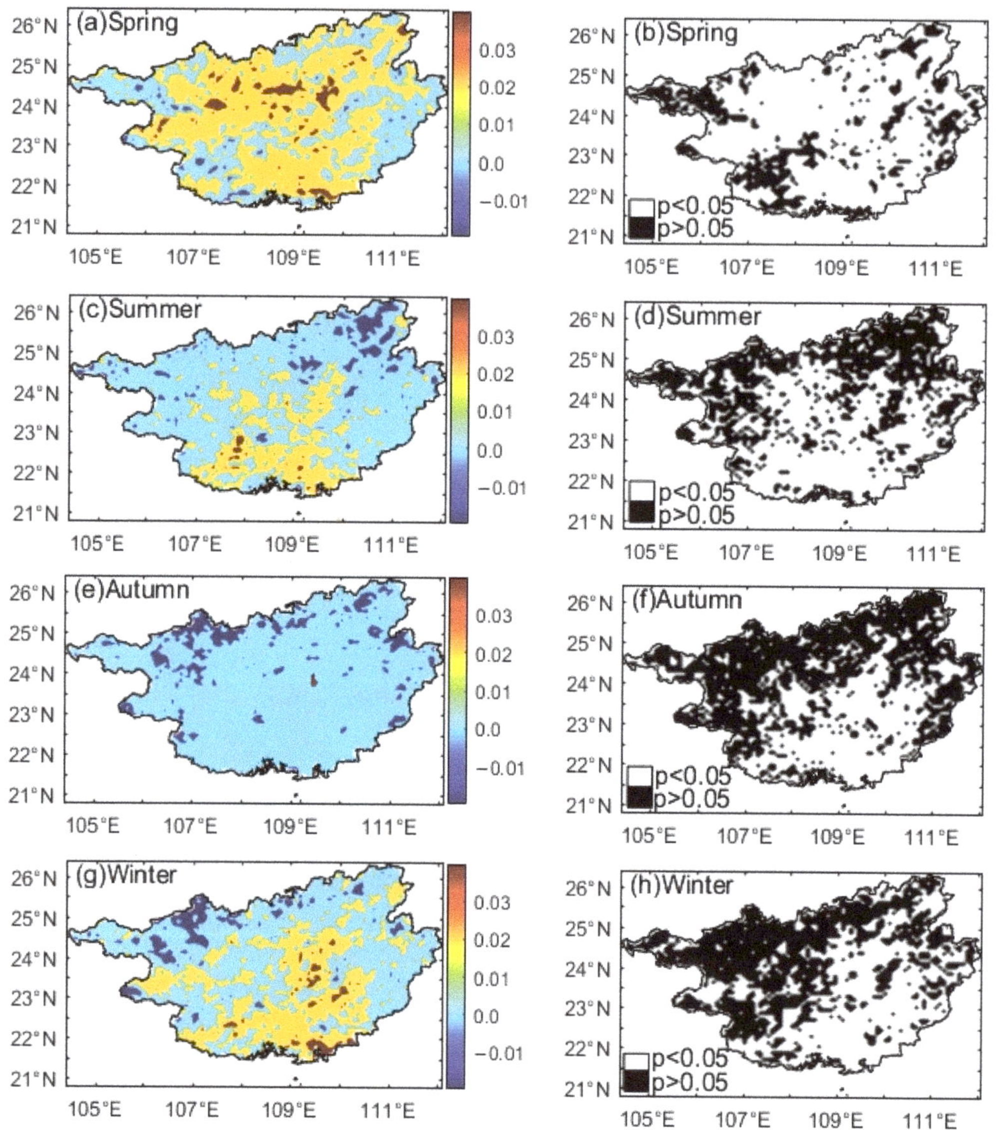

Figure 5. Spatial distributions of the trend rates of the seasonal NDVI (unit: $10 \times$ year^{-1}) (**a,c,e,g**) and their significant test (**b,d,f,h**) during 1982–2015.

However, it is hard to distinguish the difference in the greening rate among the different vegetation types based on the above spatial distributions. Table 3 provides climate inclination rates of the regional averaged NDVI for the three vegetation types. The regional mean NDVI significantly increased in each season during the past decades. As for the whole Guangxi, the highest rate was 0.0021 year^{-1} in spring, followed by winter (0.0015 year^{-1}), autumn (0.0011 year^{-1}), and summer (0.0010 year^{-1}). The farmland, forest and grassland generally showed similar trends to that of the whole Guangxi. The greening rate of farmlands was the highest in each season, whereas that of forests was the lowest.

Table 3. Trend rates (unit: year^{-1}) for the seasonal and monthly NDVI during 1982–2015.

	Guangxi	Farmlands	Forests	Grasslands
Spring	0.0021 **	0.0025 **	0.0020 **	0.0024 **
Summer	0.0010 **	0.0014 **	0.0010 **	0.0010 **
Autumn	0.0011 *	0.0015 **	0.0010 *	0.0011 *
Winter	0.0015 *	0.0019 **	0.0013	0.0013 *
January	0.0006	0.0011	0.0005	0.0003
February	0.0030 **	0.0032 **	0.0029 *	0.0032 **
March	0.0013	0.0018	0.0011	0.0014
April	0.0021 *	0.0027 **	0.0021 **	0.0030 **
May	0.0023 **	0.0028 **	0.0022 **	0.0026 **
June	0.0004	0.0008	0.0004	0.0001
July	0.0015 **	0.0021 **	0.0014 **	0.0016 **
August	0.0014 **	0.0020 **	0.0013 **	0.0017 **
September	0.0014 **	0.0017 **	0.0012 *	0.0014 *
October	0.0005	0.0007	0.0003	0.0004
November	0.0014 *	0.0020 *	0.0010	0.0013 **
December	0.0009	0.0016	0.0007	0.0010

Note: * significant at $p < 0.05$; ** significant at $p < 0.01$.

3.1.3. Trends of NDVI in Each Month

The previous analysis provided an overall picture of the NDVI variations over Guangxi. This study further analyzed the NDVI on a monthly scale to represent the variation details. The monthly mean NDVI ranged from 0.57 to 0.75, with the highest value appearing in October and the lowest appearing in March. Generally, the mean NDVI showed an increasing trend from March to October and a decreasing trend after October, which clearly represents the general changes in vegetation dynamics. During the past few decades, the monthly NDVI also exhibited significant increasing trends in many months (Table 3). As for the whole Guangxi, the highest greening rate was 0.0030 year^{-1} in February, followed by 0.0023 year^{-1} in May, 0.0021 year^{-1} in April, and 0.0015 year^{-1} in July. In August, September, and November, the NDVI showed the same increasing rate of 0.0014 year^{-1}, whereas it was relatively stable in other months. For the three vegetation types, the greening rates were also high in February, April, July, August and September. The greening rate of the farmland was relatively higher than that of the other two vegetation types in each month and, furthermore, peaked at a value of 0.0032 year^{-1} in February.

3.2. Correlations between NDVI and Climate Extremes

3.2.1. Correlations between Annual NDVI and Climate Extremes

In order to identify the main extreme indicators affecting vegetation in Guangxi, this study investigated potential connections between the annual NDVI and the extreme climatic indices by calculating the Pearson correlation coefficient (Table 4). On an annual scale, the NDVI was significantly and positively correlated with the extreme temperature indices, with the exclusion of TXn, DTR and GSL. Specifically, the correlation coefficient can be as high as 0.683, 0.641 and 0.705 for Tm, TXm and TNm, respectively. The coefficient was 0.630 for TNx, 0.605 for SU$_{25}$ and 0.562 for TR$_{25}$, whereas it was under 0.50 for TXx and TNn. The relationship between the annual NDVI and temperature extremes did not show much difference in farmlands, forests and grasslands. The correlation magnitudes were strong and comparable for the three vegetation types. However, the extreme precipitation indices, except SDII, did not show any significant connections with the NDVI in Guangxi. The correlation coefficient between SDII and NDVI was 0.472 in the whole Guangxi, and was a little higher in farmlands than in forests and grasslands. This finding indicates that the vegetation dynamics in Guangxi was not very sensitive to extreme precipitation on an annual scale.

Table 4. Correlation coefficients between the regional mean NDVI and the extreme temperature (precipitation) indices on an annual scale during 1982–2015.

	Tm	TXm	TNm	DTR	TXx	TNx	TXn	TNn	GSL	SU_{25}	TR_{25}	Rx1day	Rx5day	SDII	R20	R50	PRCPTOT
Guangxi	0.683 **	0.641 **	0.705 **	0.027	0.342 *	0.630 **	0.081	0.466 **	0.136	0.605 **	0.562 **	0.324	0.181	0.472 **	0.234	0.242	0.175
Farmlands	0.673 **	0.616 **	0.728 **	−0.029	0.322	0.674 **	0.113	0.412 *	0.092	0.615 **	0.625 **	0.392 *	0.279	0.539 **	0.264	0.253	0.197
Forests	0.680 **	0.640 **	0.687 **	0.081	0.338	0.675 **	0.117	0.450 **	0.158	0.631 **	0.551 **	0.312	0.132	0.406 *	0.160	0.132	0.092
Grasslands	0.684 **	0.634 **	0.712 **	0.025	0.331	0.626 **	0.083	0.456 **	0.195	0.617 **	0.537 **	0.368 *	0.169	0.415 *	0.184	0.195	0.126

Note: * significant at $p < 0.05$; ** significant at $p < 0.01$.

By choosing the six indices (Tm, TXm, TNm, SU_{25}, TR_{25} and SDII) that possessed strong significant correlations with the regional average NDVI, this study further investigated the spatial patterns of the potential connections between the annual NDVI and climate extremes (Figure 6). For Tm, TXm, TNm and SU_{25}, significant positive correlations with the annual NDVI were observed in more than 70% of the meteorological stations. As for Tm, TXm and TNm, the significant coefficient was 0.60 and above in many regions, whereas for SU_{25}, high coefficients of 0.40 and above occurred in central Guangxi. However, for TR_{25} and SDII, only 50% of the meteorological stations showed the significant correlations with the NDVI. In regard to TR_{25}, the significant coefficient was mostly scattered in some parts of western, eastern and southern Guangxi. With respect to SDII, only a few regions, mostly located in central-southern Guangxi, showed significant coefficients with a value around 0.40.

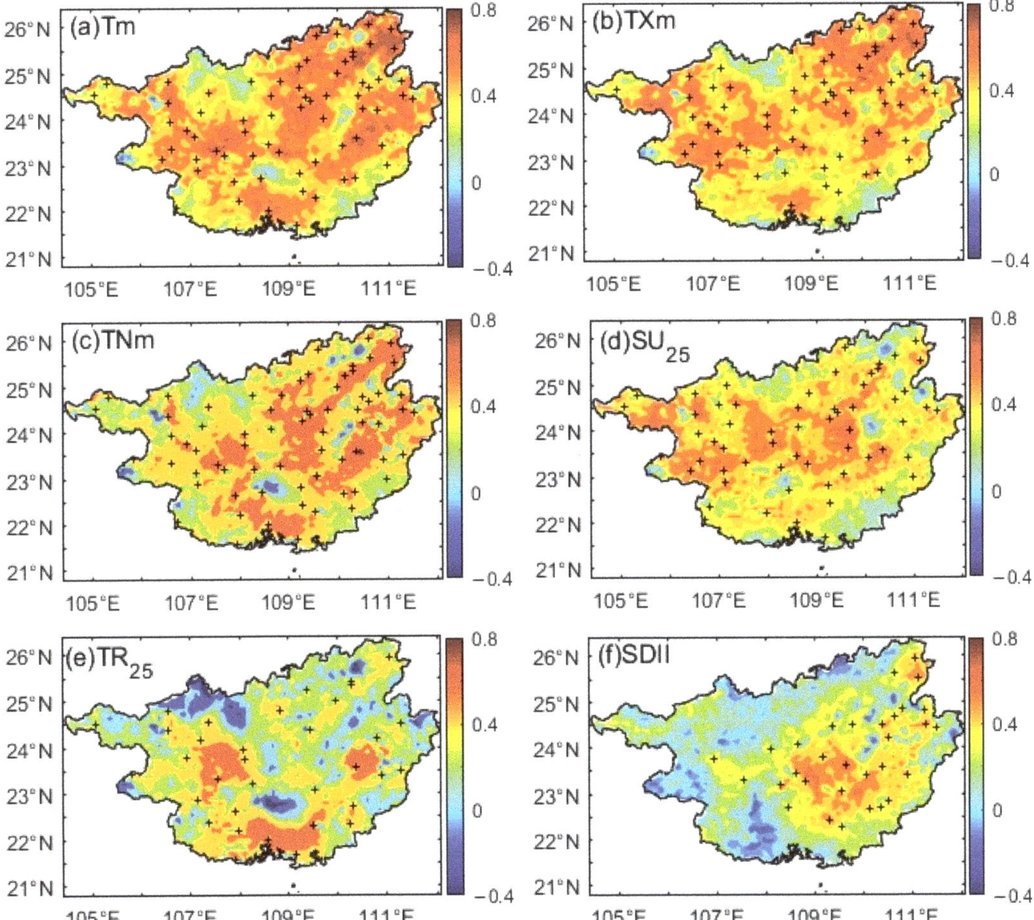

Figure 6. Spatial distributions of the correlation coefficients between the annual NDVI and each temperature extreme index during 1982–2015: (**a**) Tm; (**b**) TXm; (**c**) TNm; (**d**) SU_{25}; (**e**) TR_{25} and (**f**) SDII. Note: '+' significant at $p < 0.05$.

3.2.2. Correlations between Seasonal/Monthly NDVI and Climate Extremes

The relationships between the regional mean NDVI and the extreme climatic indices were further calculated for different seasons and months (Table 5). The correlations for the three types of vegetation did not show much difference, and they were similar to those for the whole of Guangxi. Thus, the coefficients for the three vegetation types are not presented in Table 5. The relationships indicated strong temporal-scale differences. In both spring and summer, the NDVI was significantly and positively correlated with most of the extreme temperature indices, which was also similar to that on an annual scale. In winter, the NDVI was only significantly correlated with Tm, TXm and TNx. Some coefficients were above 0.50 in spring, whereas they were mostly around 0.35–0.45 in summer and winter. Nevertheless, the NDVI did not significantly correlate with any extreme index in autumn, and it did not have any significant connections with TXn and TNn in all seasons.

Table 5. Correlation coefficients between the regional mean NDVI and the extreme temperature (precipitation) indices on a seasonal/ monthly scale during 1982–2015.

	Tm	TXm	TNm	DTR	TXx	TNx	TXn	TNn	Rx1day	Rx5day	SDII	PRCPTOT
Spring	0.526 **	0.577 **	0.477 **	0.432 *	0.226	0.547 **	0.239	0.143	0.121	0.143	0.347 *	0.145
Summer	0.425 *	0.450 **	0.363 *	0.271	0.543 **	0.341 *	0.097	0.069	−0.215	−0.240	−0.172	−0.241
Autumn	0.239	0.169	0.264	−0.096	0.119	0.245	0.021	0.195	0.105	0.094	0.094	0.083
Winter	0.387 *	0.429 *	0.327	0.301	0.238	0.405 *	0.287	0.118	−0.015	0.054	0.129	0.027
January	0.335	0.486 **	0.129	0.587 **	0.448 **	0.386 *	0.155	−0.051	−0.058	−0.001	0.052	−0.097
February	0.483 **	0.465 **	0.497 **	0.298	0.378 *	0.597 **	0.278	0.082	−0.335	−0.384 *	−0.305	−0.398 *
March	0.513 **	0.560 **	0.438 **	0.489 **	0.291	0.368 *	0.244	0.184	0.246	0.227	0.407 *	0.250
April	0.263	0.402 *	0.109	0.554 **	0.207	0.217	−0.115	−0.180	−0.089	−0.083	0.140	−0.083
May	0.129	0.273	0.063	0.315	0.174	0.332	0.211	0.035	0.006	−0.086	0.126	−0.086
June	0.651 **	0.732 **	0.255	0.585 **	0.584 **	0.163	0.304	−0.177	−0.441 **	−0.384 *	−0.399 *	−0.480 **
July	0.350 *	0.408 *	0.237	0.348 *	0.389 *	0.273	0.174	0.144	−0.185	−0.286	−0.197	−0.348 *
August	0.368 *	0.388 *	0.307	0.305	0.477 **	0.447 **	0.051	0.019	−0.236	−0.316	−0.107	−0.347 *
September	−0.056	0.032	−0.105	0.140	−0.063	0.237	−0.061	−0.246	−0.045	−0.062	0.006	−0.120
October	−0.033	0.220	−0.204	0.514 **	−0.072	−0.057	0.214	−0.144	−0.168	−0.234	−0.184	−0.243
November	−0.160	−0.106	−0.132	0.033	0.047	0.058	0.003	−0.133	−0.006	0.059	−0.009	0.068
December	−0.037	0.170	−0.155	0.278	0.025	−0.056	0.117	−0.176	−0.099	0.006	−0.015	0.048

Note: *, ** significant at $p < 0.05$, and $p < 0.01$, respectively.

On a monthly scale, the NDVI was significantly correlated with some extreme temperature indices in some months, mainly including Tm, DTR and three extreme high-temperature indices (TXm, TXx and TNx). The significant and positive correlations were mainly distributed in January, February, March, June, July and August, whereas no significant relationship was observed in May, September, November or December. Additionally, among all the indices, the NDVI was only significantly correlated with TXm and DTR in April, and with DTR in October. The NDVI did not have any significant connections with TXn and TNn in each month, which was similar to that on a seasonal scale.

As for the extreme precipitation indices, the seasonal NDVI only displayed a significant and positive correlation with SDII in spring. The NDVI also rarely had a significant relationship with the precipitation indices in each month, with the exception of a few months. In June, the NDVI was negatively correlated with all the precipitation indices, and the correlation coefficients were around −0.40. It is worth noting that the NDVI was negatively correlated with PRCPTOT in all summer months (June, July and August). Additionally, the NDVI had a significant and negative correlation with precipitation extremes in February.

3.3. Time Lags of NDVI Responses to Climate Extremes

The climatic factors may have delayed effects on vegetation. This study further investigated the time lags of the NDVI responses to extreme climatic indices based on the lag correlation analysis (Table 6). The numbers 0, 1, 2 and 3 represent the number of lag months. The correlation coefficients were much weaker if the time lags were more than three months. Thus, the coefficients with time lags more than three months are not shown in Table 6. The time lags for the three vegetation types are not presented as they were also similar to those for the whole of Guangxi.

Table 6. Correlation coefficients between the regional mean NDVI and the extreme temperature (precipitation) indices for different time lags during 1982–2015.

	Tm	TXm	TNm	DTR	TXx	TNx	TXn	TNn	Rx1day	Rx5day	SDII	PRECPTOT
0	0.614 **	0.664 **	0.576 **	0.584 **	0.539 **	0.546 **	0.646 **	0.551 **	0.311 **	0.262 **	0.384 **	0.233 **
1	0.802 **	0.805 **	0.796 **	0.267 **	0.759 **	0.786 **	0.764 **	0.760 **	0.536 **	0.487 **	0.573 **	0.482 **
2	0.765 **	0.735 **	0.780 **	0.004	0.770 **	0.787 **	0.694 **	0.752 **	0.615 **	0.579 **	0.610 **	0.588 **
3	0.504 **	0.450 **	0.538 **	−0.271 **	0.527 **	0.549 **	0.434 **	0.540 **	0.568 **	0.561 **	0.524 **	0.587 **

Note: ** significant at $p < 0.01$, respectively. '0' NDVI and extreme climate indices were collected over the same period; '1–3' NDVI lagged the extreme indices by 1–3 months. The sample size of the correlations was 408, 407, 406 and 405 for 0-month lag, 1-month lag, 2-month lag and 3-month lag, respectively.

The coefficients were as high as 0.80 when the NDVI lagged the extreme temperature indices by 1–2 months, and they were mostly around 0.60 when the NDVI lagged the extreme precipitation indices by 2–3 months. For the extreme temperature indices, with the exclusion of DTR, TXx and TNx, the time lag of the NDVI responses was one month. However, the NDVI response to DTR did not have any time lags, and the NDVI response to TXx and TNx displayed a two-month time lag. These results partly suggest that there was an unequal time lag in the NDVI responses to the extreme high-temperature indices versus those to the low-temperature indices. For the precipitation extremes, the time lags of the NDVI responses were mostly two months. Therefore, the influence of the extreme temperature indices on vegetation lasted for a shorter period than that of the precipitation indices for one month.

Figures 7 and 8 illustrate the spatial distributions of the time lag to further explore the spatial heterogeneity of the vegetation responses. For Tm, TXm, TNm, TXn and TNn (Figure 7), the vegetation responses showed a lag of one month in the most regions of Guangxi (50.0–73.0% of the total area), and sometimes showed a lag of two months in parts of the western and eastern regions (23.0–47.0% of the global grids). The delayed responses of the NDVI to DTR only occurred in a small area. With regard to the vegetation responses to the two high-temperature indices (TXx and TNx), the areas covered by a two-month delay (60.6% and 51.0%, respectively) accounted for a larger proportion than those covered by a one-month delay (37.3% and 47.3%, respectively). As for the time-lag effects of TXx and TNx, their spatial distributions were different from those of other indices, which was similar to the corresponding result on a regional mean. The unequal lagged effects of high-temperature and low-temperature extremes on terrestrial vegetation growth particularity occurred in some western and eastern regions. Nevertheless, the asymmetric responses of vegetation to the high and low temperatures were not very distinguished in Guangxi.

The delayed responses of vegetation to the precipitation indices were also spatially heterogeneous in Guangxi (Figure 8). The time lag of the NDVI responses to precipitation extremes was mostly one month in some parts of western Guangxi, whereas the lags were two or three months in the central-eastern region. For Rx1day, the grids with a lag of two months accounted for 41.9% of the global grid, those with a lag of three months accounted for 34.0%, and those with a lag of one month constituted 23.5%. For SDII, the grids with a lag of one month accounted for 40.4% of the global grid and those with a lag of two months accounted for 38.8%. For Rx5day and PRECPTOT, the grids covered by a three-month lag accounted for about 50.0% of the global grid, and those covered by a two-month lag accounted for around 35.0%. Generally speaking, most terrestrial vegetation growth was inclined to respond to Rx1day, Rx5day and PRECPTOT with time-lag effects up to 2–3 months, but to SDII with 1–2 month time-lag effects. By comparing the distributions of land use and land cover classes (Figure 1) with those of the above lag results (Figures 7 and 8), this study did not find a very obvious dependence of the time lag on land types.

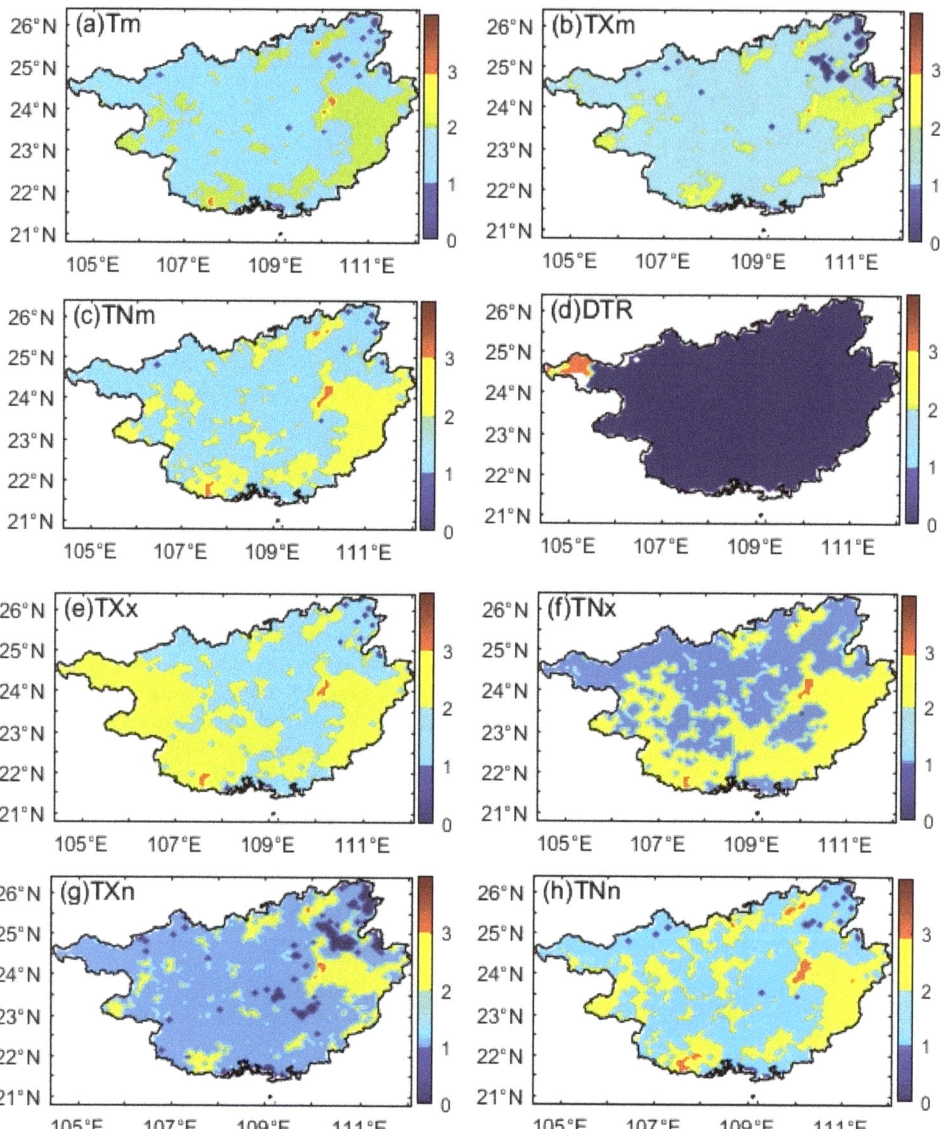

Figure 7. Spatial distributions of the time-lagged response of the NDVI to the temperature extremes: (**a**) Tm; (**b**) TXm; (**c**) TNm; (**d**) DTR; (**e**) TXx; (**f**) TNx; (**g**) TXn and (**h**) TNn. Note: '0' NDVI and extreme climate indices were collected over the same period; '1–3' NDVI lagged the extreme indices by 1–3 months. This study conducted four correlation analyses (namely, no lag, 1 month, 2 months, and 3 months), and the results of the four times were then superimposed to take the maximum value.

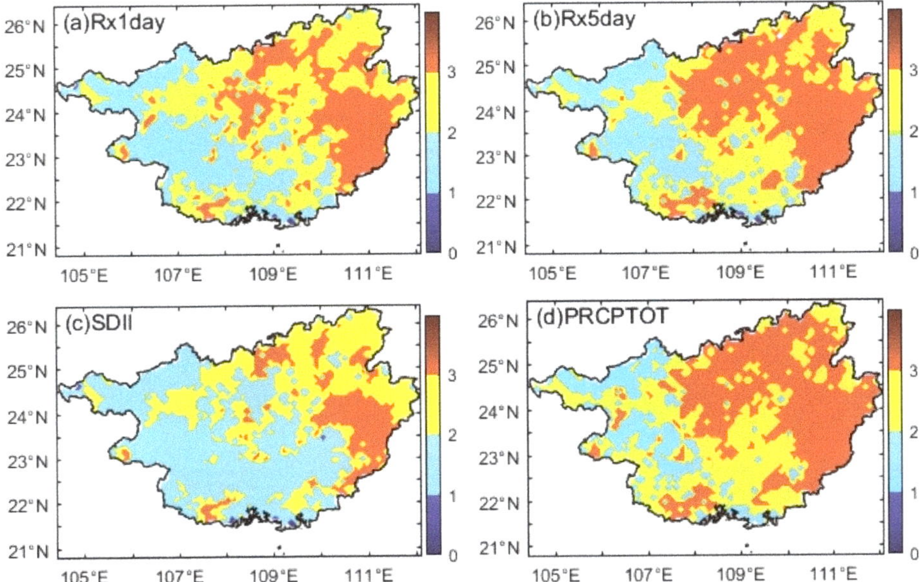

Figure 8. Spatial distributions of the time-lagged response of the NDVI to the precipitation extremes:
(**a**) Rx1day; (**b**) Rx5day; (**c**) SDII; and (**d**) PRCPTOT. Note: '0' NDVI and extreme climate indices were
collected over the same period; '1–3' NDVI lagged the extreme indices by 1–3 months. This study
conducted four correlation analyses (namely, no lag, 1 month, 2 months, and 3 months), and then
superimposed the results of the four times to take the maximum value.

4. Discussion

4.1. Variations in Vegetation Dynamics

During 1982–2015, a significant increasing trend in vegetation greenness was detected
in Guangxi not only on an annual scale, but also on seasonal and monthly scales. The
trends in the variation of terrestrial vegetation indicated obvious spatial heterogeneity, but
all types of vegetation mainly had greening trends in Guangxi, with farmland showing the
highest rate and forest showing the lowest. By comparing the spatial distributions of land
types (Figure 1) with these of the annual NDVI (Figure 9a,b) for the two key moments, it
is also clear that most vegetation areas witnessed a greening trend, especially in farming
areas. As for the annual NDVI in Guangxi, our results revealed that the increasing rate was
0.00144 year^{-1} from 1982 to 2015. The greening rate in Guangxi was much higher than that
in all of China (0.0006 year^{-1}) [20,21]. It also exceeded the increases in the Yangtze River
Basin (0.001 year^{-1}) [14,17] and Xinjiang (0.0003 year^{-1}) [22]. Conversely, the greening
rate in Guangxi was much lower than that on the Loess Plateau (0.0025 year^{-1}) [13]. The
diverse greening rate of vegetation over different regions further suggested the spatial
heterogeneity of the vegetation variation.

In Guangxi, the regional temperature indices generally had the similar increasing
trend as that of the regional averaged NDVI on an annual scale (Figure 9c,d). The MK
abrupt test for the climate indices found that many temperature indices also had turning
points around the year 2005 (Figure 9e,f: taking the minimum temperature and tropical
nights as the two examples), which was similar to the case of annual NDVI. Thus, the
extremely low NDVI in 2005 may be attributed to the negative effects of the great warming.
The regional mean NDVI showed increasing trends in each season over Guangxi. However,
according to Cui et al. (2018) [46], the NDVI showed different trends in different seasons
over some other parts of China during previous decades. This may indicate the dependence
of the NDVI variation on geographical location and spatial-scale effects. In Guangxi, the

vegetation in spring had a greater greening rate than in the other seasons, which is similar to the result on the national scale [21]. However, the spring NDVI in Guangxi showed a much higher increasing rate, compared with that in all of China [21]. The long-term variation in the monthly NDVI was also highly uneven. Therefore, our work focusing on a multiple time scale analysis is extremely helpful for understanding the time inhomogeneity of the vegetation variations.

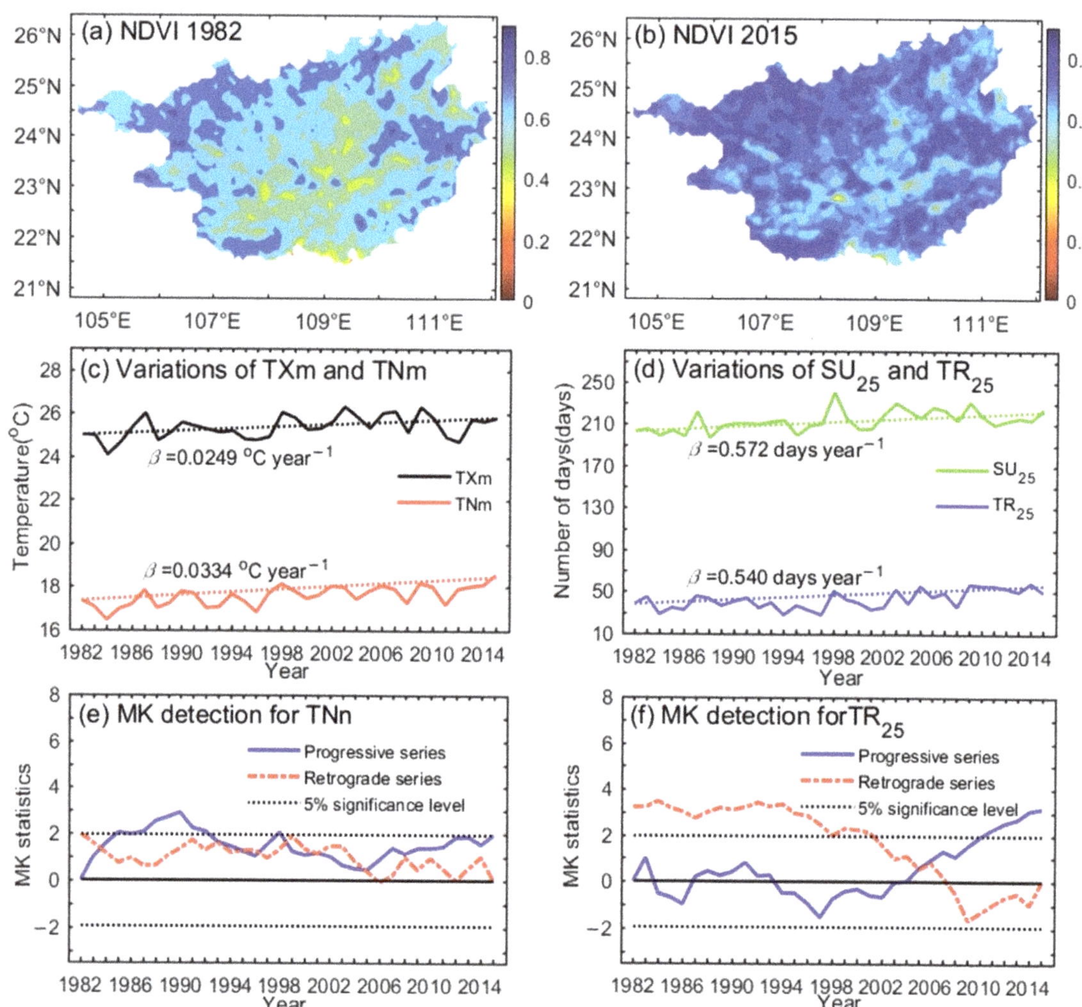

Figure 9. Spatial distributions of the annual NDVI in 1982 (**a**); spatial distributions of the annual NDVI in 2015 (**b**); inter-annual variations of TXm and TNm (**c**); inter-annual variations of SU$_{25}$ and TR$_{25}$ (**e**); the Mann–Kendall (MK) abrupt change detection for TNn (**e**); and the Mann–Kendall (MK) abrupt change detection for TR$_{25}$ (**f**). Note: in (**c,d**), the solid line represents the observed data, and the dotted line represents the fitted data.

4.2. Correlations between Vegetation and Climate Extremes

The vegetation–climate interactions are highly heterogeneous [7,26]. Knowledge of the relationship between vegetation and climate extremes would be useful for evaluating the vulnerability and resilience of vegetation to climate extremes [31,52]. From 1982 to

2015, the annual NDVI in Guangxi generally showed strong and positive correlations with most of the extreme temperature indices, which was consistent with many previous results [12,28,30,31]. Figure 10 illustrates the trend rates of some of the representative annual indices during 1982–2015. In most areas of Guangxi, the maximum and minimum temperatures increased with a rate above 0.02 °C year^{-1}, and summer days increased with a rate around 0.50 days year^{-1} (Figure 10). Wang et al. (2021) [6] found that the most recent 20-year period (after 2000) experienced greater warming than previous periods, and low-temperature extreme warming was usually greater than high-temperature extreme warming during the past 58 years. Furthermore, by comparing Figure 6 with Figure 10, it is obvious that the strong correlations between the temperature indices and vegetation mostly occurred in the areas where the warming rate was high. These indicate that the increasing extreme temperature mostly exerted a positive effect on plant growth in Guangxi.

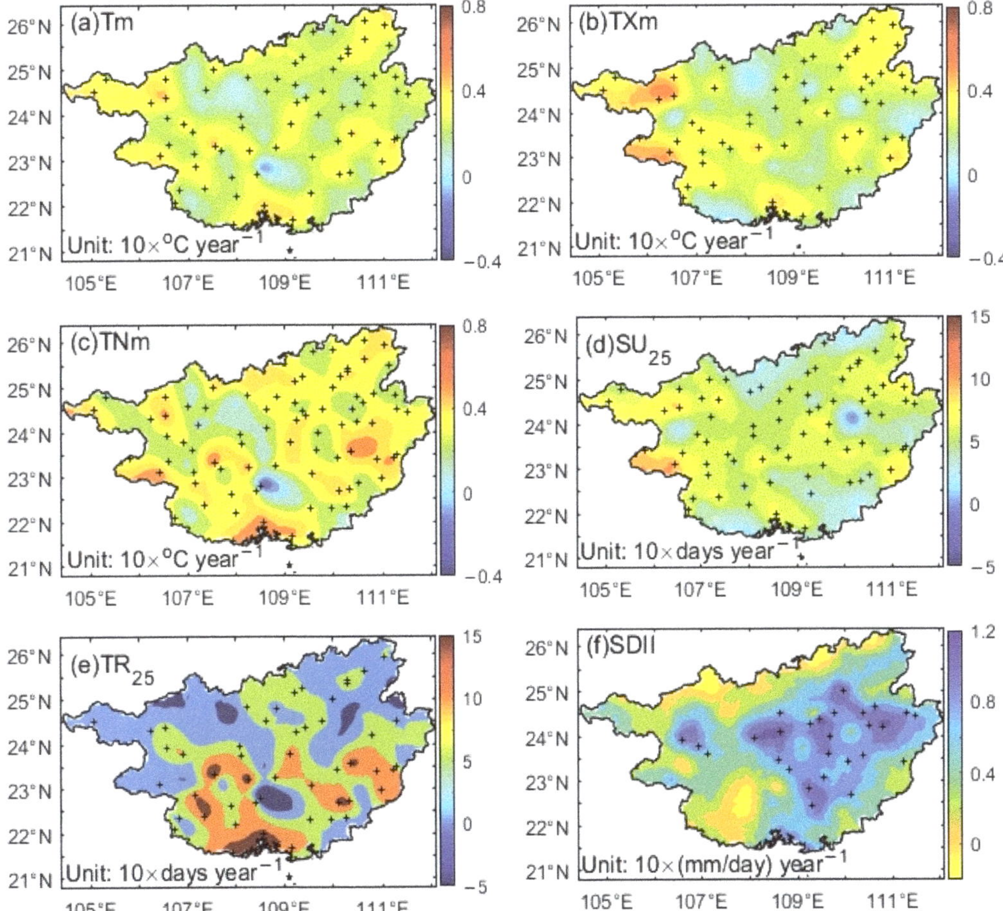

Figure 10. Spatial distributions of the trend rates of the annual climate indices during 1982–2015: (**a**) TXm, (**b**) TNm, (**c**) TXx (unit: 10 × °C year^{-1}); (**d**) SU$_{25}$, (**e**) TR$_{25}$ (unit: 10 × days year^{-1}) and (**f**) SDII (unit: 10 × (mm/day) year^{-1}). Note: '+' significant at $p < 0.05$.

Having a large area and vegetation diversity, China presented large northwest–southeast differences in its vegetation greenness (Figure 11). The multiyear mean of the annual NDVI was mostly around or above 0.5 in southeastern regions in China, such as Guangxi, whereas it was mostly under 0.3 in northwestern China, which has a low vegetation cover rate. As vegetation is a pivotal link between the atmosphere and the land's surface, the inhomogeneity of vegetation activity can enhance the spatial heterogeneity of the vegetation responses to different climate extremes. Extreme temperature mostly has a very extensive and complex effect on vegetation in China [12,18,25]. High temperature can enhance the evaporation [53] and reduce soil moisture [25,31], which increases the soil drought conditions. Thus, the increasing high temperature may tend to preclude vegetation growth in the north of China [12,18]. Nevertheless, there are relatively flourishing vegetation and abundant precipitation resources in Guangxi, and the abundant precipitation would mitigate the negative effects of increasing high-temperature indices to a certain degree.

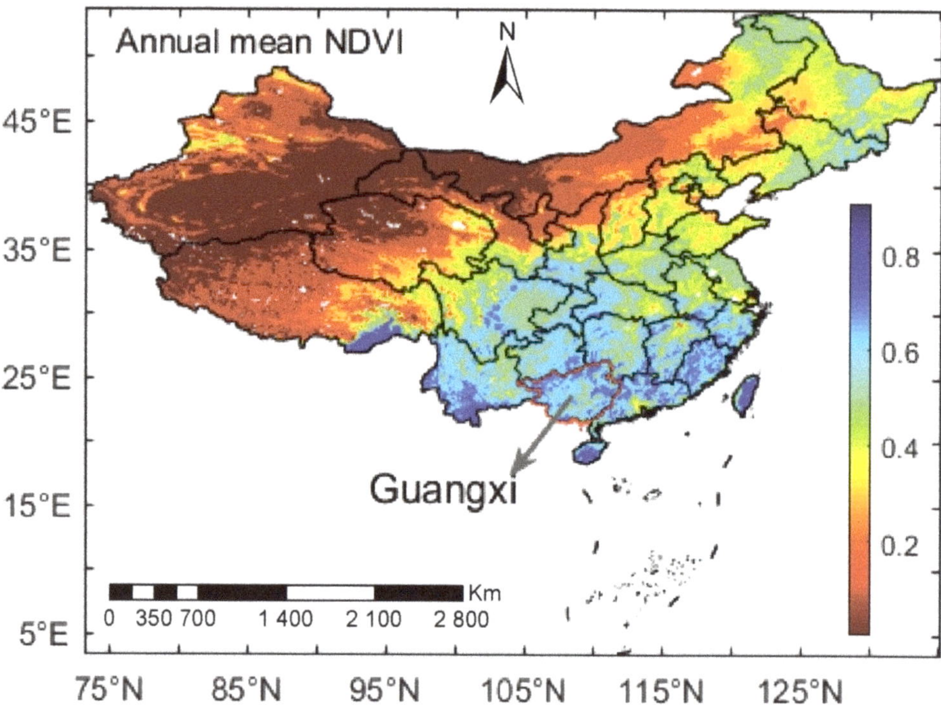

Figure 11. The mean of the annual NDVI in China during 1982–2015.

Vegetation has different mechanisms in responding to the low-temperature and high-temperature indices [25,31]. Low temperatures elongate the plant growing season [54] and enrich the soil nutrients [55]. These mechanisms might partly explain the positive correlations between the annual NDVI and the low-temperature indices over Guangxi. In addition, with obvious differences from the high-temperature indices, some low-temperature indices did not have any significant relations with the NDVI in each season and month, which further indicated the non-uniform responses of terrestrial vegetation growth to the high-temperature and low-temperature indices at the local scale. Furthermore, the correlations between the NDVI and the extreme temperature indices varied seasonally and monthly in Guangxi. The positive correlations between the NDVI and the high-temperature indices were strongest in spring, followed by summer and winter, but they were not significant at all in autumn. This seasonal distinction of the correlations in Guangxi is highly different from that at the northern hemisphere scale, where the strongest negative correlations between the mean maximum temperature and NDVI were observed in the summer of temperate dry regions [23].

At the seasonal and monthly scales, the temperature indices showed increasing trends in many months, with remarkable seasonal and monthly differences (Figure 12: taking February and June as the two representative months). This study found that February witnessed the greatest warming rate. In February, the increasing rates of the mean temperature, maximum temperature and minimum temperature were 0.0935, 0.120 and 0.0765 $°C$ $year^{-1}$, respectively, whereas they were generally under 0.050 $°C$ $year^{-1}$ in other months. High temperatures in spring provide suitable temperatures for vegetation growth. Conversely, extreme high temperatures in summer may exceed ideal growing conditions of vegetation, and lead to increased evaporation and decreased soil moisture [25,31]. All of these factors may sometimes restrain the vegetation photosynthesis to a certain degree [55], and therefore weaken the positive effects of high temperatures on the NDVI in summer. In winter, the coverage of vegetation may decrease partly due to its phenological characteristics rather than climate change. Therefore, the correlations were relatively weaker in winter. By comparing the correlation coefficients between temperature extremes and vegetation at different temporal scales, we can find that the correlations were relatively strongest on an annual scale, and were weakest on a monthly scale.

Located in southern China and close to the South Sea, Guangxi has an annual precipitation of above 1500 mm based on the multiyear mean. The extreme precipitation indices rarely changed significantly over previous decades, with the exception of the simple daily intensity index. The simple daily intensity only significantly increased in some southeastern regions accounting for 32% of Guangxi (Figure 10f). Among all the precipitation indices, the NDVI only displayed a significantly and positively weak correlation with the simple daily intensity on an annual scale, and in spring on a seasonal scale. By comparing Figure 6f with Figures 10f and 1, it can be found that the weak positive correlations mainly occurred in the area with low elevation and significant increases in precipitation. This is probably because low elevation is less prone to cause water loss and soil erosion, further promoting vegetation growth. Generally speaking, the greening vegetation was relatively non-sensitive to the extreme precipitation indices. This finding of the precipitation effects on vegetation in Guangxi differs significantly from those in northern arid and semi-arid regions where precipitation is limited and drought is a great threat to vegetation [18,21,27,56].

Analyzing the vegetation responses on a monthly scale helps to better understand the main limiting factors for vegetation growth in different growth periods [28]. The NDVI was negatively correlated with all the precipitation indices in June and with the two indices in February. Additionally, the NDVI was negatively correlated with the total wet-day precipitation in all summer months (June, July and August). This phenomenon on a monthly scale indicates that extreme precipitation will restrain vegetation growth in February and summer months, which cannot be reflected at annual and seasonal scales. Based on the analysis at a monthly scale, this study further found that the total wet-day precipitation significantly decreased at a rate of 1.325 mm $year^{-1}$ in February, whereas it significantly increased at a rate of 2.644 mm $year^{-1}$ in June from 1982 to 2015 (Figure 12e,f). Thus, the plant's growth in February was prevented by lack of water in addition to low soil moisture. Karst areas have low water-soil retaining ability, so the short-term precipitation increase in summer months can cause soil and water losses, further stunting plant growth. The influences of extreme indices on vegetation in different growing periods are better reflected at a monthly scale, which further suggests time-scale effects on the relationship between precipitation extremes and vegetation dynamics.

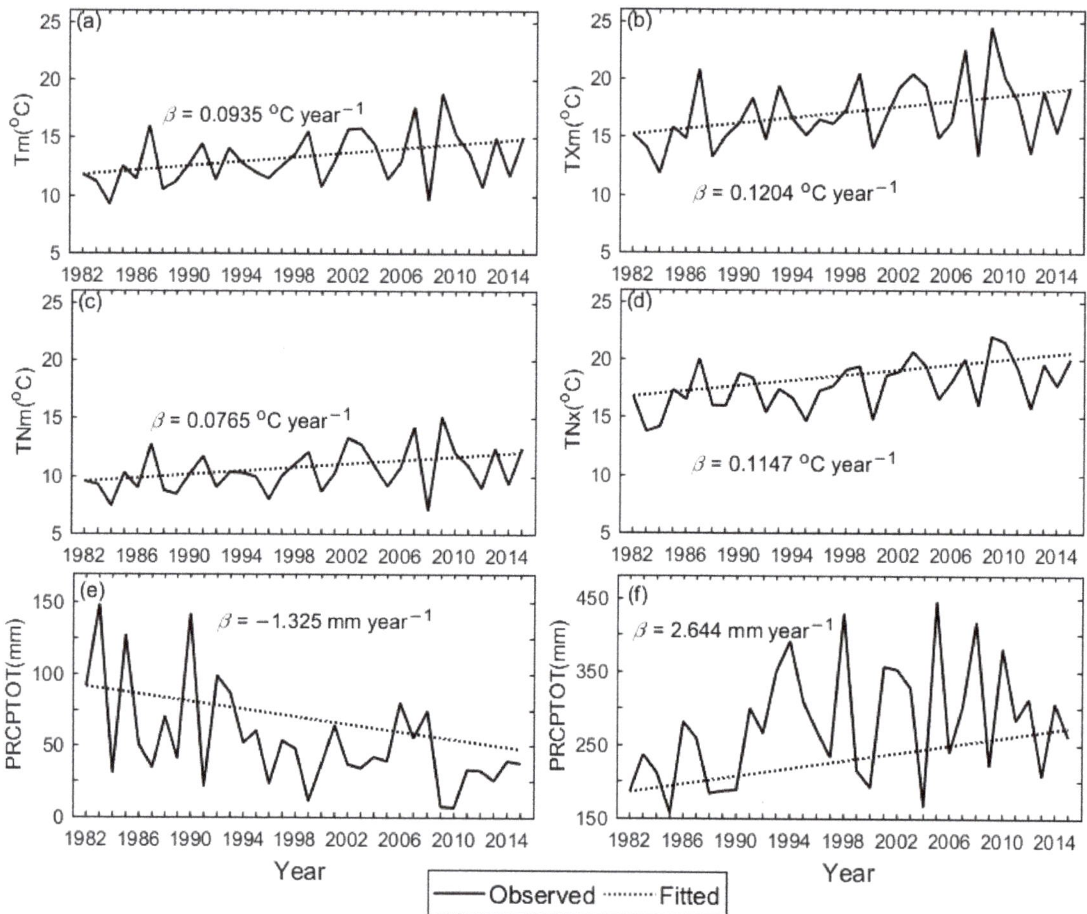

Figure 12. The interannual variation in the monthly climate indices during1982–2015: (**a**) Tm in February, (**b**) TXm in February, (**c**) TNm in February, (**d**) TNx in February, (**e**) PRCPTOT in February and (**f**) PRCPTOT in June. Note: the solid line represents the observed data, and the dotted line represents the fitted data.

The above results indicate that the variations in vegetation dynamics were mostly determined by temperature extremes rather than precipitation extremes, and the strengthened vegetation activities can be associated with enhanced extreme temperatures in Guangxi. Regarding the intensity and range of the extreme climate effects on vegetation dynamics in China, Li et al. (2021) [18] categorized these impacts into five types: the humidity-promoting type, the cold-promoting and drought-inhibiting compound type, the drought-inhibiting type, the heat-promoting and humidity-promoting compound type, and the heat-promoting and drought-inhibiting compound type. However, most of Guangxi belongs only to the heat-promoting type. Overall, although the vegetation responses to climate extremes showed great temporal heterogeneity at the different time scales, the effects of extreme climate on vegetation cannot be totally disentangled from the baseline effect of climate on a time series.

4.3. Lagged Responses of Vegetation to Extreme Climates

The influences of climate factors on vegetation often show a time lag because of lags in the adjustment of the soil moisture content and biological processes [29,30]. Many studies have shown that the antecedent ambient temperature has remarkable impacts on vegetation growth [20,30,52], which was confirmed in the present study. The lagged responses of vegetation to the extreme climatic indices differed in different regions and at different scales over Guangxi.

Generally speaking, the responses of vegetation to most temperature extremes lagged for at least one month, whereas the responses to precipitation extremes lagged for at least two months. The lagged responses of vegetation to the extreme temperature indices were generally around one month earlier than those to the precipitation indices. This result is similar to that in the Poyang Lake Basin [31], but different those in some other areas or large regions [12,28]. For example, on the Inner Mongolian Plateau, the vegetation dynamics are connected with extreme temperature indices by a time lag of at least three months, and with extreme precipitation indices by a two-month lag [12]. The impact of extreme temperatures on vegetation showed a time lag of at least one month, but the responses of maximum NDVI to extreme precipitation parameters indicated no lags in Central Asia [28]. There is a lag of approximately one month in the vegetation dynamic response to soil moisture [57], which explains the particular lag in the vegetation responses to most extreme temperature variations in Guangxi and Central Asia. The vegetation responses to some high-temperature indices displayed a two-month time lag, and had a longer time lag of one month compared to the responses to the extreme low-temperature indices. Thus, the same vegetation type sometimes showed asymmetric lag responses to the high-temperature indices versus the low-temperature indices in Guangxi, which is partly similar to the result from Wen et al. (2019) [32].

In addition, Guangxi witnessed the spatial heterogeneities in the vegetation responses to different extreme climate indices. The impacts of most of the extreme temperature indices on vegetation indicated a lag of one month in most of Guangxi, and sometimes exhibited a lag of two months. The time lag from the extreme precipitation indices were spatially more uneven, varying between one month and three months. Many studies have indicated that the effects of climate change on vegetation dynamics are spatially heterogeneous [7,13,14,17,58], and the driving factors are the geographical location, vegetation types, topography, etc. [26]. Vegetation type is an important factor in determining the time-lag effects and the strength of the correlations between vegetation dynamics and extreme climate indices in many regions [12,18,28,30,31,42,58]. However, by comparing the spatial patterns of the correlations and time lags under the different land cover types, this study found that there were no obvious differences in the correlations and time lags among farmlands, forests and grasslands, suggesting that the vegetation type itself may not have made a big difference in the vegetation responses to climate extremes in Guangxi. This is partly consistent with the results in China at a national scale [21].

Past studies were frequently performed only on one time scale. Instead, multi-time scale analysis is an effective approach to understanding how vegetation will respond to global climate change. This study can help find effective preventive measures to preserve the ecosystem in Guangxi. Local authorities should pay attention to taking adaptive mitigation measures in advance to prevent the potential negative effects of temperature extremes, as the future temperature may often exceed ideal growth conditions for vegetation if extreme temperatures continue to rise significantly. Local authorities also should formulate appropriate measures to reduce the negative effects caused by drought in February and by flood in summer months (especially in June). Drought-resistant and water-resistant vegetation should be introduced in a timely manner. With regard to agriculture, adequate irrigation and film-mulching treatments are necessary for drought control and resistance, and timely water drainage is necessary for flood resistance.

4.4. Limitations and Uncertainties

Many previous studies have validated the GIMMS NDVI data [42–44], and have pointed out that the GIMMS NDVI data are reliable for sensing the green vegetation and monitoring its long-term trends and activities [17,40,45,46]. Thus, the GIMMS NDVI data were directly used in this study without being calibrated with other independent NDVI datasets. Although the results obtained from this study were thought to be reliable, there were limitations to and uncertainties in the approach. Firstly, the GIMMS NDVI data are indirect remote sensing data simulated by the model. Their accuracy is limited to the satellite sensor sensitivity [59]. The remote sensing NDVI is susceptible to clouds, water vapor and dust aerosol pollution, and is easily saturated in areas of high-density vegetation when the vegetation cover reaches a certain high level [36,60]. This over-saturated problem leads to underestimating the greenness of vegetation in some study regions. Furthermore, despite the GIMMS NDVI data being corrected to minimize non-vegetation effects, the NDVI index itself cannot provide enough information on the vegetation and species composition [35]. There may be some errors in the modeled NDVI in karst areas dominated by hills, especially the highly fragmentized terrain with bare soil and rock outcrops. These will result in further uncertainties in the relationship between the NDVI and climatic variables.

Secondly, some uncertainties in the correlations existed because of the coarse spatial resolution of the NDVI and climate data. There were uncertainties when the climatic parameters of the meteorological stations were resampled based on the spatial resolution of the GIMMS NDVI data. The uncertainties can be minimized by field studies combined with the temporal remote sensing data of higher resolution imagery in follow-up work. In addition, the aim of this study was to systematically evaluate the influence characteristics of extreme climate on vegetation, and therefore a discussion of human influence was not included. However, in order to reduce soil erosion and improve ecological conditions, China has implemented several vegetation restoration and reforestation projects (e.g., the Natural Forest Protection Project, the Grain for Green project, and Karst Rocky Desertification Comprehensive Control and Restoration Project) initiated around 2000 [34]. These afforestation plans that have been carried out can result in a vegetation increase in some karst areas.

Long periods of time exposure to climate extremes may lead to vegetation vulnerability. Admittedly, the present study did not take into account the interaction between different climate extremes and other factors or the occurrence of different effects. Furthermore, some vegetation will become weaker or even die, whereas other vegetation may adapt to extreme climate by changing its own physiological and biochemical reactions or by modifying its own living conditions [61], and it can have feedback to the climate system [17]. This study did not strictly consider the environmental effect of the vegetation dynamics, which should be enhanced in the future for better understanding of this influence on our results. Therefore, examination of the vegetation–climate interactions still faces many challenges, and will require continued in-depth research in the future.

5. Conclusions

By means of the Normalized Difference Vegetation Index (NDVI) to capture vegetation activities, the study tried to use a multi-time scale (annual, seasonal and monthly) analysis to investigate the effects of climate extremes on vegetation from 1982 to 2015 in Guangxi. The main findings are listed as follows:

(1) The variation rates of NDVI highly differed at different time scales. The annual NDVI significantly increased at a rate of 0.00144 year^{-1}. The greening trend was strongest in spring, followed by winter, autumn and summer. On a monthly scale, the remarkable greening trends occurred in February, May and April.

(2) The effects of extreme climate on vegetation cannot be disentangled from the baseline effect of climate on a time series. The enhanced temperature extremes had positive and strong correlations with green vegetation on an annual scale. With a great seasonal and monthly heterogeneity, the significant positive correlations mostly occurred only

in January, February, March, and summer months. Precipitation extremes only had significant and negative relations with vegetation in February and summer months.

(3)   The responses of vegetation to climate extremes showed a great spatial heterogeneity, but they showed no significant differences among farmlands, forests and grasslands. The vegetation generally responded to temperature extremes with a time lag of at least one month, and there was mostly a two-month lag relative to precipitation extremes.

Nevertheless, it is essential to realize the limitations of the satellite NDVI data and the resolution gap between the NDVI and climate data. This study highlights the necessity to use a multi-time scale analysis. The detailed analysis can help better understand the response mechanism of vegetation dynamics to extreme climate and improve the investigation of vegetation–climate interactions. This study can help decision makers identify the practical issues facing regional management systems in Guangxi. Local authorities should take adaptive mitigation measures in advance to prevent the negative effects caused by enhanced temperature extremes in the future. Local authorities and farmers also should be alert to both the adverse effects of drought in February and those of flood in summer months. For example, it would be helpful to introduce drought-resistant and water-resistant vegetation in a timely manner.

Author Contributions: Conceptualization, L.W. and F.H.; methodology, L.W. and F.H.; validation, Y.M. and C.Z.; formal analysis, L.Z.; investigation, Y.M. and C.Z.; data curation, L.W. and F.H.; writing—original draft preparation, L.W. and F.H.; writing—review and editing, Y.M., C.Z., L.Z. and M.L.; visualization, F.H.; supervision, L.W.; project administration, L.W.; funding acquisition, L.W. All authors have read and agreed to the published version of the manuscript.

Funding: This research was funded by the National Natural Science Foundation of China (Grant No. 42005142).

Institutional Review Board Statement: Not applicable.

Informed Consent Statement: Not applicable.

Data Availability Statement: All the data are available in the public domain at the links provided in the texts.

Acknowledgments: Thanks to the China Meteorological Administration (CMA) for providing the meteorological data. Thanks to all editors and commenters.

Conflicts of Interest: The authors declare no conflict of interest.

References

1.   Wang, X.; Piao, S.; Ciais, P.; Li, J.; Friedlingstein, P.; Koven, C.; Chen, A. Spring temperature change and its implication in the change of vegetation growth in North America from 1982 to 2006. *Proc. Natl. Acad. Sci. USA* **2011**, *108*, 1240–1245. [CrossRef]
2.   Schlesinger, W.H.; Jasechko, S. Transpiration in the global water cycle. *Agric. For. Meteorol.* **2014**, *189*, 115–117. [CrossRef]
3.   Yin, Z.; Dekker, S.; van den Hurk, B.; Dijkstra, H. Effects of vegetation structure on biomass accumulation in a Balanced Optimality Structure Vegetation Model (BOSVM v1. 0). *Geosci. Model Dev.* **2014**, *7*, 821–845. [CrossRef]
4.   Sun, Y.; Zhang, X.B.; Zwiers, F.W.; Song, L.C.; Wan, H.; Hu, T.; Yin, H.; Ren, G. Rapid increase in the risk of extreme summer heat in eastern China. *Nat. Clim. Chang.* **2014**, *4*, 1082–1085. [CrossRef]
5.   Shi, J.; Cui, L.; Wen, K.; Tian, Z.; Wei, P.; Zhang, B. Trends in the consecutive days of temperature and precipitation extremes in China during 1961–2015. *Environ. Res.* **2018**, *161*, 381–391. [CrossRef]
6.   Wang, L.; Hu, F.; Hu, J.; Chen, C.; Liu, X.; Zhang, D.; Chen, T.; Miao, Y.; Zhang, L. Multistage spatiotemporal variability of temperature extremes over South China from 1961 to 2018. *Theor. Appl. Climatol.* **2021**, *146*, 243–256. [CrossRef]
7.   Vicente-Serrano, S.M.; Gouveia, C.; Camarero, J.J.; Beguería, S.; Trigo, R.; Lópezmoreno, J.I.; Azorínmolina, C.; Pasho, E.; Lorenzolacruz, J.; Revuelto, J. Response of vegetation to drought time-scales across global land biomes. *Proc. Natl. Acad. Sci. USA* **2013**, *110*, 52–57. [CrossRef]
8.   Rammig, A.; Wiedermann, M.; Donges, J.F.; Babst, F.; Von Bloh, W.; Frank, D.; Thonicke, K.; Mahecha, M.D. Coincidences of climate extremes and anomalous vegetation responses: Comparing tree ring patterns to simulated productivity. *Biogeosciences* **2015**, *12*, 373–385. [CrossRef]
9.   Dubovyk, O.; Landmann, T.; Dietz, A.; Menz, G. Quantifying the impacts of environmental factors on vegetation dynamics over climatic and management gradients of Central Asia. *Remote Sens.* **2016**, *8*, 600. [CrossRef]

10. Wen, Z.; Wu, S.; Chen, J.; Lü, M. NDVI indicated long-term interannual changes in vegetation activities and their responses to climatic and anthropogenic factors in the Three Gorges Reservoir Region, China. *Sci. Total Environ.* **2017**, *574*, 947–959. [CrossRef]

11. Li, C.; Leal Filho, W.; Wang, J.; Yin, J.; Fedoruk, M.; Bao, G.; Bao, Y.; Yin, S.; Yu, S.; Hu, R. An assessment of the impacts of climate extremes on the vegetation in Mongolian Plateau: Using a scenarios-based analysis to support regional adaptation and mitigation options. *Ecol. Indic.* **2018**, *95*, 805–814. [CrossRef]

12. Li, C.; Wang, J.; Hu, R.; Yin, S.; Bao, Y.; Ayal, D.Y. Relationship between vegetation change and extreme climate indices on the Inner Mongolia Plateau, China, from 1982 to 2013. *Ecol. Indic.* **2018**, *89*, 101–109. [CrossRef]

13. Zhao, L.; Dai, A.; Dong, B. Changes in global vegetation activity and its driving factors during 1982–2013. *Agric. For. Meteorol.* **2018**, *249*, 198–209. [CrossRef]

14. Zhang, W.; Wang, L.; Xiang, F.; Qin, W.; Jiang, W. Vegetation dynamics and the relations with climate change at multiple time scales in the Yangtze River and Yellow River Basin, China. *Ecol. Indic.* **2020**, *110*, 105892. [CrossRef]

15. Piao, S.L.; Nan, H.J.; Huntingford, C. Evidence for a weakening relationship between interannual temperature variability and northern vegetation activity. *Nat. Commun.* **2014**, *5*, 5058. [CrossRef]

16. Liu, Y.; Liu, X.; Hu, Y.; Li, S.; Peng, J.; Wang, Y. Analyzing nonlinear variations in terrestrial vegetation in China during 1982–2012. *Environ. Monit. Assess.* **2015**, *187*, 722. [CrossRef]

17. Cui, L.; Wang, L.; Qu, S.; Singh, R.P.; Lai, Z.; Yao, R. Spatiotemporal extremes of temperature and precipitation during 1960–2015 in the Yangtze River Basin (China) and impacts on vegetation dynamics. *Theor. Appl. Climatol.* **2019**, *136*, 675–692. [CrossRef]

18. Li, S.; Wei, F.; Wang, Z.; Shen, J.; Liang, Z.; Wang, H.; Li, S. Spatial heterogeneity and complexity of the impact of extreme climate on vegetation in China. *Sustainability* **2021**, *13*, 5748. [CrossRef]

19. Schuldt, B.; Buras, A.; Arend, M.; Vitasse, Y.; Beierkuhnlein, C.; Damm, A.; Gharun, M.; Grams, T.E.; Hauck, M.; Hajek, P.; et al. A first assessment of the impact of the extreme 2018 summer drought on Central European forests. *Basic Appl. Ecol.* **2020**, *45*, 86–103. [CrossRef]

20. Xu, G.; Zhang, H.; Chen, B.; Zhang, H.; Innes, J.L.; Wang, G.; Yan, J.; Zheng, Y.; Zhu, Z.; Myneni, R.B. Changes in vegetation growth dynamics and relations with climate over China landmass from 1982 to 2011. *Remote Sens.* **2014**, *6*, 3263–3283. [CrossRef]

21. Liu, Y.; Lei, H. Responses of natural vegetation dynamics to climate drivers in China from 1982 to 2011. *Remote Sens.* **2015**, *7*, 10243–10268. [CrossRef]

22. Du, J.; Shu, J.; Yin, J.; Yuan, X.; Jiaerheng, A.; Xiong, S.; He, P.; Liu, W. Analysis on spatio-temporal trends and drivers in vegetation growth during recent decades in Xinjiang, China. *Int. J. Appl. Earth Obs.* **2015**, *38*, 216–228. [CrossRef]

23. Tan, J.; Piao, S.; Chen, A.; Zeng, Z.; Ciais, P.; Janssens, I.A.; Mao, J.; Myneni, R.; Peng, S.; Peñuelas, J.; et al. Seasonally different response of photosynthetic activity to daytime and night-time warming in the Northern Hemisphere. *Glob. Chang. Biol.* **2014**, *21*, 377. [CrossRef]

24. John, R.; Chen, J.; Ou-Yang, Z.-T.; Xiao, J.; Becker, R.; Samanta, A.; Ganguly, S.; Yuan, W.; Batkhishig, O. Vegetation response to extreme climate events on the Mongolian Plateau from 2000 to 2010. *Environ. Res. Lett.* **2013**, *8*, 035033. [CrossRef]

25. Peng, S.; Piao, S.; Ciais, P.; Myneni, R.; Chen, A.; Chevallier, F.; Dolman, A.J.; Janssens, I.A.; Peñuelas, J.; Zhang, G.; et al. Asymmetric effects of daytime and night-time warming on Northern Hemisphere vegetation. *Nature* **2013**, *501*, 88–94. [CrossRef]

26. Hua, W.; Chen, H.; Zhou, L.; Xie, Z.; Qin, M.; Li, X.; Ma, H.; Huang, Q.; Sun, S. Observational Quantification of Climatic and Human Influences on Vegetation Greening in China. *Remote Sens.* **2017**, *9*, 425. [CrossRef]

27. Zhang, Q.; Kong, D.D.; Singh, V.P.; Shi, P.J. Response of vegetation to different time-scales drought across China: Spatiotemporal patterns, causes and implications. *Global Planet. Chang.* **2017**, *152*, 1–11. [CrossRef]

28. Luo, M.; Sa, C.; Meng, F.; Duan, Y.; Liu, T.; Bao, Y. Assessing extreme climatic changes on a monthly scale and their implications for vegetation in Central Asia. *J. Clean. Prod.* **2020**, *271*, 122396. [CrossRef]

29. Piao, S.; Mohammat, A.; Fang, J.; Cai, Q.; Feng, J. NDVI-based increase in growth of temperate grasslands and its responses to climate changes in China. *Glob. Environ. Chang.* **2006**, *16*, 340–348. [CrossRef]

30. Wu, D.; Zhao, X.; Liang, S.; Zhou, T.; Huang, K.; Tang, B.; Zhao, W. Time-lag effects of global vegetation responses to climate change. *Glob. Chang. Biol.* **2015**, *21*, 3520–3531. [CrossRef]

31. Tan, Z.Q.; Tao, H.; Jiang, J.H.; Zhang, Q. Influence of climate extremes on NDVI (Normalized Difference Vegetation Index) in the poyang lake basin, China. *Wetlands* **2015**, *35*, 1033–1042. [CrossRef]

32. Wen, Y.; Liu, X.; Yang, J.; Lin, K.; Du, G. NDVI indicated inter-seasonal non-uniform time-lag responses of terrestrial vegetation growth to daily maximum and minimum temperature. *Global Planet. Chang.* **2019**, *177*, 27–38. [CrossRef]

33. Pan, N.; Feng, X.; Fu, B.; Wang, S.; Ji, F.; Pan, S. Increasing global vegetation browning hidden in overall vegetation greening: Insights from time-varying trends. *Remote Sens. Environ.* **2018**, *214*, 59–72. [CrossRef]

34. Wu, L.; Wang, S.; Bai, X.; Tian, Y.; Luo, G.; Wang, J.; Li, Q.; Chen, F.; Deng, Y.; Yang, Y.; et al. Climate change weakens the positive effect of human activities on karst vegetation productivity restoration in southern China. *Ecol. Indic.* **2020**, *115*, 106392. [CrossRef]

35. Tong, X.; Wang, K.; Yue, Y.; Brandt, M.; Liu, B.; Zhang, C.; Liao, C.; Fensholt, R. Quantifying the effectiveness of ecological restoration projects on long-term vegetation dynamics in the karst regions of Southwest China. *Int. J. Appl. Earth Obs. Geoinf.* **2017**, *54*, 105–113. [CrossRef]

36. Richardson, A.J. Distinguishing vegetation from soil background information. *Photogramm. Eng. Remote Sens.* **1977**, *43*, 1541–1552.

37. Tucker, C.J. Red and photographic infrared linear combinations for monitoring vegetation. *Remote Sens. Environ.* **1979**, *8*, 127–150. [CrossRef]

38. Zhang, C.; Lu, D.; Chen, X.; Zhang, Y.; Maisupova, B.; Tao, Y. The spatiotemporal patterns of vegetation coverage and biomass of the temperate deserts in Central Asia and their relationships with climate controls. *Remote Sens. Environ.* **2016**, *175*, 271–281. [CrossRef]
39. Barichivich, J.; Bria, K.R.; Myneni, R.B.; Osborn, T.J.; Melvin, T.M.; Philippe, C.; Shilong, P.; Compton, T. Large-scale variations in the vegetation growing season and annual cycle of atmospheric CO2 at high northern latitudes from 1950 to 2011. *Glob. Chang. Biol.* **2013**, *19*, 3167–3183. [CrossRef]
40. Zhu, Z.; Bi, J.; Pan, Y.; Ganguly, S.; Anav, A.; Xu, L.; Samanta, A.; Piao, S.; Nemani, R.R.; Myneni, R.B. Global Data Sets of Vegetation Leaf Area Index (LAI)3g and Fraction of Photosynthetically Active Radiation (FPAR)3g Derived from Global Inventory Modeling and Mapping Studies (GIMMS) Normalized Difference Vegetation Index (NDVI3g) for the Period 1981 to 2011. *Remote Sens.* **2013**, *5*, 927–948.
41. Tucker, C.J.; Pinzon, J.E.; Brown, M.E.; Slayback, D.A.; Pak, E.W.; Mahoney, R.; Vermote, E.; Saleous, N.E. An extended AVHRR 8-km NDVDI dataset compatible with MODIS and SPOT vegetation NDVI data. *Int. J. Remote Sens.* **2005**, *26*, 4485–4498. [CrossRef]
42. Zhou, Y.; Fengsong, P.; Yan, X.; Wu, C.; Zhong, R.; Wang, K.; Wang, H.; Cao, Y. Assessing the Impacts of Extreme Climate Events on Vegetation Activity in the North South Transect of Eastern China (NSTEC). *Water* **2019**, *11*, 2291. [CrossRef]
43. Fensholt, R.; Proud, S.R. Evaluation of Earth Observation based global long term vegetation Trends-Comparing GIMMS and MODIS global NDVI time series. *Remote Sens. Environ.* **2012**, *119*, 131–147. [CrossRef]
44. Sarmah, S.; Jia, G.; Zhang, A.; Singha, M. Assessing seasonal trends and variability of vegetation growth from NDVI3g, MODIS NDVI and EVI over South Asia. *Remote Sens. Lett.* **2018**, *9*, 1195–1204. [CrossRef]
45. Zeng, F.W.; Collatz, G.J.; Pinzon, J.E.; Ivanoff, A. Evaluating and quantifying the climate-driven interannual variability in Global Inventory Modeling and Mapping Studies (GIMMS) Normalized Difference Vegetation Index (NDVI3g) at Global Scales. *Remote Sens.* **2013**, *5*, 3918–3950. [CrossRef]
46. Cui, L.; Wang, L.; Singh, R.P.; Lai, Z.; Jiang, L.; Yao, R. Association analysis between spatiotemporal variation of vegetation greenness and precipitation/temperature in the Yangtze River Basin (China). *Environ. Sci. Pollut. Res.* **2018**, *25*, 21867–21878. [CrossRef]
47. Easterling, D.R.; Alexander, L.V.; Mokssit, A.; Detemmerman, V. CCI/CLIVAR workshop to develop priority climate indices. *Bull. Amer. Meteor. Soc.* **2003**, *84*, 1403–1407.
48. Sen, P.K. Estimates of the regression coefficient based on Kendall's tau. *J. Am. Stat. Assoc.* **1968**, *63*, 1379–1389. [CrossRef]
49. Mann, H.B. Nonparametric tests against trend. *Econom. J. Econom. Soc.* **1945**, *13*, 245–259. [CrossRef]
50. Zhou, Y.; Zhang, L.; Fensholt, R.; Wang, K.; Vitkovskaya, I.; Tian, F. Climate contributions to vegetation variations in Central Asian drylands: Pre- and post-USSR collapse. *Remote Sens.* **2015**, *7*, 2449–2470. [CrossRef]
51. Zhao, J.; Huang, S.; Huang, Q.; Wang, H.; Leng, G.; Fang, W. Time-lagged response of vegetation dynamics to climatic and teleconnection factors. *Catena* **2020**, *189*, 104474. [CrossRef]
52. Mulder, C.P.H.; Iles, D.T.; Rockwell, R.F. Increased variance in temperature and lag effects alter phenological responses to rapid warming in a subarctic plant community. *Glob. Chang. Biol.* **2016**, *23*, 801–814. [CrossRef] [PubMed]
53. Yan, H.; Yu, Q.; Zhu, Z.; Myneni, R.B.; Yan, H.; Wang, S.; Shugart, H.H. Diagnostic analysis of interannual variation of global land evapotranspiration over 1982–2011: Assessing the impact of ENSO. *J. Geophys. Res. Atmos.* **2013**, *118*, 8969–8983. [CrossRef]
54. Piao, S.; Friedlingstein, P.; Ciais, P.; Viovy, N.; Demarty, J. Growing season extension and its impact on terrestrial carbon cycle in the Northern Hemisphere over the past 2 decades. *Glob. Biogeochem. Cycles* **2007**, *21*, GB3018. [CrossRef]
55. Zhang, T.; Yang, S.; Guo, R.; Guo, J. Correction: Warming and nitrogen addition alter photosynthetic pigments, sugars and nutrients in a temperate meadow ecosystem. *PLoS ONE* **2016**, *11*, e0158249. [CrossRef] [PubMed]
56. Zhang, Q.Y.; Wu, S.H.; Zhao, D.S.; Dai, E.F. Responses of growing season vegetation changes to climatic factors in inner mongolia grassland. *J. Nat. Res.* **2013**, *28*, 754–764.
57. Chen, T.; De Jeu, R.; Liu, Y.; Van der Werf, G.; Dolman, A. Using satellite based soil moisture to quantify the water driven variability in NDVI: A case study over mainland Australia. *Remote Sens. Environ.* **2014**, *140*, 330–338. [CrossRef]
58. Dong, Y.; Yin, D.; Li, X.; Huang, J.; Su, W.; Li, X.; Wang, H. Spatial-temporal evolution of vegetation NDVI in association with climatic, environmental and anthropogenic factors in the Loess Plateau, China during 2000–2015: Quantitative analysis based on Geographical Detector Model. *Remote Sens.* **2021**, *13*, 4380. [CrossRef]
59. Julien, Y.; Sobrino, J.A. Comparison of cloud-reconstruction methods for time series of composite NDVI data. *Remote Sens. Environ.* **2010**, *114*, 618–625. [CrossRef]
60. Brantley, S.T.; Zinnert, J.C.; Young, D.R. Application of hyperspectral vegetation indices to detect variations in high leaf area index temperate shrub thicket canopies. *Remote Sens. Environ.* **2011**, *115*, 514–523. [CrossRef]
61. Lloret, F.; Escudero, A.; Lriondo, J.M.; Martínez-Vilalta, J.; Valladares, F. Extreme climate events and vegetation: The role of stabilizing processes. *Glob. Chang. Biol.* **2012**, *18*, 797–805. [CrossRef]

remote sensing

Article

Heatwaves Significantly Slow the Vegetation Growth Rate on the Tibetan Plateau

Caixia Dong [1,2], Xufeng Wang [1,*], Youhua Ran [1] and Zain Nawaz [1]

1 Key Laboratory of Remote Sensing of Gansu Province, Heihe Remote Sensing Experimental Research Station, Northwest Institute of Eco-Environment and Resources, Chinese Academy of Sciences, Lanzhou 730000, China; dongcaixia@nieer.ac.cn (C.D.); ranyh@lzb.ac.cn (Y.R.); zain-nawaz@lzb.ac.cn (Z.N.)
2 University of Chinese Academy of Sciences, Beijing 100049, China
* Correspondence: wangxufeng@lzb.ac.cn; Tel.: +86-931-4967724

Abstract: In recent years, heatwaves have been reported frequently by literature and the media on the Tibetan Plateau. However, it is unclear how alpine vegetation responds to the heatwaves on the Tibetan Plateau. This study aimed to identify the heatwaves using long-term meteorological data and examine the impact of heatwaves on vegetation growth rate with remote sensing data. The results indicated that heatwaves frequently occur in June, July, and August on the Tibetan Plateau. The average frequency of heatwaves had no statistically significant trends from 2000 to 2020 for the entire Tibetan Plateau. On a monthly scale, the average frequency of heatwaves increased significantly ($p < 0.1$) in August, while no significant trends were in June and July. The intensity of heatwaves indicated a negative correlation with the vegetation growth rate anomaly (ΔVGR) calculated from the normalized difference vegetation index (NDVI) (r = -0.74, $p < 0.05$) and the enhanced vegetation index (EVI) (r = -0.61, $p < 0.1$) on the Tibetan Plateau, respectively. Both NDVI and EVI consistently demonstrate that the heatwaves slow the vegetation growth rate. This study outlines the importance of heatwaves to vegetation growth to enrich our understanding of alpine vegetation response to increasing extreme weather events under the background of climate change.

Keywords: heatwave; alpine vegetation; Tibetan Plateau; remote sensing; extreme climate events

Citation: Dong, C.; Wang, X.; Ran, Y.; Nawaz, Z. Heatwaves Significantly Slow the Vegetation Growth Rate on the Tibetan Plateau. *Remote Sens.* **2022**, *14*, 2402. https://doi.org/10.3390/rs14102402

Academic Editor: Miaogen Shen

Received: 8 April 2022
Accepted: 13 May 2022
Published: 17 May 2022

Publisher's Note: MDPI stays neutral with regard to jurisdictional claims in published maps and institutional affiliations.

1. Introduction

A heatwave is defined as a period with sustained high-temperature anomalies resulting in strong impacts on human health, the ecological environment, and socioeconomic development [1]. A recent study has indicated that heatwaves have increased in prevalence significantly since the 1950s [2]. The heatwave has received growing attention in global change ecology study because of its remarkable effects on carbon, water, and energy exchange between the land surface and atmosphere [3]. Much evidence from crop yield, tree ring, and manipulative experiments has demonstrated that the occurrence of extreme high-temperature events can trigger significant impacts on terrestrial ecosystems and human society [4–11]. The heatwave also significantly impacts vegetation from regional to global scales, which has been witnessed by the satellite-derived normalized difference vegetation index (NDVI)-based studies [12,13]. These studies have mainly focused on tropical and temperate regions but have ignored cold regions.

With global warming, cold regions have been experiencing increased intensity, frequency, and duration of heatwaves in the past several decades [14–16]. As the "Third Pole", the Tibetan Plateau is warming twice as fast as the global average warming rate [17]. In situ meteorological observations and model projections indicated that extreme high-temperature events have happened frequently on the Tibetan Plateau. Due to the low intensity of anthropogenic activities [18], the Tibetan Plateau is an ideal region for studying the responses of alpine vegetation to extreme temperature events. The alpine vegetation on the Tibetan Plateau is very sensitive to high-temperature events due to the heat-limiting environment [19]. Many recent studies have begun to examine the impacts of extreme temperature on vegetation (e.g., productivity, phenology) on the Tibetan Plateau [20,21]. As a specific type of extreme high–temperature event, the heatwave has been reported recently by literature [22] on the Tibetan Plateau. However, very few studies have been conducted on the ecological effects of the heatwave on the Tibetan Plateau. We only found one site scale study, which reported that the heatwave can substantially increase alpine ecosystem respiration on the Tibetan Plateau [20]. Therefore, the response of alpine vegetation on the Tibetan Plateau to heatwaves is poorly understood. It is necessary to evaluate the heatwave effect on vegetation more widely and provide valuable information to address the climate change in this region.

Meanwhile, the MODIS bidirectional reflectance distribution function (BRDF)-adjusted daily reflectance product has made it possible to detect vegetation change in a very short period. On the Tibetan Plateau, heatwaves are usually of short duration and take place at a small part of the plateau. Some small heatwaves may be ignored if using the 16-day or monthly composite vegetation index. The BRDR-corrected daily snow–free MODIS reflectance product and long-term daily meteorological observations make it possible to examine the alpine vegetation response to heatwaves at a regional scale on the Tibetan Plateau.

We aim to fill the knowledge gap about the response mechanism of the alpine vegetation to heatwave on the Tibetan Plateau. The objectives of this study include: (1) identify the heatwaves from 2000 to 2020 on the Tibetan Plateau and analyze their temporal trends; (2) examine the response of vegetation growth to heatwaves intensity and duration.

2. Materials and Methods

2.1. Study Area

The Tibetan Plateau, with an area of 2.5 million km^2 and an average altitude over 4000 m above sea level [23], is the largest and highest plateau in the world [4]. The geographical range of the Tibetan Plateau is 26–40°N, 73–105°E. It spans six provinces, namely the Tibet and Xinjiang Uygur Autonomous Regions and Qinghai, Yunnan, Sichuan, and Gansu Provinces [15]. The characteristics of the climate on the Tibetan Plateau are strong solar radiation, low air temperature, and large day–night temperature difference [18,24]. This climatic pattern determines the general distribution of the vegetation [25]. The dominant vegetation type is alpine grasslands, which accounts for about 60% of the entire plateau area [26]. Figure 1 depicts the spatial distribution of the meteorological stations on the Tibetan Plateau.

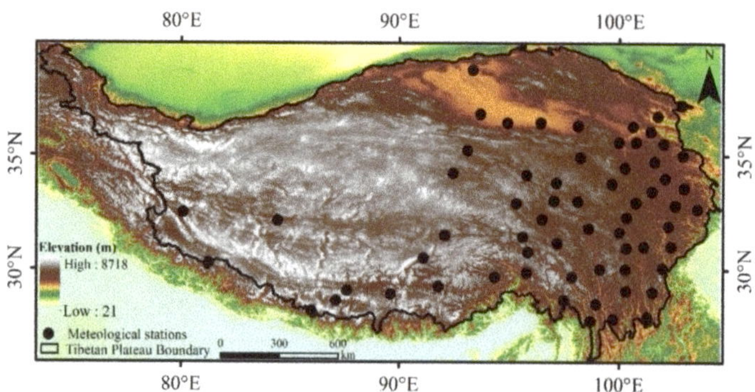

Figure 1. The distribution of the meteorological stations on the Tibetan Plateau.

2.2. Datasets

The daily maximum temperature (T$_{max}$) and precipitation data used in this study were provided by the Climatic Data Center, National Meteorological Information Center, China

Meteorological Administration (http://data.cma.cn/, accessed on 21 August 2021). In the dataset, there are 86 meteorological stations across the Tibetan Plateau. Because heatwave detection requires long-term temperature data, we excluded the sites with data gaps from 1980 to 2020. Finally, 64 stations were selected in this study and are depicted in Figure 1. Table A1 in Appendix A depicts information about each meteorological station (WMO code, name, latitude, longitude, and elevation.

Digital elevation model data for the Tibetan Plateau was obtained from Shuttle Radar Topography Mission (https://earthexplorer.usgs.gov/, accessed on 21 August 2021) and its spatial resolution is 30 m.

The MODIS Nadir Bidirectional Reflectance Distribution Function Adjusted Reflectance (NBAR) product (MCD43A4) was used to monitor the vegetation growth in this work. MCD43A4 provides daily surface reflectance by combining Terra and Aqua MODIS data at a 500–m spatial resolution. The surface reflectance was normalized to nadir using the bidirectional reflectance distribution function for the solar angle at local noontime. This product has removed view angle effects, and masks cloud cover and snow contamination [27]. Collection 6 of MCD43A4 from 2000 to 2020 on the Tibetan Plateau was obtained from the Google Earth Engine (https://earthengine.google.com/, accessed on 21 August 2021). The two vegetation indices (Vis) have been widely used in ecological studies, whereas the NDVI is chlorophyll-sensitive, the EVI is more responsive to canopy structural variations [28,29]. A combination of the NDVI and EVI to complement each other examined the robustness and comparability of the vegetation growth rate changes [30]. In this study, we used the surface reflectance from the MCD43A4 product to calculate daily Vis, including the NDVI and EVI.

The Land Surface Soil Moisture Dataset over the Tibetan Plateau was downloaded from National Tibetan Plateau Data Center [31]. This dataset is daily surface soil moisture with a spatial resolution of $0.25°$, retrieved from passive microwave brightness temperature data. The dataset synthesized microwave brightness temperature measurements from SMMR, SSM/I, SSMIS, AMSR-E, AMSR2, SMAP, and FY3B to produce a long-term soil moisture product [32]. We used Land Surface Soil Moisture Dataset data from 2002 to 2020 to examine the soil moisture difference before and during the heatwave.

2.3. Methods

In this study, we used a percentile-based thresholds method to identify the heatwaves in June, July, and August at each station on the Tibetan Plateau. [33]. The heatwaves were identified following the definition of a heatwave in the literature [34] with a few modifications in this study. The 90th percentile values of daily maximum temperature during the climatological baseline period (1980–2020) are used as the threshold to identify hot days. A period with at least five consecutive hot days (the maximum temperature is greater than the threshold) was identified as a heatwave event at a single station. The frequency, duration, and intensity are characteristics of a heatwave event. The frequency of heatwaves is the sum of the heatwaves at 64 sites on the Tibetan Plateau in a year. The number of stations experiencing heatwaves is also counted each year. Each site is counted only once per year. The duration and temperature anomaly above the threshold are introduced as two dimensions to quantify the severity of the heatwaves. The duration is the length of a heatwave event, and the temperature anomaly is the average temperature difference between the daily maximum temperature and the 90th percentile threshold. The accumulative intensity of a heatwave is the sum of temperature differences above the threshold during the heatwave. The average intensity of heatwaves is the average temperature difference during the heatwave. To explore the impact of heatwaves on vegetation, we only focused on the heatwaves occurring in June, July, and August.

VIs were used to calculate the vegetation growth rate, which is the change of the value of VIs before and after a heatwave. The NDVI and EVI [30] are calculated with the Equations (1) and (2), respectively. In addition to correct inferior values in VIs, a time series reconstruction algorithm, the Savitzky–Golay filter (Equation (3)), is applied to long-term daily VIs in this study [35]. We excluded these heatwaves in the analysis when MCD43A4 data was missing to make our result reliable. Vegetation growth rate (VGR) was calculated as the difference between the vegetation index before and during the heatwave (Equation (4)). Then, the vegetation growth rate anomaly (ΔVGR) (Equation (5)) was calculated from the VGR during the heatwave and the multi–year average VGR in the corresponding period. The vegetation index and vegetation growth rate anomaly were used as vegetation proxies to examine the heatwave impacts on vegetation. The formulas are expressed as follows:

$$\text{NDVI} = \frac{\rho_{nir} - \rho_{red}}{\rho_{nir} + \rho_{red}} \tag{1}$$

$$\text{EVI} = \frac{G \times (\rho_{nir} - \rho_{red})}{\rho_{nir} + (C_1 \times \rho_{nir} - C_2 \times \rho_{blue}) + L} \tag{2}$$

where $G = 2.5$, $C_1 = 6$, $C_2 = 7.5$, and $L = 1$; ρ_{blue}, ρ_{red}, and ρ_{nir} are the reflectance of the blue, red, and near-infrared bands, respectively.

$$Y_j^* = (2m + 1)^{-1} \sum_{i=-m}^{i=m} C_i Y_{j+i} \tag{3}$$

where Y represents the original time–series data, Y^* is the reconstructed time-series data, C_i is the weight of the filter window, and $2m + 1$ is the size of the filter window. The window size and polynomial order in the Savitzky–Golay filter were set to 31 and 4, respectively [36].

$$\text{VGR} = VIs_{dur} - VIs_{bf} \tag{4}$$

$$\Delta\text{VGR} = VGR - VGR_{baseline} \tag{5}$$

where VIs_{bf} and VIs_{dur} are the average of VIs before the heatwave and during the heatwave, respectively. $VGR_{baseline}$ represents the multi-year average VGR during the reference period (2000–2020). ΔVGR is the difference between VGR and $VGR_{baseline}$.

The linear trend of the number of sites with heatwaves was analyzed using the Mann–Kendall methods [37,38]. The Mann–Kendall method is a nonparametric test for monotonic trends. This method does not assume a specific distribution for the data and is not sensitive to outliers. The Theil–Sen method was used to calculate the slope of the linear trend [39]. The slope of the trend measures the number of the heatwaves' change rate over time. To explore the impact of heatwaves on vegetation growth, we calculated the correlation coefficients between the heatwaves (the intensity of the heatwaves) and vegetation growth rate (change rate anomaly of NDVI and EVI) using the Pearson correlation method.

3. Results

3.1. Trends of Heatwaves Frequency

Based on the 64 meteorological stations on the Tibetan Plateau, we first identified the heatwaves and calculated the duration and intensity of the heatwaves for each station. The interannual variation of heatwaves frequency at these stations is depicted in Figure 2. From 2000 to 2020, the heatwaves frequently occurred in June, July, and August. Overall, from 2000 to 2020, the frequency of heatwaves in the growing season had no statistically significant trends for the entire Tibetan Plateau in Figure 2. The occurring frequencies of the heatwaves are different among June, July, and August. The heatwaves happened more frequently in August than in June and July. The heatwave frequency increased significantly ($p < 0.1$) in August, while no significant trends occurred in June and July.

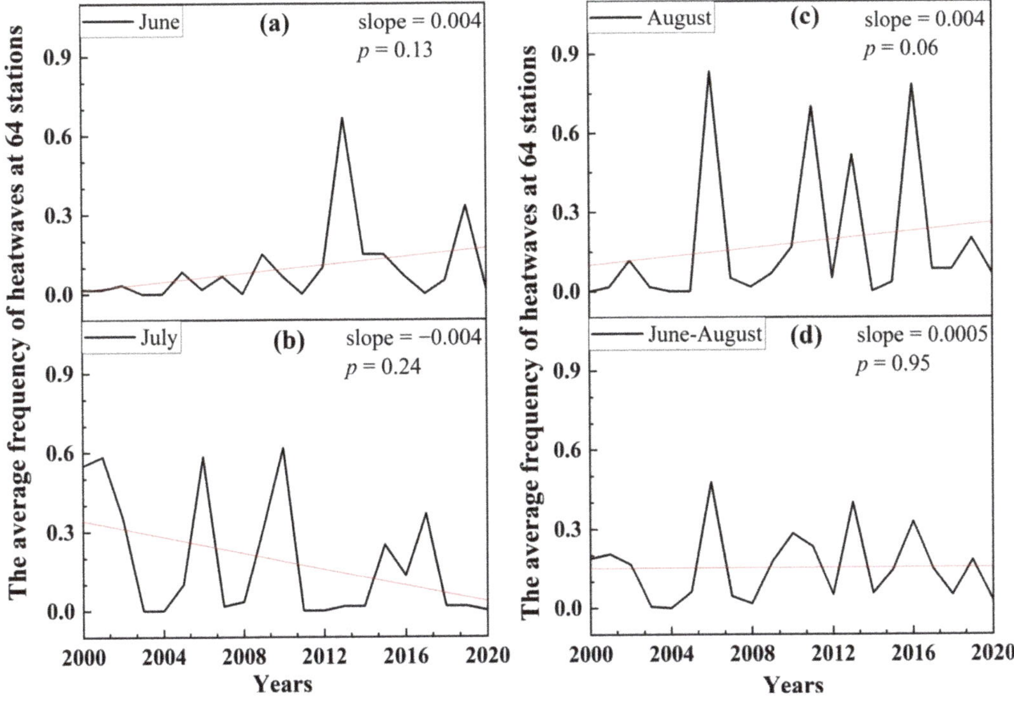

Figure 2. The average frequency of heatwaves at 64 stations in June (**a**), July (**b**), and August (**c**), and June–August (**d**) on the Tibetan Plateau from 2000 to 2020.

Figure 3 depicts the interannual dynamic of the duration and intensity of heatwaves from 2000 to 2020. The color changes from yellow to red indicate that the heatwave severity varies from weak to strong. Overall, the duration and intensity of heatwaves ranged from 5.00 to 11.57 days and from 0.67 to 2.55 °C/d, respectively. Most of the heatwaves are short with a duration of about 6 days. The longest duration appeared in August. Among June, July, and August, heatwaves in August lasted longer than in other months. The intensity of the heatwaves has neither obvious monthly patterns nor evident temporal changing trends. Through analyzing the intensity and duration of heatwaves, we found that heatwaves with long durations may have low intensity (average high–temperature anomaly). The heatwaves occurred most frequently in recent years, such as 2006, 2013, 2016, etc.

To explore the extent of the heatwaves, a matrix heatmap is used to depict the number of stations where heatwaves happened in June, July, and August from 2000 to 2020 (Figure 4). Generally, as expected, there are no widespread heatwave occurrences in most years on the Tibetan Plateau due to the cold environment. From 2000 to 2020, heatwaves occurred at more than half of the total stations on the Tibetan Plateau in 2000, 2001, 2006, 2010, 2013, and 2016. The most two severe heatwave events detected from the daily maximum temperature happened in August 2016 with 47 sites and in June 2013 with 38 sites.

Figure 3. Matrix heatmap for heatwave duration and intensity in June, July, and August on the Tibetan Plateau from 2000 to 2020. The matrix heatmap (**a**) refers to the temporal change of the heatwave duration; the matrix heatmap (**b**) refers to the temporal change of the heatwave intensity. Each grid cell represents the average duration and intensity of a heatwave at all sites in a given month of a year. The blank grid cell represents no heatwave, and the value is NaN. The color of the matrix heatmap represents the size of the value.

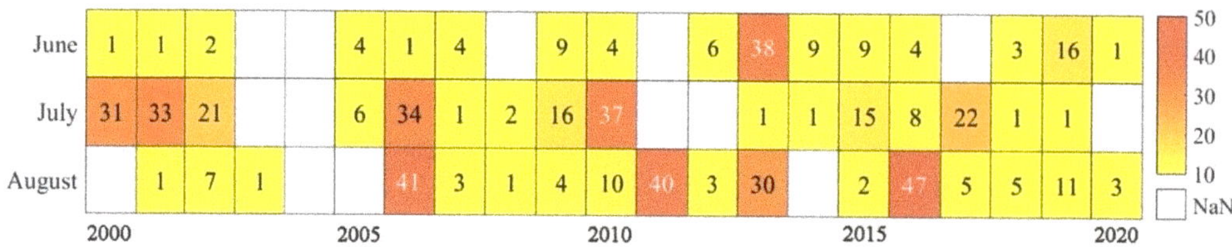

Figure 4. Matrix heatmap of the number of sites with heatwaves from June to August on the Tibetan Plateau from 2000 to 2020. Each grid cell represents the number of sites with heatwaves in a given month of a year. The blank grid represents no heatwave, and the value is NaN. The color of the matrix heatmap represents the size of the value.

3.2. Effects of Heatwaves on Vegetation

The deviation analysis was used to calculate the ΔVGR, which can reflect the VGR change caused by the heatwave. A positive anomaly indicates an increase in the VGR, and a negative anomaly indicates a decrease in VGR. Figure 5 depicts the average ΔVGR calculated from NDVI (a) and EVI (b) in June, July, and August from 2000 to 2020. The ΔVGR ranging from positive to negative was expressed as the color changing from green to yellow. The range of the ΔVGR calculated by NDVI and EVI is from −0.0088 to 0.057 and from −0.0095 to 0.023, respectively. Overall, the ΔVGR calculated from NDVI is consistent with the ΔVGR calculated from EVI.

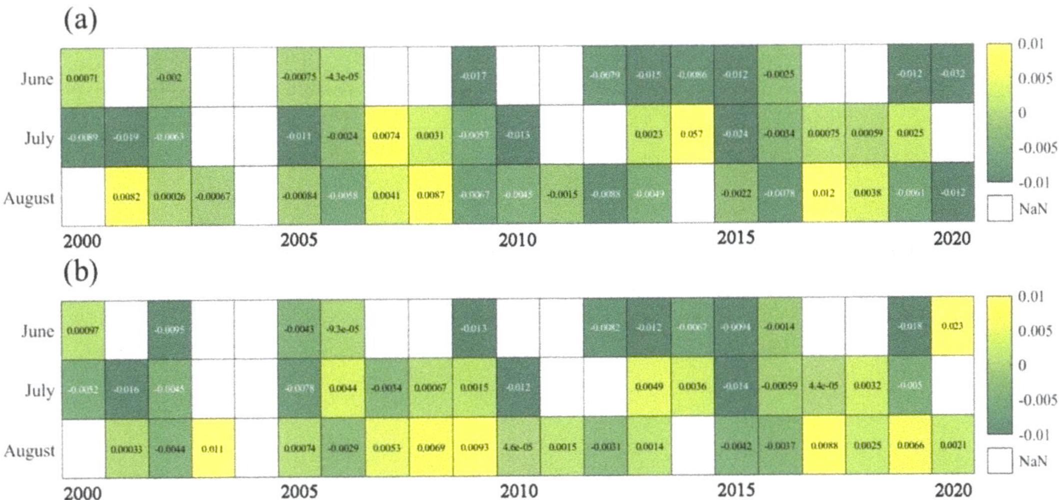

Figure 5. Matrix heatmap of the ΔVGR from June to August on the Tibetan Plateau from 2000 to 2020. The matrix heatmap (**a**) refers to the ΔVGR calculated from NDVI; the matrix heatmap (**b**) refers to the ΔVGR calculated from EVI. Each grid cell represents the average vegetation growth rate anomaly at all sites in a given month of a year. The blank grid cell represents no heatwave happened, and the value is NaN. The color of the matrix heatmap represents the size of the value.

The monthly variation of ΔVGR corresponding to heatwaves on the Tibetan Plateau in June, July, and August is displayed in Figure 6. The ΔVGR was negative when the heatwave happened in June, July, and August. That means the VGR during the heatwaves is lower than the multi-year mean on the Tibetan Plateau. Compared to other months, the minimum value of the ΔVGR was found in June, indicating that vegetation growth in June was more sensitive to the heatwave than in July and August. The value of the ΔVGR calculated by NDVI and EVI corresponding to the heatwaves is more consistent in June and July than in August. Overall, heatwaves in the growing season significantly slow the VGR on the Tibetan Plateau. The analysis indicates that the ΔVGR can capture the vegetation growth response to heatwaves on the Tibetan Plateau.

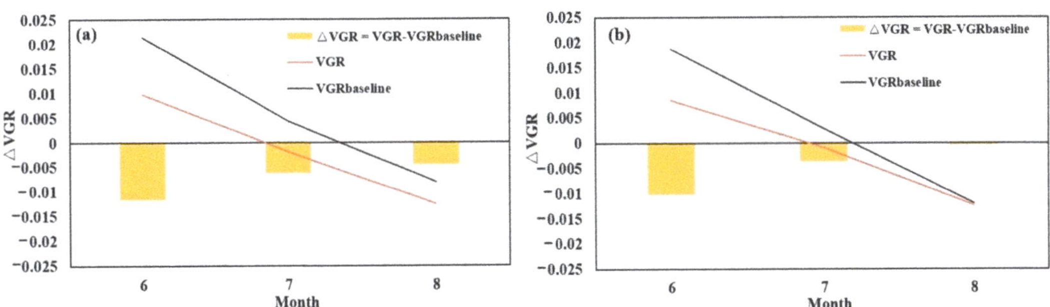

Figure 6. The ΔVGR in each month in June, July, and August on the Tibetan Plateau from 2000 to 2020. The ΔVGR is calculated from NDVI (**a**) and EVI (**b**) on the Tibetan Plateau from June to August. The black lines represent the monthly multiyear mean of each variable between 2000 and 2020; the red lines indicate the seasonal evolution during extremely high temperatures.

The correlation relationships were analyzed between the ΔVGR and the intensity of the heatwaves. The intensity of the heatwaves was calculated as temperature anomaly during heatwaves multiplied by heatwave duration. The intensity was grouped with a step of 2.5 °C×d to make the analysis clearer. As depicted in Figure 7a,c, the accumulative intensity (°C×d) of heatwaves indicates a negative correlation with the ΔVGR calculated from NDVI ($r = -0.74$, $p < 0.05$) and EVI ($r = -0.61$, $p < 0.1$) on the Tibetan Plateau, respectively. The average intensity (°C/d) of heatwaves indicates a negative correlation with the ΔVGR calculated by NDVI ($r = -0.77$, $p < 0.05$) and EVI ($r = -0.66$, $p < 0.1$) on the Tibetan Plateau (Figure 7b,d), respectively. Vegetation growth is strongly affected by the heatwaves in June, July, and August on the Tibetan Plateau. With the intensity increase of heatwaves, the VGR decreases linearly.

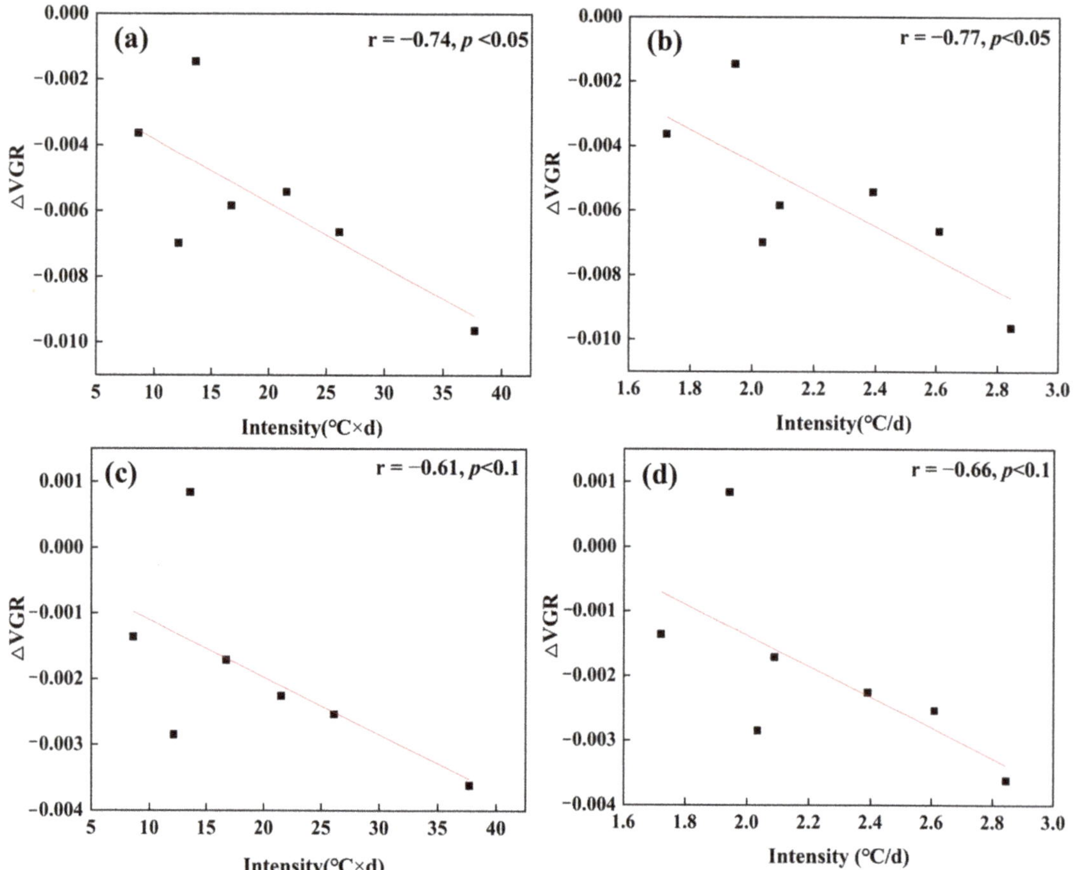

Figure 7. Pearson's correlation coefficient between the intensity of the heatwave and the ΔVGR from June to August during the growing season for all 64 sites on the Tibetan Plateau. (**a,c**) the accumulative intensity of heatwaves versus the ΔVGR. (**b,d**) the average intensity of heatwaves versus the ΔVGR. The ΔVGR in (**a,b**) is calculated from NDVI, and in (**c,d**) is calculated from EVI.

To corroborate our findings, we select the year 2013 and 2016, when widespread heatwaves occurred, to specially study the anomaly of VGR and the anomaly of climate factors before and after the occurrence of the heatwaves. The spatial distribution of the ΔVGR on the Tibetan Plateau is displayed in Figure 8a for June 2013 and Figure 8c for August 2016. The average VGRs in June 2013 and August 2016 were significantly lower than the multi-year average VGR. In June 2013, the ΔVGRs were negative at 33 of the 38 sites where the heatwave occurred. In August 2016, the ΔVGRs were negative at 30 of the 47 sites where the heatwave occurred. An obvious decrease in vegetation growth rate can be found in most sites where the heatwave occurred. It is suggested that the heatwaves in 2013 and 2016 significantly slowed down the VGR. During the selected two heatwave events, the temperature was significantly higher than the multi−year average in the corresponding period, including 5.8 °C above the multi−year average in 2016 and 4.9 °C above the multi−year average in 2013 (Figure 9).

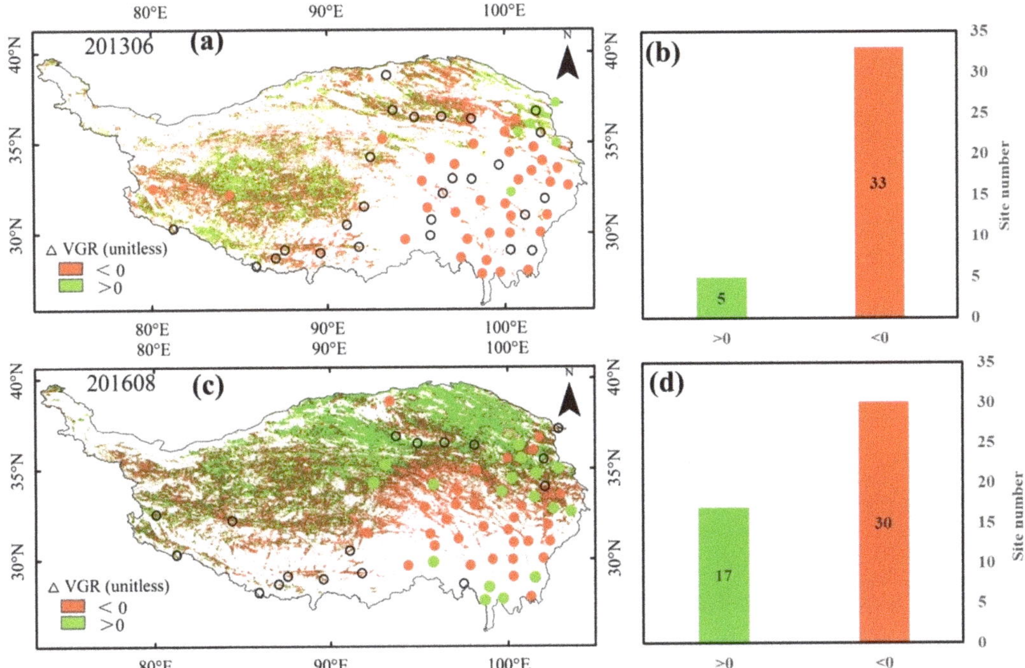

Figure 8. ΔVGR response to heatwaves in June 2013 (**a**) and August 2016 (**c**). The blank circle represents the site without a heatwave, while the filled circle represents the site with a heatwave; (**b**,**d**) represent the site number in June 2013 and August 2016, respectively; the red-filled circle represents the negative ΔVGR; the green−filled circle represents the positive ΔVGR. Positive ΔVGR stands for an increase in the VGR; negative ΔVGR stands for a decrease in the VGR. The right plots illustrate the number of sites with negative and positive VGR.

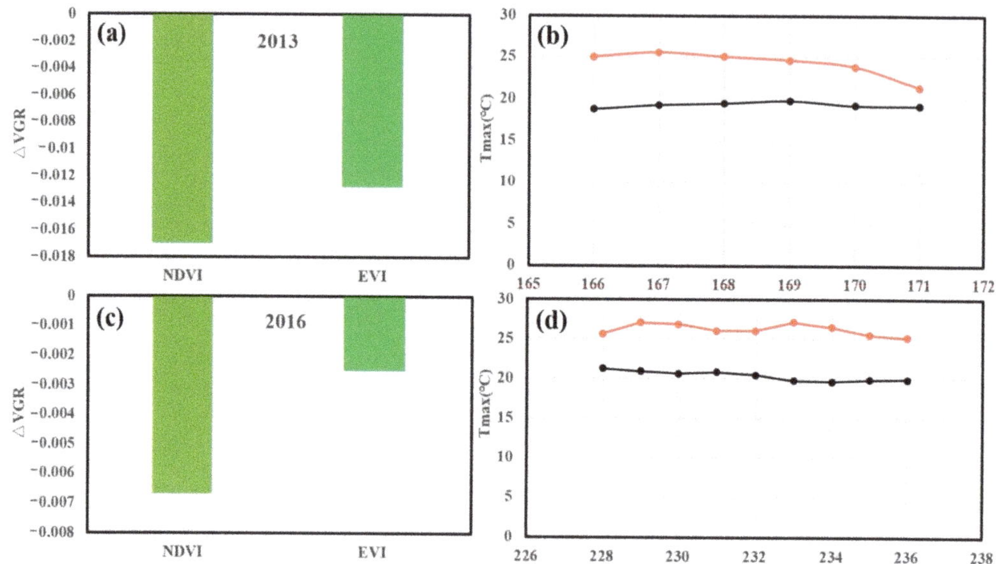

Figure 9. The ΔVGR and daily maximum temperatures during the heatwaves in June 2013 and August 2016. (**a,c**) represent the ΔVGR in June 2013 and August 2016, respectively; (**b,d**) represent the average daily maximum temperature (T_{max}) versus the multi-year average T_{max} during the heatwave occurrence in June 2013 (**b**) and August 2016 (**d**). The red line indicates the T_{max} (15 June 2013–20 June 2013 DOY: 166–171 and 16 August 2016–24 August 2016 DOY: 228–236) during a heatwave in one severe year. The black line indicates the multi-year average T_{max} during heatwaves occurrence in June 2013 and August 2016.

4. Discussion

To our knowledge, very few studies have focused on heatwaves on the Tibetan Plateau. Previous studies mainly used the traditional extreme high-temperature indices to explore their effects on alpine vegetation, such as TX90p (percentage of days when TX > 90th percentile), WSDI (warm spell duration index), etc. The traditional extreme high-temperature indices are usually calculated at a monthly or annual scale, which is too coarse to capture the short climate disturbance on vegetation. A recent study examined the trends of extreme temperature events using 71 meteorological stations from 1961 to 2005 and found that there were statistically significant increasing trends for extreme high-temperature indices [15]. He et al. [21] analyzed the spatial pattern and long-term trend in extreme high-temperature indices in the Kobresia meadow region from 1961 to 2008, and found a significant increase in the warmest daytime temperature. However, in this study, we found that heatwaves have no significant increasing trend from 2000 to 2020, and the trends vary greatly among June, July, and August. This is inconsistent with the extreme high−temperature indices trend reported previously on the Tibetan Plateau [40]. This is caused by the different definitions between heatwave and extreme high-temperature events. The inconsistency in trends between heatwave and extreme high-temperature indices could also be attributed to the different periods and datasets. Liu et al. [40] reported the characteristic of extreme high-temperature events from 2001 to 2015 based on the monthly gridded datasets, but this study explored the heatwaves based on the meteorological station data. The heatwaves on the Tibetan Plateau mainly occurred locally, only a few widespread heatwaves were detected, and no heatwaves occurred for the entire Plateau or all sites simultaneously. Due to the sparse and non-uniformly distributed weather stations, it is difficult to accurately extract the heatwave spatial extent [41]. However, the evolution of spatial extent is essential to better understand the varying mechanism of heatwaves on the Tibetan Plateau. Thus, future studies on better understanding the dynamic of heatwaves will be benefited from the high-resolution and reliable grid meteorological dataset.

Heatwaves can limit vegetation photosynthesis by pushing the ambient temperature to exceed the optimal photosynthetic temperature, increasing the vapor pressure deficit (VPD), and reducing the soil moisture. High temperatures over the optimal photosynthetic temperature could constrain the activity of Rubisco, increase photorespiration, and lead to a decline in net photosynthesis [42]. An ambient temperature lower or higher than the optimal photosynthetic temperature will inhibit vegetation growth [43]. The slowdown of vegetation growth rate during heatwave occurrence indicates that the extreme temperatures of heatwaves overpassed the optimum photosynthesis temperature for alpine vegetation on the Tibetan Plateau. Additionally, summer heatwaves affect photosynthesis primarily due to the physiological response to water deficit and high temperatures, including reductions in enzymatic activity and stomatal conductance to prevent water loss [44]. The stress effects are increased by water deficits [45]. In general, the occurrence of heatwaves was frequently accompanied by a decline in precipitation and a decrease in air relative humidity [5]. To validate this phenomenon, we examined the soil moisture and precipitation for the two selected heatwaves in June 2013 and August 2016. The soil moisture and precipitation during heatwave periods are obviously lower than the multi−year average in the corresponding periods (Figure 10). This demonstrates that heatwaves affect alpine vegetation by combining temperature stress and water limitation on the Tibetan Plateau. However, comparing the soil moisture and precipitation before and during the heatwave in August 2016, soil moisture and precipitation showed different change patterns. Maybe, this resulted from data noises in soil moisture product. The microwave–based soil moisture used in this work is retrieved using in−situ at five pixel-scale fields and 25km microwave remote sensing [32]. But, it is not widely validated on the Tibetan Plateau due to no widespread in-situ measurements. Moreover, the alpine vegetation responds differently to heatwaves in different phenology stages. The ΔVGR in July is more significant than that of July and August. It is indicated that the alpine vegetation is more sensitive to heatwaves in the early growing season than in the later growing season. Vegetation is fragile and sensitive to the environment in the early growing season [26,45,46], for example, spring phenology is more sensitive to environmental factors than autumn phenology [47]. Meanwhile, vegetation usually grows faster in the early growing season than in the later growing season; therefore, the growth rate may be more sensitive to environmental factors in the early growing season [46]. In this study, it is indicated that ΔVGR decreased linearly with heatwave intensity. This is partly because most heatwaves are weak on the Tibetan Plateau, and alpine vegetation can recover from these disturbances. Different grasslands on the Tibetan Plateau exhibited different response patterns to climate changes [26]. Interestingly, the alpine meadow is more sensitive to heatwaves than the alpine steppe in June (Figure 11), but the opposite is true in July and August. This may result from the different coverage between the two types. Moreover, the growth rate changing mechanism is complex; more factors should be considered [48–51]. Further research is needed to clarify the detailed mechanism of these changes. On the Tibetan Plateau, the vegetation that has adapted to the cold, alpine environment might differ from other ecosystems concerning vegetation responses to temperature extremes [26,40]. Given that the Tibetan Plateau will continue warming in the future [52,53], heatwaves will happen more widely and intensely and will lead to abrupt changes by approaching the temperature threshold of alpine vegetation.

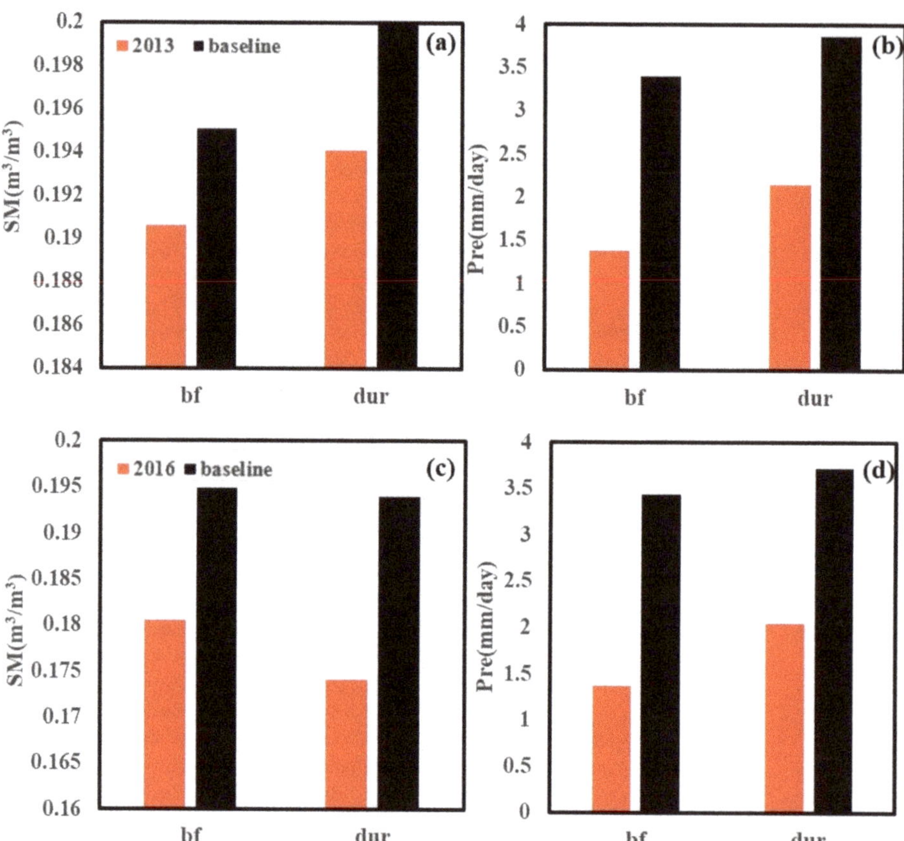

Figure 10. Precipitation and soil moisture comparison before and during the heatwave occurrence in June 2013 and August 2016. (**a,c**) represent the average soil moisture before/during two selected heatwave periods versus multi-year average soil moisture in the corresponding periods. (**b,d**) represent the average daily precipitation before/during two selected heatwave periods versus multi-year average soil moisture in the corresponding periods. The red bars indicate the heatwave years. The black bars indicate the baseline period (soil moisture: 2002–2019, pre: 2000–2020). The "bf" represents the period before the heatwaves (9 June 2013–14 June 2013 DOY: 160–165 and 9 August 2016–15 August 2016 DOY: 221–227). The "dur" indicates the heatwave periods (15 June 2013–20 June 2013 DOY: 166–170 and 16 August 2016–24 August 2016DOY: 228–236).

There are some uncertainties involved in this study. Firstly, meteorological observations have a relatively sparse and uneven distribution, resulting in the low representativeness of the identification of heatwaves [41]. Secondly, the quality of the remote sensing dataset is low on the Tibetan Plateau due to the contamination of snow, clouds, and complex terrain. Thirdly, the reconstructed vegetation index value can result in uncertainties in the analysis. The different spatial representativeness among station data, MODIS data, and coarse resolution soil moisture data can also lead to uncertainties in the study. Fourthly, the definition of a heatwave is uncertain due to the unique natural conditions on the Tibetan Plateau. In some heatwave definitions in the tropic or temperate region, a fixed high-temperature threshold is usually used by combining the temperature 90% percentiles.

In this work, only the 90% percentile was used and may result in uncertainties when comparing heatwaves on the Tibetan Plateau with other regions.

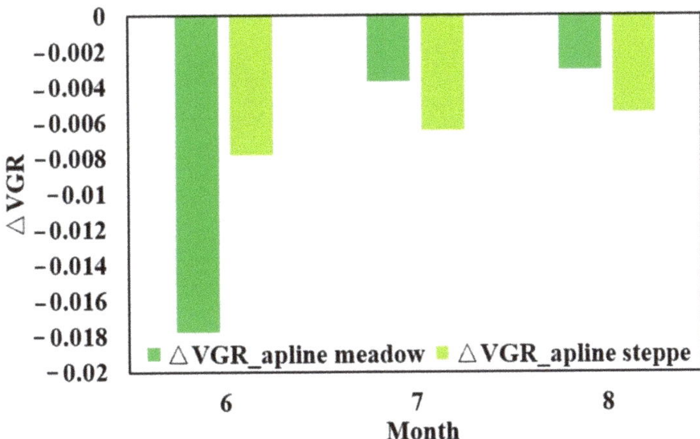

Figure 11. ΔVGR in each month in June, July, and August on the Tibetan Plateau from 2000 to 2020. The dark green column represents the ΔVGR of the alpine meadow; the light green column indicates the ΔVGR of the alpine steppe.

5. Conclusions

In this study, heatwaves were detected on the Tibetan Plateau by using long-term meteorological station data. The characteristics of heatwaves were explored at these meteorological stations. By combining the remotely sensed vegetation indices, the alpine vegetation response mechanism to heatwaves was examined on the Tibetan Plateau. The results indicate that: (1) With rapid warming, heatwaves frequently occur in June, July, and August on the Tibetan Plateau. The heatwaves have no significant increasing trend from 2000 to 2020; (2) The correlation between heatwave intensity and vegetation growth rate anomalies was significantly negative on the Tibetan Plateau. The vegetation growth rate estimated from NDVI and EVI consistently indicates that heatwaves slow vegetation growth. The alpine vegetation growth rate is more sensitive to the heatwave in June than in July and August. This study outlines the importance of heatwaves to vegetation growth to enrich our understanding of alpine vegetation response to increasing extreme weather events under the background of climate change.

Author Contributions: C.D. and X.W. Data collection, conceptualization, writing—original draft preparation, software, methodology, formal analysis; X.W. funding acquisition, writing review and editing, supervision; Y.R. and Z.N. writing—reviewing and editing the paper. All authors have read and agreed to the published version of the manuscript.

Funding: This work was supported by the National Natural Science Foundation of China (Grant No. 41771466), the Youth Innovation Promotion Association CAS to X.W. (NO. 2020422), and the National Key R&D Program of China (Grant No. 2017YFA0604801).

Data Availability Statement: The meteorological data is available at the Climatic Data Center, National Meteorological Information Center, China Meteorological Administration (http://data.cma.cn/, accessed on 7 April 2022). The DEM (Digital Elevation Model) data is available at http://srtm.csi.cgiar.org/, accessed on 7 April 2022. The Collection 6 of MCD43A4 is available in the Google Earth Engine (https://earthengine.google.com/, accessed on 7 April 2022). The soil moisture data is available at the National Tibetan Plateau Data Center (http://data.tpdc.ac.cn/, accessed on 7 April 2022).

Acknowledgments: The authors would like to thank TPDC for providing the data and anonymous reviewers for their valuable comments.

Conflicts of Interest: The authors declare no competing interest.

Appendix A

Table A1. Information for meteorological stations used in this study.

WMO Code	Station Name	Latitude	Longitude	Elevation (m)
52602	Lenghu	38.75	93.33	2770
52707	Xiaozaohuo	36.80	93.68	2767
52787	Wushaoling	37.20	102.87	3045.1
52818	Golmud	36.42	94.90	2807.6
52825	Nuomuhong	36.43	96.42	2790.4
52836	Doulan	36.30	98.10	3191.1
52856	Gonghe	36.27	100.62	2835
52866	Xining	36.72	101.75	2295.2
52868	Guide	36.03	101.43	2237.1
52908	Wudaoliang	35.22	93.08	4612.2
52943	Xinghai	35.58	99.98	3323.2
52955	Guihan	35.58	100.75	3202.9
52974	Tongren	35.52	102.02	2491.4
55228	Shiquanhe	32.50	80.08	4278
55248	Gaize	32.15	84.42	4414.9
55299	Naqu	31.48	92.07	4507
55437	Pulan	30.28	81.25	3900
55493	Dangxiong	30.48	91.10	4200
55569	Lazi	29.08	87.60	4000
55598	Shannan	29.25	91.77	3551.7
55655	Nielaer	28.18	85.97	3810
55664	Dingri	28.63	87.08	4300
55680	Jiangzi	28.92	89.60	4040
56004	Tuotuohe	34.22	92.43	4533.1
56018	Zaduo	32.90	95.30	4066.4
56021	Qumalai	34.13	95.78	4175
56029	Yushu	33.02	97.02	3681.2
56033	Maduo	34.92	98.22	4272.3
56034	Qingshuihe	33.80	97.13	4415.4
56038	Shiqu	32.98	98.10	4200
56043	Guoluo	34.47	100.25	3719
56046	Dari	33.75	99.65	3967.5
56065	Henan	34.73	101.60	3500
56067	Jiuzhi	33.43	101.48	3628.5
56074	Maqu	34.00	102.08	3471.4
56079	Ruoergai	33.58	102.97	3439.6

Table A1. *Cont.*

WMO Code	Station Name	Latitude	Longitude	Elevation (m)
56080	Hezuo	35.00	102.90	2910
56116	Qingqing	31.42	95.60	3873.1
56125	Nangqian	32.20	96.48	3643.7
56137	Changdu	31.15	97.17	3306
56144	Dege	31.80	98.58	3184
56146	Ganzi	31.62	100.00	3393.5
56151	Banma	32.93	100.75	3530
56152	Seda	32.28	100.33	3893.9
56167	Daodu	30.98	101.12	2957.2
56172	Maerkang	31.90	102.23	2664.4
56173	Hongyuan	32.80	102.55	3491.6
56178	Xiaojing	31.00	102.35	2369.2
56182	Songpan	32.65	103.57	2850.7
56223	Luolong	30.75	95.83	3640
56227	Bomi	29.87	95.77	2736
56247	Batang	30.00	99.10	2589.2
56251	Xinlong	30.93	100.32	3000
56257	Litang	30.00	100.27	3948.9
56312	Linzhi	29.67	94.33	2991.8
56331	Zuogong	29.67	97.83	3780
56357	Daocheng	29.05	100.30	3727.7
56374	Kangding	30.05	101.97	2615.7
56434	Chayu	28.65	97.47	2327.6
56444	Deqin	28.48	98.92	3319
56459	Muli	27.93	101.27	2426.5
56462	Jiulong	29.00	101.50	2987.3
56533	Gongshan	27.75	98.67	1583.3
56543	Zhongdian	27.83	99.70	3276.1

References

1. Lavaysse, C.; Cammalleri, C.; Dosio, A.; van der Schrier, G.; Toreti, A.; Vogt, J. Towards a monitoring system of temperature extremes in Europe. *Nat. Hazards Earth Syst. Sci.* **2018**, *18*, 91–104. [CrossRef]
2. Perkins-Kirkpatrick, S.E.; Lewis, S.C. Increasing trends in regional heatwaves. *Nat. Commun.* **2020**, *11*, 3357. [CrossRef]
3. Reichstein, M.; Bahn, M.; Ciais, P.; Frank, D.; Mahecha, M.D.; Seneviratne, S.I.; Zscheischler, J.; Beer, C.; Buchmann, N.; Frank, D.C.; et al. Climate extremes and the carbon cycle. *Nature* **2013**, *500*, 287–295. [CrossRef]
4. Wang, R.; He, M.; Niu, Z. Responses of Alpine Wetlands to Climate Changes on the Qinghai-Tibetan Plateau Based on Remote Sensing. *Chin. Geogr. Sci.* **2020**, *30*, 189–201. [CrossRef]
5. Piao, S.; Zhang, X.; Chen, A.; Liu, Q.; Lian, X.; Wang, X.; Peng, S.; Wu, X. The impacts of climate extremes on the terrestrial carbon cycle: A review. *Sci. China Earth Sci.* **2019**, *62*, 1551–1563. [CrossRef]
6. Yan, Y.; Jeong, S.; Park, C.-E.; Mueller, N.D.; Piao, S.; Park, H.; Joo, J.; Chen, X.; Wang, X.; Liu, J.; et al. Effects of extreme temperature on China's tea production. *Environ. Res. Lett.* **2021**, *16*, 044040. [CrossRef]
7. Lobell, D.B.; Bänziger, M.; Magorokosho, C.; Vivek, B. Nonlinear heat effects on African maize as evidenced by historical yield trials. *Nat. Clim. Chang.* **2011**, *1*, 42–45. [CrossRef]

8. Klesse, S.; Babst, F.; Lienert, S.; Spahni, R.; Joos, F.; Bouriaud, O.; Carrer, M.; Di Filippo, A.; Poulter, B.; Trotsiuk, V.; et al. A Combined Tree Ring and Vegetation Model Assessment of European Forest Growth Sensitivity to Interannual Climate Variability. *Glob. Biogeochem. Cycles* **2018**, *32*, 1226–1240. [CrossRef]

9. Bastos, A.; Gouveia, C.M.; Trigo, R.M.; Running, S.W. Analysing the spatio-temporal impacts of the 2003 and 2010 extreme heatwaves on plant productivity in Europe. *Biogeosciences* **2014**, *11*, 3421–3435. [CrossRef]

10. Mou, C.; Sun, G.; Luo, P.; Wang, Z.; Luo, G. Flowering Responses of Alpine Meadow Plant in the Qinghai-Tibetan Plateau to Extreme Drought Imposed in Different Periods. *Chin. J. Appl. Environ. Biol.* **2013**, *19*, 272–279. [CrossRef]

11. Qin, C.; Yang, B.; Bräuning, A.; Sonechkin, D.M.; Huang, K. Regional extreme climate events on the northeastern Tibetan Plateau since AD 1450 inferred from tree rings. *Glob. Planet Chang.* **2011**, *75*, 143–154. [CrossRef]

12. Zhou, Y.; Pei, F.; Xia, Y.; Wu, C.; Zhong, R.; Wang, K.; Wang, H.; Cao, Y. Assessing the Impacts of Extreme Climate Events on Vegetation Activity in the North South Transect of Eastern China (NSTEC). *Water* **2019**, *11*, 2291. [CrossRef]

13. Zaitchik, B.F.; Macalady, A.K.; Bonneau, L.R.; Smith, R.B. Europe's 2003 heat wave: A satellite view of impacts and land–atmosphere feedbacks. *Int. J. Climatol.* **2006**, *26*, 743–769. [CrossRef]

14. Baldwin, J.W.; Dessy, J.B.; Vecchi, G.A.; Oppenheimer, M. Temporally Compound Heat Wave Events and Global Warming: An Emerging Hazard. *Earth's Future* **2019**, *7*, 411–427. [CrossRef]

15. You, Q.; Kang, S.; Aguilar, E.; Yan, Y. Changes in daily climate extremes in the eastern and central Tibetan Plateau during 1961–2005. *J. Geophys. Res.* **2008**, *113*, D07101. [CrossRef]

16. Li, Z.; Guo, X.; Yang, Y.; Hong, Y.; Wang, Z.; You, L. Heatwave Trends and the Population Exposure Over China in the 21st Century as Well as Under 1.5 °C and 2.0 °C Global Warmer Future Scenarios. *Sustainability* **2019**, *11*, 3318. [CrossRef]

17. You, Q.; Kang, S.; Pepin, N.; Flügel, W.-A.; Sanchez-Lorenzo, A.; Yan, Y.; Zhang, Y. Climate warming and associated changes in atmospheric circulation in the eastern and central Tibetan Plateau from a homogenized dataset. *Glob. Planet Chang.* **2010**, *72*, 11–24. [CrossRef]

18. Zhang, L.; Guo, H.; Ji, L.; Lei, L.; Wang, C.; Yan, D.; Li, B.; Li, J. Vegetation greenness trend (2000 to 2009) and the climate controls in the Qinghai-Tibetan Plateau. *J. Appl. Remote Sens.* **2013**, *7*, 073572. [CrossRef]

19. Hua, T.; Wang, X. Temporal and Spatial Variations in the Climate Controls of Vegetation Dynamics on the Tibetan Plateau during 1982–2011. *Adv. Atmos. Sci.* **2018**, *35*, 1337–1346. [CrossRef]

20. Zhang, F.; Cao, G. Resilience of Energy and CO_2 Exchange to a Summer Heatwave in an Alpine Humid Grassland on the Qinghai-Tibetan Plateau. *Pol. J. Environ. Stud.* **2017**, *26*, 385–394. [CrossRef]

21. He, S.; Richards, K.; Zhao, Z. Climate extremes in the Kobresia meadow area of the Qinghai-Tibetan Plateau, 1961–2008. *Environ. Earth Sci.* **2015**, *75*, 60. [CrossRef]

22. Yang, L.; Wang, H.; Lu, T.; Liu, L.; Fu, W.; Wei, D. Characteristics of Air Temperature Variation in Lhasa City over the Past 49 Years. *Earth Environ.* **2021**, *49*, 492–503. [CrossRef]

23. Li, C.; Wulf, H.; Schmid, B.; He, J.; Schaepman, M.E. Estimating Plant Traits of Alpine Grasslands on the Qinghai-Tibetan Plateau Using Remote Sensing. *IEEE J. Sel. Top. Appl. Earth Obs. Remote Sens.* **2018**, *11*, 2263–2275. [CrossRef]

24. Yao, T.; Xue, Y.; Chen, D.; Chen, F.; Thompson, L.; Cui, P.; Koike, T.; Lau, W.K.M.; Lettenmaier, D.; Mosbrugger, V.; et al. Recent Third Pole's Rapid Warming Accompanies Cryospheric Melt and Water Cycle Intensification and Interactions between Monsoon and Environment: Multidisciplinary Approach with Observations, Modeling, and Analysis. *Bull. Am. Meteorol. Soc.* **2019**, *100*, 423–444. [CrossRef]

25. Chen, F.; Zhang, J.; Liu, J.; Cao, X.; Hou, J.; Zhu, L.; Xu, X.; Liu, X.; Wang, M.; Wu, D.; et al. Climate change, vegetation history, and landscape responses on the Tibetan Plateau during the Holocene: A comprehensive review. *Quat. Sci. Rev.* **2020**, *243*, 106444. [CrossRef]

26. Bhattarai, P.; Zheng, Z.; Bhatta, K.P.; Adhikari, Y.P.; Zhang, Y. Climate-Driven Plant Response and Resilience on the Tibetan Plateau in Space and Time: A Review. *Plants* **2021**, *10*, 480. [CrossRef]

27. Lucht, W.; Lewis, P. Theoretical noise sensitivity of BRDF and albedo retrieval from the EOS-MODIS and MISR sensors with respect to angular sampling. *Int. J. Remote Sens.* **2010**, *21*, 81–98. [CrossRef]

28. Gao, X.; Huete, A.R.; Ni, W.; Miura, T. Optical–Biophysical Relationships of Vegetation Spectra without Background Contamination. *Remote Sens. Environ.* **2000**, *74*, 609–620. [CrossRef]

29. Tucker, C.J. Red and Photographic Infrared Linear Combinations for Monitoring Vegetation. *Remote Sens. Environ.* **1979**, *8*, 127–150. [CrossRef]

30. Huete, A.; Didan, K.; Miura, T.; Rodriguez, E.P.; Gao, X.; Ferreira, L.G. Overview of the radiometric and biophysical performance of the MODIS vegetation indices. *Remote Sens. Environ.* **2002**, *83*, 195–213. [CrossRef]

31. Chai, L.; Zhu, Z.; Liu, S. Land Surface Soil Moisture Dataset of SMAP Time-Expanded Daily 0.25° × 0.25° over Qinghai-Tibet Plateau Area (SMsmapTE, V1). *Natl. Tibet. Plateau Data Cent.* **2020**. [CrossRef]

32. Qu, Y.; Zhu, Z.; Chai, L.; Liu, S.; Montzka, C.; Liu, J.; Yang, X.; Lu, Z.; Jin, R.; Li, X.; et al. Rebuilding a Microwave Soil Moisture Product Using Random Forest Adopting AMSR-E/AMSR2 Brightness Temperature and SMAP over the Qinghai–Tibet Plateau, China. *Remote Sens.* **2019**, *11*, 683. [CrossRef]

33. You, Q.; Jiang, Z.; Kong, L.; Wu, Z.; Bao, Y.; Kang, S.; Pepin, N. A comparison of heat wave climatologies and trends in China based on multiple definitions. *Clim. Dyn.* **2016**, *48*, 3975–3989. [CrossRef]

34. Li, X.; Ren, G.; Wang, S.; You, Q.; Sun, Y.; Ma, Y.; Wang, D.; Zhang, W. Change in the heatwave statistical characteristics over China during the climate warming slowdown. *Atmos. Res.* **2021**, *247*, 105152. [CrossRef]

35. Ma, M.; Veroustraete, F. Reconstructing pathfinder AVHRR land NDVI time-series data for the Northwest of China. *Adv. Space Res.* **2006**, *37*, 835–840. [CrossRef]

36. Zhang, H.; Ren, Z. Comparison and Application Analysis of Several NDVI Time-Series Reconstruction Methods. *Sci. Agric. Sin.* **2014**, *47*, 2998–3008. [CrossRef]

37. Yan, B.; Lu, Z.; Li, J.; Huo, J. Analysis of Runoff Characteristics in Dry Season at Datong Station on the Mainstream of the Yangtze River. *Environ. Earth Sci.* **2021**, *768*, 012066. [CrossRef]

38. Wang, X.; Xiao, J.; Li, X.; Cheng, G.; Ma, M.; Zhu, G.; Altaf Arain, M.; Andrew Black, T.; Jassal, R.S. No trends in spring and autumn phenology during the global warming hiatus. *Nat. Commun.* **2019**, *10*, 2389. [CrossRef]

39. Sen, P.K. Estimates of the Regression Coefficient Based on Kendall's Tau. *J. Am. Stat. Assoc.* **1968**, *63*, 1379–1389. [CrossRef]

40. Liu, D.; Wang, T.; Yang, T.; Yan, Z.; Liu, Y.; Zhao, Y.; Piao, S. Deciphering impacts of climate extremes on Tibetan grasslands in the last fifteen years. *Sci. Bull.* **2019**, *64*, 446–454. [CrossRef]

41. You, Q.; Fraedrich, K.; Ren, G.; Pepin, N.; Kang, S. Variability of temperature in the Tibetan Plateau based on homogenized surface stations and reanalysis data. *Int. J. Climatol.* **2013**, *33*, 1337–1347. [CrossRef]

42. Hozain, M.I.; Salvucci, M.E.; Fokar, M.; Holaday, A.S. The differential response of photosynthesis to high temperature for a boreal and temperate Populus species relates to differences in Rubisco activation and Rubisco activase properties. *Tree Physiol.* **2010**, *30*, 32–44. [CrossRef]

43. Chen, A.; Huang, L.; Liu, Q.; Piao, S. Optimal temperature of vegetation productivity and its linkage with climate and elevation on the Tibetan Plateau. *Glob. Chang. Biol.* **2021**, *27*, 1942–1951. [CrossRef]

44. Fu, Z.; Ciais, P.; Bastos, A.; Stoy, P.C.; Yang, H.; Green, J.K.; Wang, B.; Yu, K.; Huang, Y.; Knohl, A.; et al. Sensitivity of gross primary productivity to climatic drivers during the summer drought of 2018 in Europe. *Philos. Trans. R. Soc. Lond. B Biol. Sci.* **2020**, *375*, 20190747. [CrossRef]

45. Hatfield, J.L.; Prueger, J.H. Temperature extremes: Effect on plant growth and development. *Weather Clim. Extrem.* **2015**, *10*, 4–10. [CrossRef]

46. Yu, H.; Xu, J.; Okuto, E.; Luedeling, E. Seasonal response of grasslands to climate change on the Tibetan Plateau. *PLoS ONE* **2012**, *7*, e49230. [CrossRef]

47. Yu, H.; Luedeling, E.; Xu, J. Winter and spring warming result in delayed spring phenology on the Tibetan Plateau. *Proc. Natl. Acad. Sci. USA* **2010**, *107*, 22151–22156. [CrossRef]

48. Cheng, G.; Wu, T. Responses of permafrost to climate change and their environmental significance, Qinghai-Tibet Plateau. *J. Geophys. Res.* **2007**, *112*, F2. [CrossRef]

49. Yi, S.; Zhou, Z.; Ren, S.; Xu, M.; Qin, Y.; Chen, S.; Ye, B. Effects of permafrost degradation on alpine grassland in a semi-arid basin on the Qinghai–Tibetan Plateau. *Environ. Res. Lett.* **2011**, *6*, 045403. [CrossRef]

50. Piao, S.; Cui, M.; Chen, A.; Wang, X.; Ciais, P.; Liu, J.; Tang, Y. Altitude and temperature dependence of change in the spring vegetation green-up date from 1982 to 2006 in the Qinghai-Xizang Plateau. *Agric. For. Meteorol.* **2011**, *151*, 1599–1608. [CrossRef]

51. Ganjurjav, H.; Gornish, E.S.; Hu, G.; Schwartz, M.W.; Wan, Y.; Li, Y.; Gao, Q. Warming and precipitation addition interact to affect plant spring phenology in alpine meadows on the central Qinghai-Tibetan Plateau. *Agric. For. Meteorol.* **2020**, *287*, 107943. [CrossRef]

52. An, W.; Hou, S.; Hu, Y.; Wu, S. Delayed warming hiatus over the Tibetan Plateau. *Earth Space Sci.* **2017**, *4*, 128–137. [CrossRef]

53. Liu, X.; Chen, B. Climatic warming in the Tibetan Plateau during recent decades. *Int. J. Climatol.* **2000**, *20*, 1729–1742. [CrossRef]

Article

Quantitative Analysis of Natural and Anthropogenic Factors Influencing Vegetation NDVI Changes in Temperate Drylands from a Spatial Stratified Heterogeneity Perspective: A Case Study of Inner Mongolia Grasslands, China

Shengkun Li [1,2], Xiaobing Li [1,2,*], Jirui Gong [1,2], Dongliang Dang [2], Huashun Dou [2] and Xin Lyu [2]

[1] State Key Laboratory of Earth Surface Process and Resource Ecology, Beijing Normal University, Beijing 100875, China; sklee@mail.bnu.edu.cn (S.L.); jrgong@bnu.edu.cn (J.G.)
[2] Faculty of Geographical Science, Beijing Normal University, Beijing 100875, China; 201621190009@mail.bnu.edu.cn (D.D.); hsdou@mail.bnu.edu.cn (H.D.); lyuxin@bnu.edu.cn (X.L.)
* Correspondence: xbli@bnu.edu.cn; Tel.: +86-010-58807212

Citation: Li, S.; Li, X.; Gong, J.; Dang, D.; Dou, H.; Lyu, X. Quantitative Analysis of Natural and Anthropogenic Factors Influencing Vegetation NDVI Changes in Temperate Drylands from a Spatial Stratified Heterogeneity Perspective: A Case Study of Inner Mongolia Grasslands, China. *Remote Sens.* **2022**, *14*, 3320. https://doi.org/10.3390/rs14143320

Academic Editors: Tinghai Ou, Wenxin Zhang, Youhua Ran and Xuejia Wang

Received: 27 May 2022
Accepted: 5 July 2022
Published: 10 July 2022

Publisher's Note: MDPI stays neutral with regard to jurisdictional claims in published maps and institutional affiliations.

Abstract: The detection and attribution of vegetation dynamics in drylands is an important step for the development of effective adaptation and mitigation strategies to combat the challenges posed by human activities and climate change. However, due to the spatial heterogeneity and interactive influences of various factors, quantifying the contributions of driving forces on vegetation change remains challenging. In this study, using the normalized difference vegetation index (NDVI) as a proxy of vegetation growth status and coverage, we analyzed the temporal and spatial characteristics of the NDVI in China's Inner Mongolian grasslands using Theil–Sen slope statistics and Mann–Kendall trend test methods. In addition, using the GeoDetector method, a spatially-based statistical technique, we assessed the individual and interactive influences of natural factors and human activities on vegetation-NDVI change. The results show that the growing season average NDVI exhibited a fluctuating upward trend of 0.003 per year from 2000 to 2018. The areas with significant increases in NDVI ($p < 0.05$) accounted for 45.63% of the entire region, and they were mainly distributed in the eastern part of the Mu Us sandy land and the eastern areas of the Greater Khingan Range. The regions with a decline in the NDVI were mainly distributed in the central and western regions of the study area. The GeoDetector results revealed that both natural and human factors had significant impacts on changes in the NDVI ($p < 0.001$). Precipitation, livestock density, wind speed, and population density were the dominant factors affecting NDVI changes in the Inner Mongolian grasslands, explaining more than 15% of the variability, while the contributions of the two topography factors (terrain slope and slope aspect) were relatively low (less than 2%). Furthermore, NDVI changes responded to the changes in the level of specific influencing factors in a nonlinear way, and the interaction of two factors enhanced the effect of each singular factor. The interaction between precipitation and temperature was the highest among all factors, accounting for 39.3% of NDVI variations. Findings from our study may aid policymakers in better understanding the relative importance of various factors and the impacts of the interactions between factors on vegetation change, which has important implications for preventing and mitigating land degradation and achieving sustainable pasture use in dryland ecosystems.

Keywords: NDVI; vegetation dynamics; influencing factors; spatial stratified heterogeneity; geographical detector method

1. Introduction

Drylands, covering about 41.30% of the Earth's terrestrial surface and supporting more than 38% of the global population [1], are characterized by a lack of water, infertile soil, and high climate variability. They are highly susceptible to climate fluctuations and human activities [2,3]. Because of the limitations imposed by water resource availability and challenging climate change effects, drylands fall victim to persistent land degradation problems that have led to the desertification of 3.6 billion hectares worldwide and have threatened the lives and livelihoods of the local people [3,4]. Monitoring land degradation and identifying its potential causes are of great significance to sustainable land use. As the primary producer in the ecosystem, the ground vegetation links the carbon–water cycle and the energy flow within the hydrosphere, pedosphere, and atmosphere [5,6], and it plays a fundamental role in providing ecosystem goods and maintaining terrestrial ecosystem functions [7]. The vegetation conditions of degraded land have always been used as a proxy to quantitatively detect ecosystem processes at both local and regional scales [8–11]. With the help of satellite remote sensing images, the detection and attribution of vegetation greening and browning trends have emerged as a popular subject in the scientific community over the past several decades [12], and the relation between the normalized difference vegetation index (NDVI) and vegetation growth status and coverage has been well established. Due to the spatial heterogeneity and the combined effects of the driving factors [13,14], quantifying the contributions of the main drivers of vegetation change remains challenging. It is urgent that techniques be developed to help disentangle the contributions of factors to variations in vegetation for the development of strategies for vegetation restoration and desertification prevention in drylands.

In general, vegetation change was influenced by intertwined natural and human-induced factors. The impact of global climate change on vegetation growth is a major research priority. Numerous studies have been carried out related to the response of the NDVI to variations in climatic factors (e.g., air temperature, solar radiation, and precipitation) at different spatiotemporal scales [5,6,15], aimed at improving our knowledge of the mechanistic link between the effects of climate change on vegetation activity. Over the last decades, human activities became diverse and intensive, exerting greater pressure on terrestrial ecosystems [3,16]. Anthropogenic factors manifest primarily in land-use change (LUCC) or changes in management measures [17,18]. Urbanization, characterized by the occupation of vegetation-covered surfaces by impervious ground, may lead to vegetation degradation [19]. Overgrazing, cultivation of arable land, and deforestation have resulted in bare ground and soil erosion, which may result in vegetation degradation [20], while the enclosure management of degraded rangeland may promote vegetation restoration [21].

The residual trend method (RESTREND), ecosystem modeling methods, and various mathematical models are widely used for quantifying the influences of driving factors on changes in vegetation growth. The RESTREND method distinguishes between human-induced and climate-driven vegetation changes based on the trend of NDVI residues (defined as the differences between the actual and predicted NDVI values) [22], and it is predominantly useful in studies of regions where water is limiting [9,23,24]. However, the RESTREND method is associated with some uncertainties [25]. The results of the RESTREND method may vary considerably with the time employed to compute the NDVI-precipitation regression and the trends of its residuals [23,26]. Moreover, this method attributes the residuals to the total effects of all human disturbances, making it difficult to disentangle and compare the contributions of different human activities on vegetation variations [18]. The mathematical models mainly include regression, correlation analysis, and the structural equation modeling method [6,18,19,27]. Most of the mathematical models detect the impacts of the environmental variables on the vegetation dynamics using a linear hypothesis [28]. However, theory and empirical evidence suggest that the trajectory of the responses of the vegetation index to the influencing factors is often nonlinear [2,29,30], so the linearity assumptions may result in erroneous conclusions and misleading interpretations. Many process-based ecosystem models have been developed to quantify the responses of environmental variables to key ecological processes in a nonlinear way [28,31,32], overcoming the deficiencies of the mathematical models. However, ecosystem modes usually require a large number of inputs and parameter settings, and there are uncertainties in the models' structures and parameter choices [33], which may lead to inconsistent model results [15]. The GeoDetector method, which was developed by Wang in 2010 [34], quantifies the impacts of factors on geographical phenomena or attributes from a spatially stratified heterogeneity perspective [34,35]. The GeoDetector method does not involve complex parameter settings, nor does it follow the restrictive assumptions of traditional statistical methods. This technique has been used to evaluate the influences of factors in the eco-environmental and social science fields [15,36–40]. The GeoDetector method can be a promising tool for exploring the associations between various impact factors and vegetation changes in drylands.

With climate change and increasing anthropogenic activities, the vulnerable ecosystems of the drylands in northern China have been degraded to varying degrees, posing severe ecological and environmental problems [41–43]. In order to reverse the environmental degradation trend, particularly in the ecologically fragile regions, several ecological conservation programs were carried out in the late 1990s [44,45]. Since then, land-use patterns have changed substantially [45]. An in-depth understanding of the spatial features and the changes in and underlying the driving mechanisms of vegetation activity is important to improve policymakers' understanding of the sustainable use of vegetation resources and for the development of reasonable strategies for ongoing ecological restoration. At present, scholars have mainly focused on the relationships between vegetation variations and climatic factors at different time and spatial scales in the drylands of northern China [15,24,26,46–49], but they have paid little attention to different human activities (e.g., grazing pressure, land use conversions). In addition, few studies have considered the potential of interactive effects between the factors impacting vegetation changes. If the interactions between factors are not taken into consideration, the results may be biased [19].

In this study, the Inner Mongolian grasslands were selected as the study area. The Inner Mongolian grasslands act as an ecological protective belt for eastern China and provide plenty of ecosystem services (e.g., food supply, grass production, climate regulation, carbon sequestration, soil erosion control, and cultural heritage) [50]. The objectives of this work were twofold: (1) to investigate the temporal and spatial characteristics of the NDVI in the Inner Mongolian grasslands from 2000–2018; and (2) to examine the individual contributions and interactive effects of natural factors and human activities on vegetation changes using the GeoDetector method. This work aims to provide a scientific foundation for detecting the underlying mechanism of vegetation changes in temperate grasslands.

2. Materials and Methods

2.1. Study Area

The Inner Mongolian grasslands, located in the drylands of northern China (105°18′–125°15′E, 37°38′–50°50′N; Figure 1a), cover an area of approximately 78.2×10^4 km^2. This region is mainly arid and semi-arid [48]. It is ecologically fragile and is vulnerable to climate variations and human activities, but it plays a critical role in safeguarding the ecological security of northern China's agricultural plain and metropolitan regions [44]. Dominated by a temperate continental climate, the Inner Mongolian grasslands have annual precipitation of 120–520 mm and an annual mean temperature of −2 to 10 °C. More than 80% of the annual total precipitation is concentrated between July and September, which coincides with the period of high temperatures [44]. The climate has distinct seasonal characteristics: (1) windy and dry in spring, with strong evaporation; (2) warm and hot in summer, with an uneven precipitation distribution; (3) a short autumn, with early frost and snow and large day–night temperature differences; and (4) a cold and long winter. The vegetation across the Inner Mongolian grasslands has obvious east–west zonal distribution characteristics. From west to east, the precipitation and soil fertility gradually increase, and the solar radiation gradually decreases, forming three different types of temperate grasslands (Figure 1d). There are also nonzonal vegetation types, including saline meadows and marshes along riverbanks and large tracts of sandy lands, which are closely associated with site-specific geographical characteristics (e.g., water bodies, topography, and salinization). The study area is dominated by high plains and low mountains, which are characterized by high elevations in the central and western areas and low elevations in the southeastern and northern areas, with altitudes ranging from 90 m to 2300 m (Figure 1b).

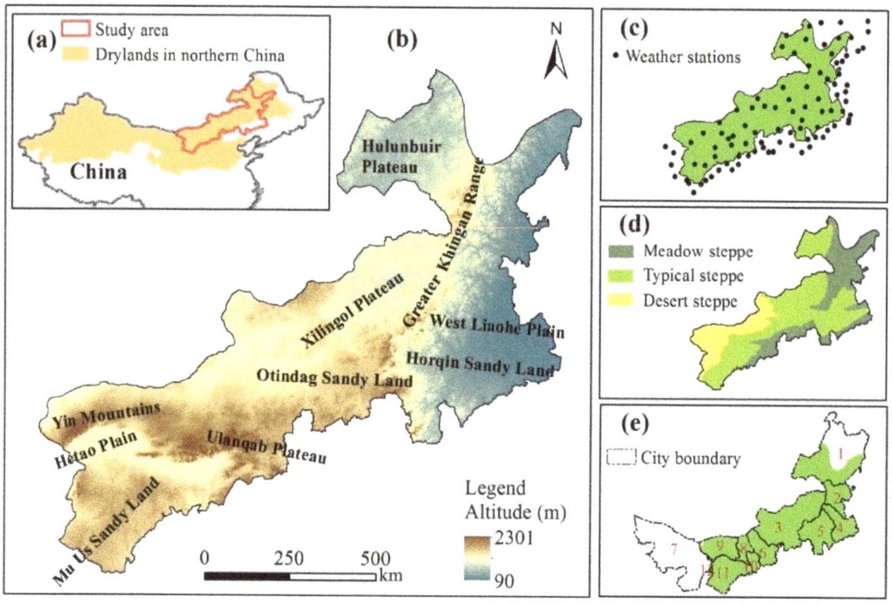

Figure 1. (a) Location of the study area, (b) topographical conditions, (c) weather station distribution, (d) ecoregions of the study area, and (e) city boundaries. The numbers in (e) represent the prefecture-level cities (1—Hulunbuir; 2—Hinggan; 3—Xilingol; 4—Tongliao; 5—Chifeng; 6—Ulanqab; 7—Alxa; 8—Baotou; 9—Bayannur; 10—Hohhot; 11—Erdos; 12—Wuhai).

2.2. Data Acquisition and Processing

The NDVI values were extracted from the MOD13A2 product (Version 6, 1000 m resolution, 16-days composite). The MOD13A2 product was geographically projected using the MODIS Reprojection Tool (MRT) software; then, the maximum value composite method (MVC) was used to obtain monthly NDVI data to reduce the effects of clouds and image noise. The MVC method still cannot guarantee that all pixels of an image are cloud-free. In this study, we used the variable weight filtering method proposed by Zhu to reconstruct a set of high-quality NDVI time-series data; the reconstructed vegetation index time-series data can enhance the application capability of vegetation index time-series data in the study of vegetation–climate factor interactions [51]. Given that most of the plants withered and stopped growing during the winter, we used the growing season (defined as April to October) NDVI to detect the inter-annual variations in the vegetation activity [48,49].

The meteorological datasets covering the period of 2000–2018 were compiled from ninety-six weather stations (Figure 1c). The datasets included daily values of the mean temperature, precipitation, sunshine duration, relative air humidity, and mean wind speed, which were obtained from the National Meteorological Information Center of China. The Solar Energy Resource Evaluation method (QX/T 89-2008), which was developed by the China Meteorological Administration, was used to estimate solar radiation. The spatial distribution results for meteorological station data at a spatial resolution of 1000 m were obtained via spatial interpolation using ArcGIS 10.2.

The 2000 and 2015 land cover type data (1000 m resolution) were retrieved from the Resource and Environment Science and Data Center. The dataset was interpreted visually based on Landsat thematic mapper images and unmanned aerial vehicles, which are characterized as being highly accurate via random sampling checks and field surveys. The data contain 26 secondary categories with a comprehensive evaluation accuracy of >90% [52].

The soil type data were extracted from the soil map of China (1:1,000,000), which was provided by the Chinese Soil Census Office. The data were compiled by soil generation classification standards.

The topographic data consisted of altitude, slope, and aspect data. Through image mosaicking, we obtained a DEM of the study area with a spatial resolution of 90 m from the Geospatial Data Cloud site (http://www.gscloud.cn, accessed on 25 March 2022).

The administrative boundaries, roads, and settlements (1:250,000) vector data were obtained from the National Catalogue Service for Geographic Information, Ministry of Natural Resources of China.

The statistical data, including the total population, gross regional product, agricultural mechanical power, fertilizer applied for agriculture, grain production, oil production, and quantity of livestock (including goats, sheep, horses, cattle, and camels), at the county level, were obtained from the Inner Mongolia Statistical Yearbook (https://data.cnki.net/, accessed on 20 March 2022). According to prior research, we used an equivalent unit of grazing (i.e., "sheep unit") to normalize the grazing intensity among different species [53]. Using the empirical formula [10,53], we set the transition factor for large livestock (e.g., cattle, donkeys, camels, and horses) to 6, whereas the transition factor was set to 1 for goats and sheep.

2.3. Mann-Kendall Trend Test and Sen's Slope Estimator

In this work, the Sen trend analysis and the Mann–Kendall test [54–56] were used to detect the trend slopes and significance of trends in the NDVI time series, respectively. The procedure for the nonparametric Mann–Kendall trend test [55,56] is as follows:

$$S = \sum_{i=1}^{n-1} \sum_{j=i+1}^{n} \text{sgn}(x_j - x_i) \tag{1}$$

In Equation (1), S denotes the standardized test statistic value, x_i and x_j are data values at time i and j, respectively; n is the length of time series; and $\text{sgn}(x_j - x_i)$ is the sign function, which is calculated as follows:

$$\text{sgn}(x_j - x_i) = \begin{cases} -1, & if\ x_j - x_i < 0 \\ 0, & if\ x_j - x_i = 0 \\ +1, & if\ x_j - x_i > 0 \end{cases} \tag{2}$$

In this study, the length of time series n = 19, and the trend test were conducted using the Z_S value, which is defined as follows:

$$Z_S = \begin{cases} \dfrac{S+1}{\sqrt{\text{Var}(S)}}, & if\ S < 0 \\ 0, & if\ S = 0 \\ \dfrac{S-1}{\sqrt{\text{Var}(S)}}, & if\ S > 0 \end{cases} \tag{3}$$

In Equation (3), the variance Var(S) is computed as:

$$\text{Var}(S) = \frac{n(n-1)(2n+5) - \sum_{i=1}^{m} t_i(t_i - 1)(2t_i + 5)}{18} \tag{4}$$

In Equation (4), m is the number of tied groups in the time series and t_i is the width of the tied groups. In this study, a significance level of $\alpha = 0.05$ was used. It is assumed that, for null hypothesis, the data are arranged with no significant trend. When $|Z| > Z_{1-\alpha/2}$, the null hypothesis is rejected and the trend of the change in the series data is considered to be significant.

The Sen slope calculation is carried out as:

$$\beta = \text{Median}\left(\frac{x_j - x_i}{j - i}\right) \tag{5}$$

where Median() denotes the median function of the requested series; β is the slope of the time series x; and a negative β value indicates a decreasing trend in the series.

2.4. GeoDetector Method

Spatial stratified heterogeneity (SSH), referring to a within sub-region variance of less than that between the sub-regions [35], is ubiquitous in ecological phenomena, such as soil types, land use types, and climate zones. The GeoDetector comprises a series of spatial statistical methods, and it is frequently used to detect the SSH of the dependent variables without linear assumptions and reveal the driving forces behind a phenomenon by quantifying the impact of associated factors. The GeoDetector assumes that if an independent variable (e.g., precipitation) has a certain degree of influence on a dependent variable (i.e., NDVI changes), then the spatial patterns of the two variables have high similarity. Figure 2 illustrates the principle of the GeoDetector; spatial variances within each sub-region and among the different sub-regions are compared to identify the explanatory powers of the potential explanatory variables [34,39,57].

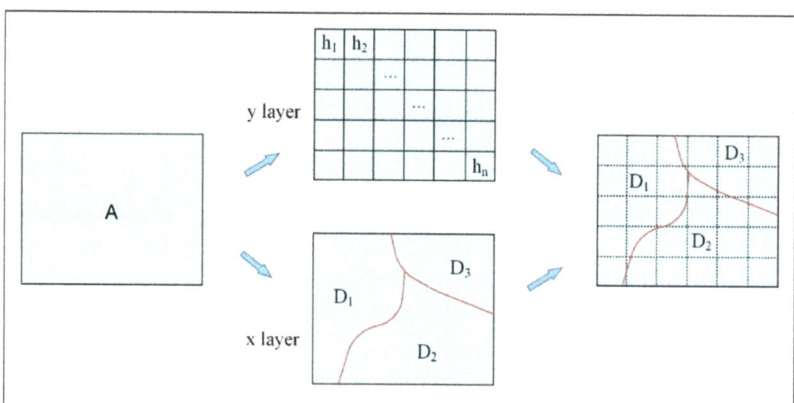

Figure 2. Illustration of the principle of the GeoDetector model. The study area A, the grid system H = {h$_i$, i = 1, 2, … , n}, and the sub-region of the potential factor D = {D$_i$, i = 1, 2, 3}.

2.4.1. Single Factor Influence Detection

The impact of an individual factor on changes in NDVI can be measured using the q-statistic [34,35]:

$$q = 1 - \frac{\sum_{h=1}^{L} N_h \sigma_h^2}{N \sigma^2} \tag{6}$$

where q is the measurement index of the factor. The range of q-statistic is [0, 1]. Based on the model principle, the larger the q-statistic is, the stronger the independent variable represents the heterogeneity of the dependent variable. L refers to the number of stratifications of factor X; N_h and N are the numbers of units in sub-region h and over the whole study region, respectively; and σ^2 and σ_h^2 represent the variances of variable Y over the entire study region and in sub-region h, respectively.

2.4.2. Interaction Detection of Pairwise Factors

The interactive impact of two explanatory factors (X1 and X2) on NDVI change also can be quantified by q-statistic. The module of interaction detection quantifies the interaction between two factors by comparing q(X1∩X2) with q(X1) and q(X2) to assess whether the factors weaken or enhance one another or are independent of each other, in which $q(X1 \cap X2)$ indicates the explanatory power of a new factor created by overlaying the layer of the two variables in GIS tools (Supplementary Figure S1). Generally, the results of the interaction detector encompass five categories (Figure 3).

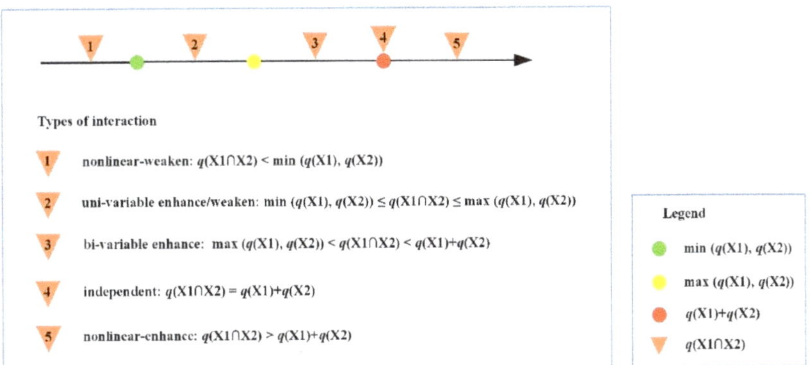

Figure 3. Judgment for interaction types between explanatory variables. Note: max() and min() denote the maximum and minimum functions, respectively. $q(X1 \cap X2)$ represents the interaction between factors X1 and X2. Modified from prior research [38,57,58].

2.4.3. Selection of Factors

In this work, we chose the slope of the change in the NDVI from 2000–2018 as the dependent variable and selected 15 potential natural and human factors (Table 1, Figure 4). Specifically, in addition to the climatic factors, we included soil and topography as fundamental environmental factors, which have been demonstrated to be critical to inter-annual variations of vegetation [9,59–61]. Additionally, six factors (road impact, geographical location, population pressure, grazing pressure, land use/cover change, and economic development) were selected to reflect the magnitude of anthropogenic influences [19,59]. In this study, we reclassified land cover types into six types, and the land cover maps (water bodies were excluded) for 2000 and 2015 were superimposed to generate a land use/cover change (LUCC) map. The spatial distribution of the grades for all driving factors can be found in Figure 4.

Table 1. Potential driving factors of vegetation variation in the study area.

Category	Index	Abbreviation	Unit
Climate	Annual precipitation	Pre	mm
	Annual mean temperature	Tem	°C
	Annual solar radiation	SR	$MJ \cdot m^{-2}$
	Annual mean wind speed	WS	$m \cdot s^{-1}$
	Annual mean relative air humidity	RH	%
Topography	Altitude	Alt	m
	Terrain slope	Slopd	°
	Slope aspect	Slopa	°
Soil	Soil type	Soilt	categorical
Human activity	Distance to the nearest road	DNR	km
	Distance to the nearest county centers	DNC	km
	Population density	Popd	$person \cdot km^{-2}$
	Per capita gross regional product	GRP	$10,000 \ yuan \ person^{-1}$
	Livestock density	Livstd	$sheep \cdot km^{-2}$
	Land use/cover change	LUCC	categorical

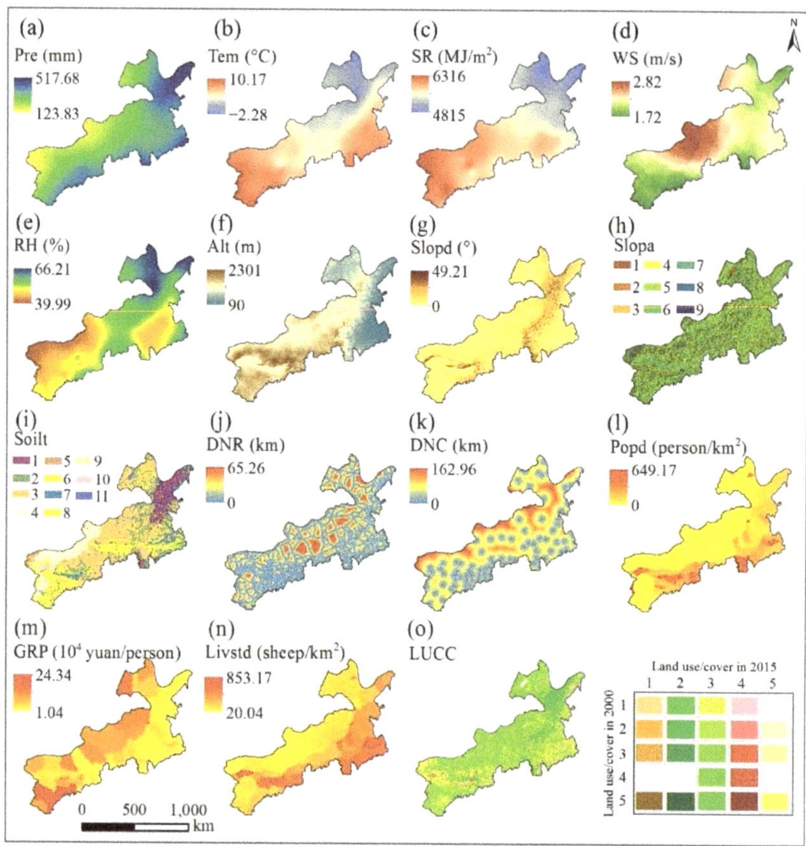

Figure 4. The spatial distributions of all factors. The numbers in the legend of (**h**) represent (1) Flat ground, (2) North slope, (3) Northeast slope, (4) East slope, (5) Southeast slope, (6) South slope, (7) Southwest slope, (8) West slope, and (9) Northwest slope. The numbers in the legend of (**i**) represent (1) Luvisols, (2) Semi-luvisols, (3) Caliche soils, (4) Arid soils, (5) Desert soils, (6) Skeletol primitive soils, (7) Semi-hydromorphic soils, (8) Hydromorphic soils, (9) Saline soils, (10) Anthrosols, and (11) Others. The numbers in the legend of (**o**) represent (1) Cropland, (2) Forest, (3) Grassland, (4) Construction land, and (5) Unused land. Pre: precipitation; Tem: air temperature; SR: solar radiation; WS: wind speed; RH: relative air humidity; Alt: altitude; Slopd: terrain slope; Slopa: slope aspect; Soilt: soil type; DNR: distance to the nearest road; DNC: distance to the nearest county centers; Popd: population density; GRP: Per capita gross regional product; Livstd: livestock density; LUCC: land use/cover change.

2.4.4. Factor Grading Optimization in the GeoDetector Method

Since the GeoDetector method is only suitable for dealing with discrete or categorical variables, all the continuous predictor variables should be discretized using appropriate discretization methods before modeling [38,62]. In this study, the twelve factors, namely, five post-interpolation meteorological factors, two topography factors (altitude and terrain slope), and five anthropogenic factors (distance to the nearest road, distance to the nearest county centers, population density, per capita gross regional product, and livestock density), are continuous variables. We converted the twelve continuous variables into discrete ones.

To reduce the subjectivity of user-defined discretization and ensure the best-quality modelling results, the optimal discretization methods were determined from five types of unsupervised discretization methods, including geometrical interval (GI), natural breaks

(NB), equal interval (EI), quantile (QU), and standard deviation (SD) methods [38,62,63]. The procedures used for the factor grading optimization are as follows (Figure 5). First, we classified each continuous variable based on the five discretization methods and fourteen levels (stratification numbers of 2 to 15). Then, we extracted the values of the NDVI tendency layer and all of the classification layers. Finally, we calculated the q-statistics of each continuous predictor variable in all of the classification cases and plotted them to show their changes (Figure 6).

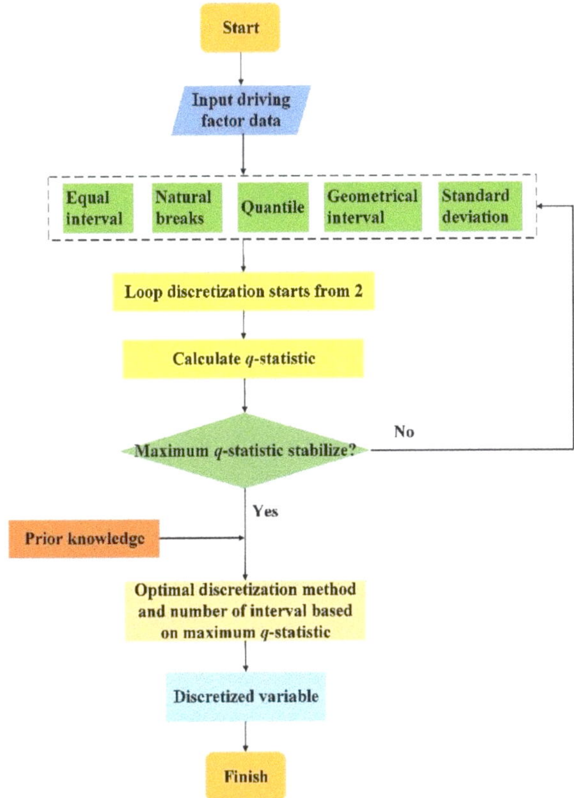

Figure 5. Flowchart illustrating the process of determining the optimal discretization method and stratification number.

A combination of classification algorithms with prior knowledge was needed to classify the continuous variable when using the GeoDetector method [34,57]. As shown in Figure 6, the maximum q-statistics of the factors generally increased as the number of stratifications increased. When the number of stratifications reached a certain value, the maximum q-statistic stabilized. This certain value was defined as the stable value. When the stratification number was bigger than the stable value, the characterization identified by GeoDetector remained unchanged, implying that more discretization intervals do not mine the information of the continuous variables. Considering the maximum stratification numbers (7, 8, 7, and 9, respectively) used in relevant studies [38,64–66] and the stable value observed in this study (approximately 10 in Figure 6), we limited the maximum number of stratifications to 10. The largest q-statistic value indicates the optimal discretization method and stratification number [38,62,63]. Based on this principle, we determined the optimal discretization methods and stratification numbers of each predictor variable. The impact factors with the optimal discretization methods and stratification numbers can be found in Table 2.

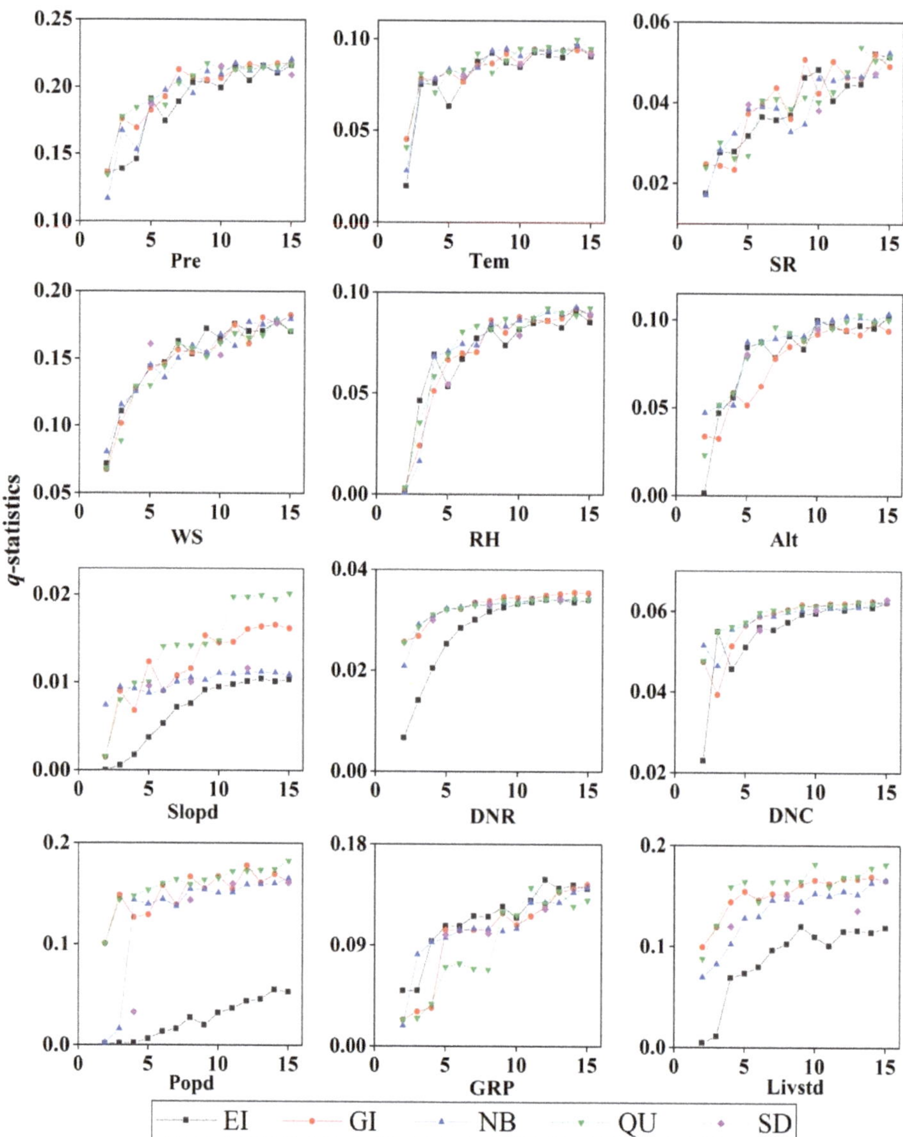

Figure 6. Comparison of the *q*-statistics under the different discretization methods and stratification number combinations. The five types of discretization methods include equal interval (EI), geometrical interval (GI), natural break (NB), quantile (QU), and standard deviation (SD). Pre: precipitation; Tem: air temperature; SR: solar radiation; WS: wind speed; RH: relative air humidity; Alt: altitude; Slopd: slope; DNR: distance to the nearest road; DNC: distance to the nearest county centers; Popd: population density; GRP: Per capita gross regional product; and Livstd: livestock density.

Table 2. The classification of potential driving factors. The units of Pre, Tem, SR, WS, RH, Alt, Slopd, DNR, DNC, Popd, GRP, and Livstd are mm, °C, MJ·m^{-2}, m·s^{-1}, %, m, °, km, km, person·km^{-2}, 10,000 yuan person^{-1}, and sheep·km^{-2}, respectively. In the parentheses of the table header, QU, EI, and GI correspond to three discretization methods of the quantile, equal interval, and geometrical interval, respectively, and the numbers represent the number of stratifications.

Category/Factor	Pre (QU-9)	Tem (QU-10)	SR (GI-9)	WS (EI-9)	RH (GI-10)	Alt (EI-10)	Slopd (GI-9)
1	123.8–194.9	−2.28 to 0.74	4815–5130	1.72–1.85	40.0–43.7	90–311	0–0.17
2	194.9–233.5	0.74–2.11	5130–5371	1.85–1.97	43.7–46.5	311–532	0.17–0.24
3	233.5–269.0	2.11–3.18	5371–5555	1.97–2.09	46.5–48.5	532–753	0.24–0.40
4	269.0–306.1	3.18–3.97	5555–5695	2.09–2.21	48.5–50.0	753–974	0.40–0.79
5	306.1–332.3	3.97–4.94	5695–5802	2.21–2.33	50.0–51.1	974–1195	0.79–1.71
6	332.3–352.4	4.94–5.72	5802–5884	2.33–2.45	51.1–52.7	1195–1416	1.71–3.88
7	352.4–378.7	5.72–6.89	5884–5991	2.45–2.58	52.7–54.7	1416–1638	3.88–8.97
8	378.7–418.8	6.89–7.48	5991–6131	2.58–2.70	54.7–57.5	1638–1859	8.97–20.92
9	418.8–517.7	7.48–8.07	6131–6316	2.70–2.82	57.5–61.2	1859–2080	20.92–49.21
10		8.07–10.17			61.2–66.2	2080–2301	

Category\Factor	Slopa	Soilt	DNR (GI-9)	DNC (GI-9)	Popd (GI-10)	GRP (EI-9)	Livstd (QU-10)
1	Flat ground	Luvisols	0–1.44	0–19.5	0.98–1.95	1.04–3.63	20.05–29.85
2	North	Semi-luvisols	1.44–2.35	19.5–31.9	1.95–3.79	3.63–6.22	29.85–52.72
3	Northeast	Caliche soils	2.35–3.79	31.9–39.8	3.79–7.28	6.22–8.81	52.72–55.99
4	East	Arid soils	3.79–6.11	39.8–44.8	7.28–13.89	8.81–11.39	55.99–69.06
5	Southeast	Desert soils	6.11–9.83	44.8–52.8	13.89–26.42	11.39–13.99	69.06–78.86
6	South	Skeletol primitive soils	9.83–15.79	52.8–65.2	26.42–50.17	13.99–16.57	78.86–111.53
7	Southwest	Semi-hydromorphic soils	15.79–25.35	65.2–84.6	50.17–95.20	16.57–19.16	111.53–157.27
8	West	Hydromorphic soils	25.35–40.68	84.6–115.1	95.20–180.56	19.16–21.75	157.27–186.67
9	Northwest	Saline soils	40.68–65.26	115.1–163.0	180.56–342.39	21.75–24.34	186.67–261.82
10		Anthrosols			342.39–649.17		261.82–853.17
11		Others					

3. Results

3.1. Spatio-Temporal Variability of NDVI

The areas with mean growing season NDVI values of greater than 0.6 accounted for 5.34% of the entire area from 2000–2018 (Figure 7a), indicating the generally inferior nature of the vegetation cover in the Inner Mongolian grasslands. The spatial pattern of the multi-year mean NDVI during the growing season exhibited an increasing trend from south to north and from west to east (Figure 7a), which was highly consistent with the distribution pattern of water and heat resources. The northeastern part of the Inner Mongolian grasslands is located in the transitional zone between the Greater Khingan Range forest region and the Inner Mongolia temperate grasslands, and high NDVI values (>0.6) are concentrated in this area (Figure 7a). The topography of the central part of the region is dominated by high plains and low mountains, with good forage quality and NDVI values from 0.3–0.5. The western part of the region is subject to low rainfall and is home to xerophytic vegetation, which is mainly composed of xerophytic bunch grass mixed with semi-shrubs and allium plants. This area also has widely distributed low NDVI values (<0.2), indicating poor vegetation coverage (Figure 7a).

The growing season average NDVI of the entire study area ranged from 0.289 to 0.365 during the period of 2000 to 2018 (Supplementary Figure S2) and exhibited a significant increase at a rate of 0.003 a^{-1} ($p < 0.05$). Figure 7b shows the NDVI changing trends at the pixel scale in the Inner Mongolian grasslands, which indicates that the NDVI mainly increased, with the areas of increase and the areas of decrease being 71.90×10^4 and 6.30×10^4 km^2, accounting for 91.94% and 8.06% of the entire region, respectively. The areas with significant increases in NDVI ($p < 0.05$) accounted for 45.63% of the entire region, and they were mainly distributed in the eastern part of the Mu Us sandy land (i.e., Erdos and Hohhot) and in the eastern areas of the Greater Khingan Range (i.e., Hinggan City, Tongliao City, and Chifeng City). Although the vegetation conditions improved in general, different degrees of degradation were also observed across the study area. The regions with a decline in NDVI were mainly distributed in the central and western regions of the study area (Figure 7b), especially in four prefecture-level cities (i.e., Bayannur, Baotou, northern Ulanqab, and western Xilingol). The areas with significant decreases in NDVI were small, accounting for 0.44% of the study region, and they were relatively scattered.

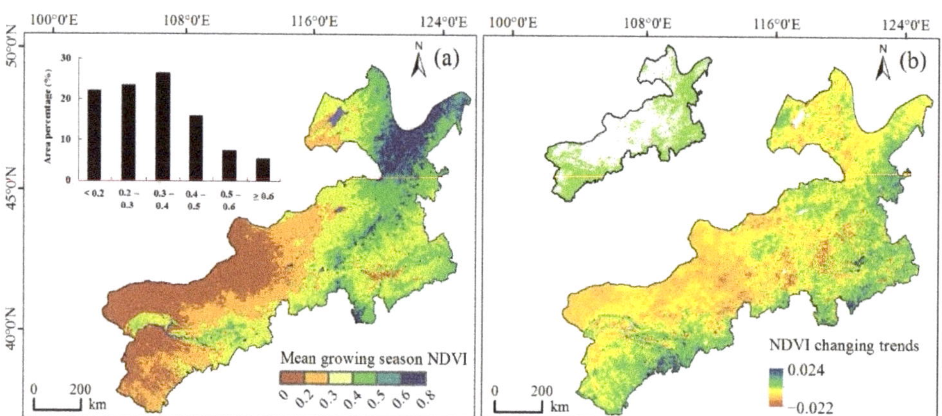

Figure 7. The (**a**) spatial distribution of the growing season average NDVI and (**b**) the slope of the change in the NDVI from 2000–2018. The inset graph in (**a**) is a statistical histogram; the inset map in (**b**) shows the significant decreases (red) and increases (green) in the NDVI at the 95% confidence level.

3.2. Impacts of Natural and Human Factors on NDVI Changes

3.2.1. Impacts of the 15 Factors on NDVI Changes

The q-statistics of all of the influencing factors passed the significance test ($p < 0.001$) (Supplementary Figure S3). The q-statistic values of the factors exhibited a marked difference that can be ranked as follows: Pre (0.217) > Livstd (0.182) > WS (0.173) > Popd (0.167) > GRP (0.126) > Alt (0.100) > Tem (0.096) > RH (0.088) > Soilt (0.067) > DNC (0.062) > SR (0.051) > LUCC (0.049) > DNR (0.035) > Slopd (0.015) > Slopa (0.001) (Supplementary Figure S3). These results indicate that the precipitation, which had the highest q-statistic value, predominantly explains the spatial heterogeneity of NDVI changes. The next most important factors were the livestock density, wind speed, and population density, with contributions of greater than 15%, while the impacts of the two topography factors (terrain slope and slope aspect) were relatively low, with q-statistic values of less than 0.02. Therefore, both the natural and human factors were identified as important factors influencing the vegetation NDVI changes in the Inner Mongolian grasslands.

3.2.2. Interactions between the 15 Factors

Two types of interaction relationships (i.e., nonlinear enhancement and bivariate enhancement) were identified among the 105 cases. For 55 cases, the q-statistics of the pairwise factor interactions were larger than the sum of the q-statistics of the two involved factors (Figure 8), which implies a nonlinear enhancement effect. The top five interactive q-statistics decreased in the following order: Pre∩Tem (0.393) > Tem∩Popd (0.336) > Pre∩Livstd (0.334) > Tem∩SR (0.332) > Tem∩RH (0.331). This indicates that the interactions between the climatic factors, population density, and livestock density had the greatest impacts on vegetation changes. These results show that the q-statistic of any pair of interacting factors was greater than the q-statistics of the single factors in the pair, implying that no factor influenced the vegetation changes in an independent manner but rather through interactions with the other factors.

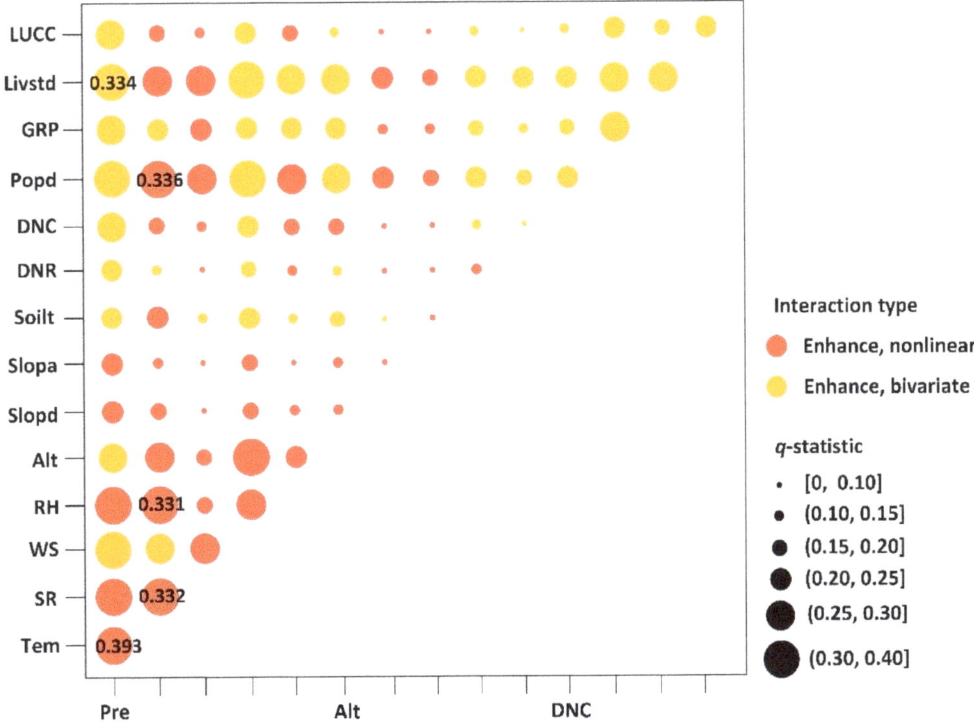

Figure 8. Influences of the interactions between two factors. Pre: precipitation; Tem: air temperature; SR: solar radiation; WS: wind speed; RH: relative air humidity; Alt: altitude; Slopd: slope; Slopa: slope aspect; Soilt: soil type; DNR: distance to the nearest road; DNC: distance to the nearest county centers; Popd: population density; GRP: Per capita gross regional product; Livstd: livestock density; and LUCC: land use/cover change type.

3.2.3. Effects of the Different Grades of the Factors

The rate at which the NDVI increased varied substantially with the different levels of the factors (Figure 9). Specifically, as the precipitation, population density, per capita gross regional product, and livestock density increased, the magnitude of the increase in the NDVI generally increased. As the wind speed increased, the rate of the increase in the NDVI generally decreased. The altitude, distance to the nearest road, and distance to the nearest county centers showed characteristics similar to those of the wind speed. As the relative air humidity increased, the rate of increase of the NDVI continued to increase and reached a maximum, and then it fluctuated slightly. The rate of increase of the NDVI fluctuated for different ranges or types of temperature, solar radiation, slope, aspect, and soil type (Figure 9). As shown in Table 3, most of the types of land use conversion led to an increase in the NDVI. The land use conversion from grasslands to croplands caused the largest increasing rate of NDVI. There were two types of land use conversions (from cropland to construction land and from unused land to construction land) that lead to a decrease in the NDVI.

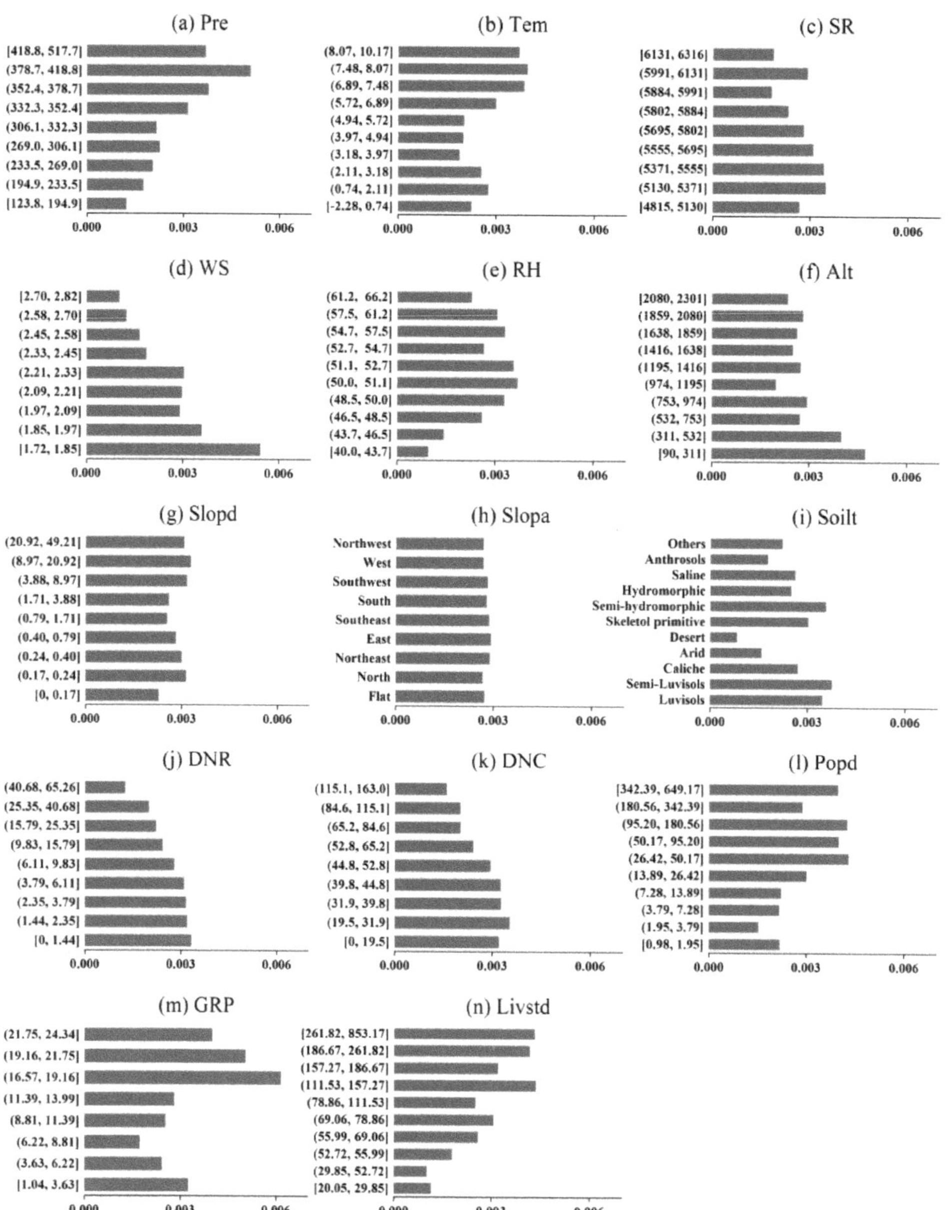

Figure 9. Influences of the factors' different grades on the magnitude of the increase in the NDVI. The units of Pre (precipitation), Tem (air temperature), SR (solar radiation), WS (wind speed), RH (relative air humidity), Alt (altitude), Slopd (slope), Slopa (slope aspect), DNR (distance to the nearest road), DNC (distance to the nearest county center), Popd (population density), GRP (per capita gross regional product), and Livstd (livestock density) are mm, °C, MJ·m^{-2}, m·s^{-1}, %, m, °, °, km, km, person·km^{-2}, 10,000 yuan person^{-1}, and sheep·km^{-2}, respectively. Soilt represents the soil type.

Table 3. Impacts of land use/cover change types on NDVI variations. The numbers in parentheses are the percentage of specific land use/cover change to the total area (%).

2000/ 2015	Cropland	Forest	Grassland	Construction Land	Unused Land
Cropland	0.0039 (13.649)	0.0033 (0.080)	0.0038 (0.229)	−0.0009 (0.106)	0.0039 (0.020)
Forest	0.0047 (0.036)	0.0031 (9.461)	0.0039 (0.059)	0.0039 (0.013)	0.0030 (0.014)
Grassland	0.0057 (0.364)	0.0037 (0.151)	0.0025 (60.929)	0.0010 (0.238)	0.0026 (0.423)
Construction land			0.0045 (0.006)	0.0032 (1.381)	
Unused land	0.0039 (0.048)	0.0049 (0.038)	0.0025 (0.474)	−0.0007 (0.047)	0.0025 (12.227)

4. Discussion

4.1. Applicability and Limitation of the GeoDetector Method

In relation to vegetation variations, numerous studies have explored the separation of natural and human factors. It should be noted that the commonly used methods (e.g., RESTREND, statistical correlation, or regression analysis) suffer from potential limitations. Specifically, RESTREND analysis cannot differentiate anthropogenic impacts from different aspects of human activities [14,18]. Statistical methods of evaluating the factors influencing vegetation changes mainly include correlation analysis [48], regression analysis [27], factor analysis [66], and geographically weighted regression [18]. However, these statistical methods involve assumptions regarding the data, fail to reveal the interactions between factors, or are hindered by the multicollinearities among the influencing factors [67]. The GeoDetector method was employed in this work, and it has three distinct advantages. (1) The GeoDetector method is not based on linear hypotheses, thus, it provides easier data preparation and wider applicability. (2) Working with both categorical and continuous variables, the GeoDetector method is not limited by data type. (3) The GeoDetector method can be used to determine how the interaction of explanatory variables affects the dependent variables without the restriction of multicollinearities [37,39,68,69]. Our study demonstrates that the GeoDetector method is an efficient technique for quantifying the impacts of driving factors and their interactions on vegetation changes.

Different grading standards (involving the discretization method and the stratification number) have certain impacts on the GeoDetector results [38,62,68]. However, the selection of the discretization methods and stratification numbers in prior model applications were subject to weaknesses such as randomness and subjectivity [39,59,64,70,71], which may introduce uncertainties and lead to misleading interpretations. In this study, an optimal discretization method was obtained based on the five types of unsupervised discretization methods (Figure 5). In addition, on the basis of a changing curve of the degree of influence (of influencing factors) with different numbers of stratifications, an optimal stratification number was also determined for each predictor variable (Figure 7). The optimization of factor-grading improves the accuracy and effectiveness of the modeling [38,63,72].

In this study, the socio-economic data are obtained from the statistical yearbooks at the county scale. It should be noted that the lack of spatial information of these socio-economic indicators forces them to be uniformly distributed within administrative divisions. This involves the spatial scale effect, which may have critical influences on the spatially stratified heterogeneity analysis. However, it has not been fully investigated and integrated in the GeoDetector method [38].

4.2. Effects of Factors on Vegetation Changes

4.2.1. Effects of the Main Natural Drivers

Our results indicate that precipitation was the dominant factor influencing the changes in the NDVI. This finding is consistent with the results of prior studies, which have indicated that vegetation growth in dryland ecosystems is very sensitive to precipitation changes [15,23,41,48]. As shown in Figure 9a, as the precipitation increased, the increase in the NDVI initially kept rising and reached a peak, and then it decelerated. A possible explanation for this is that the long-term cloud cover due to the excess precipitation may have resulted in reductions in temperature and solar radiation [48], which are not conducive to the improvement of the productivity of the grassland vegetation. In a similar vein, prior research reported that as precipitation increases, there is a threshold for the response of vegetation NDVI to precipitation, beyond which the magnitude of increase in NDVI driven by precipitation will decrease [73].

The rate of increase of the NDVI decreased substantially as the wind speed increased (Figure 9d) for two main reasons. First, the high wind speed increased evaporation and decreased the surface moisture, resulting in adverse effects on vegetation growth. Second, the Inner Mongolian grasslands are located in the sandstorm source region of northern China [46], with frequent strong winds in spring. In aeolian desertification areas, vegetation growth has been found to be constrained by burial and abrasion, the loss of surface soil resources, and the interruption of nutrient accumulation [74]. Zou and Zhai reported that vegetation coverage, as indicated by NDVI, was significantly negatively correlated with the occurrence frequency of spring dust storms in Inner Mongolia [75].

Overall, the magnitude of the increase in the NDVI values on the east-facing slopes was larger than that on the west-facing slopes (Figure 9h). One possible reason was that the study area is located in a marginal zone of the East Asian summer monsoon. Compared with the east-facing slopes, the west-facing slopes receive less precipitation and more solar radiation and thus are characterized by drier and hotter microclimates, which are harsher environments for vegetation growth.

The areas with semi-luvisols showed significant increases in the NDVI, while the desert soil had little effect on the increase in the NDVI (Figure 9i). Luvisol soils form under temperate forest and grassland vegetation. Due to the high accumulation of dead forest leaves and herbaceous debris and the cool climate, microbial decomposition is limited to a certain extent, leading to the formation of a humus layer with high fertility that can serve as important agricultural and forest soil resources. Desert soils, which develop under the temperate desert grassland vegetation in the northwestern marginal areas of the study region (Figure 4i), have little humus accumulation, a low organic matter content, and harsh environments (e.g., low precipitation, strong winds, and solar radiation), all of which are unfavorable for vegetation growth.

4.2.2. Effects of Human Activities

Livestock grazing is the main form of grassland utilization in Inner Mongolia [23,43]. Our results show that livestock density was the most influential anthropogenic factor affecting NDVI changes. Prior studies in arid Inner Mongolia were predicated on the belief that overgrazing substantially decreases vegetation cover and biomass production [76–78]. However, in this study, with an increase in livestock density, the magnitude of NDVI increase generally rose (Figure 9n), apparently suggesting, paradoxically, that grazing intensity, as indicated by livestock density, improved grassland vegetation. One possible reason for this is that the Inner Mongolia government has actively promoted the policy of herdsmen settlements and livestock pen-raising in recent decades [44]. According to the statistical yearbook of Inner Mongolia, in the past 19 years, the total area of livestock sheds has significantly increased at a rate of 9.31 million square meters per year ($p < 0.05$) (Figure 10), and the number of livestock in captivity has significantly increased at a rate of 5.37 million sheep units per year ($p < 0.05$) (Figure 10). During the same period, the total number of livestock in Inner Mongolia has increased at a rate of 2.07 million sheep units per year, and the growth rate of livestock in captivity is much higher than that of the total amount of livestock. The intensive mode of livestock production accounts for the increasing proportion of animal husbandry production in pastoral areas. Due to the strong implementation of an ecological restoration policy and this intensive livestock production mode, forage sources are more dependent on external imports, and less damage is caused to the local vegetation. In addition, the livestock density is inherently high in areas with high vegetation coverage. Hence, with the increase in livestock density, grassland vegetation conditions still improved.

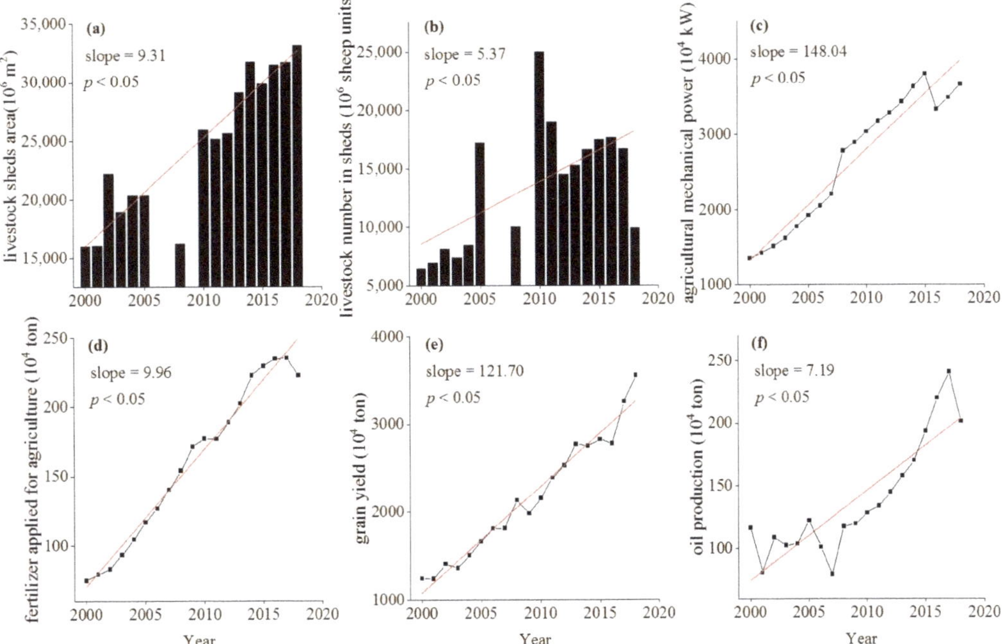

Figure 10. Inter-annual variations in (**a**) livestock shed area, (**b**) livestock numbers in sheds, (**c**) the total agricultural mechanical power, (**d**) the amount of fertilizer applied for agriculture, (**e**) the grain yield, and (**f**) the oil production in Inner Mongolia from 2000–2018. Note: The data were obtained from the Inner Mongolia Statistical Yearbook [79]. In (**a,b**), the data for livestock shed area and livestock numbers in sheds for 2006, 2007, and 2009 are not recorded in the Inner Mongolia Statistical Yearbook.

Although the GeoDetector method excludes the influence of multcollinearity among independent variables [37,39], the results show that the q-statistics and the ranking order of the livestock density, population density, and per capita gross regional product are very close (Supplementary Figure S3). This implied that the relationships between the human factors are closer than those between the natural factors. Actually, the interference of the anthropogenic factors in Inner Mongolia is not as complicated as those in developed areas. Grazing is the primary human activity affecting the ecological environment in the study area, and animal husbandry is the main source of income for local herders [41,44]. Where there are large numbers of livestock, there is often a high population density and social productivity (Figure 4j–n).

Land use/cover change is a manifestation of human activity [52]. Most of the types of land use conversion had positive influences on vegetation change in this study (Table 3). With the technological improvement and update of the industrial structures in agricultural sectors, a 0.0039 a^{-1} increase in the NDVI was observed in the unchanged cropland, which is corroborated by the fact that from 2000 to 2018, the agricultural mechanical power and fertilizer used for agriculture in Inner Mongolia increased by 2.71 and 2.98 times, respectively; the grain production increased from 12.42×10^6 t to 35.53×10^6 t, and the oil production increased from 1.16×10^6 t to 2.02×10^6 t [79] (Figure 10). The conversions of grassland and unused land into cropland increased the NDVI, indicating that reasonable reclamation has a positive effect on vegetation recovery. Through a series of ecological restoration measures, such as grazing prohibitions with grassland closures, restoration of cropland to grassland, and reforestation with hillside closures, the NDVI values of the unchanged grassland and forest land increased at rates of 0.0025 a^{-1} and 0.0031 a^{-1}, respectively. For the unused land that was converted into grassland and forest, the NDVI values increased at rates of 0.0025 a^{-1} and 0.0049 a^{-1}, respectively. For the cropland that was converted into grassland and forest, the NDVI values increased at rates of 0.0038 a^{-1} and 0.0033 a^{-1}, respectively. Prior research also reported that the implementation of ecological restoration programs was beneficial to the improvement of vegetation coverage in Inner Mongolia [44,80]. However, urban expansion has caused decreases in the NDVI of 0.0009 a^{-1} due to the conversion of cropland into construction land and 0.0007 a^{-1} due to the conversion of unused land into construction land (Table 3). Therefore, more attention should be paid to ensuring the development of green infrastructures, such as parks and green spaces, during urban expansion.

The interactions among factors can greatly enhance the effect of a single factor (Figure 8). Although the distance to the nearest road, slope, and aspect did not contribute ideally to NDVI changes, their explanatory powers were enhanced when they interacted with other factors, especially precipitation and livestock density. Natural factors such as soil type (q(Soilt∩Livstd)) > q(Livstd)), slope (q(Slopd∩Livstd)) > q(Livstd)), and aspect (q(Slopa∩Livstd)) > q(Livstd)) tended to enhance the influence of human activities on vegetation changes.

4.3. Limitations and Future Research Directions

This study had certain limitations which can be improved in future research. First, the spatial differentiation of the relationships between NDVI variations and the driving factors was not taken into consideration. For example, most of the degraded vegetation from 2000 to 2018 was in the central and western regions of the Inner Mongolian grasslands (Figure 7b), and the interactions between the precipitation and temperature had the greatest impact on vegetation changes. Thus, with the high surface evaporation potential and low soil moisture due to a relatively small increase in precipitation and large increases in temperature in the central and western regions (Figure 11), the increase in the water stress level was probably the main reason for vegetation degradation. In further studies, the introduction of spatial statistical methods (e.g., a geographically weighted regression model), which can reflect the spatial nonstationarity of the parameters in different spaces, may improve our understanding of the spatial heterogeneity of the relationship between vegetation change and its driving forces [18,66]. Secondly, the drylands in northern China are regions with diverse land uses (mainly deserts and grasslands) and substantial seasonal climatic differences [81]. More evidence showed that soil moisture, which exhibits significant spatial and temporal variability [82], is crucial in regulating vegetation productivity. In future studies, a multiple time and spatial scale analysis can contribute to a better understanding of the drivers of vegetation growth change in order to develop suitable management schemes that are regionally and temporally specific. Thirdly, prior research observed the NDVI asymptotically saturating in high biomass regions [83]. Regarding this issue, the EVI (enhanced vegetation index) and SAVI (soil-adjusted vegetation index) were developed to make up for some of the shortcomings of the NDVI (e.g., atmospheric noise, soil background, saturation). Due to the limited spatial resolution of MODIS NDVI, it is difficult to meet the requirements of fine mapping. Combining the process with the Sentinel dataset or other vegetation indexes (e.g., EVI, SAVI) may help to obtain more precise estimates of vegetation dynamics. Last but not least, if breakpoints, which indicate a shift in the mechanism of influence on the time series under certain circumstances, are neglected, the results of the trend analysis may lead to a misjudging of the factors that influence vegetation changes [40]. In future studies, the times at which breakpoints occurred should be first identified, noting points at which the time series was split into sub-series. Then, the trends and significance levels of the sub-series would be quantified separately to obtain more accurate conclusions regarding the driving forces of vegetation changes.

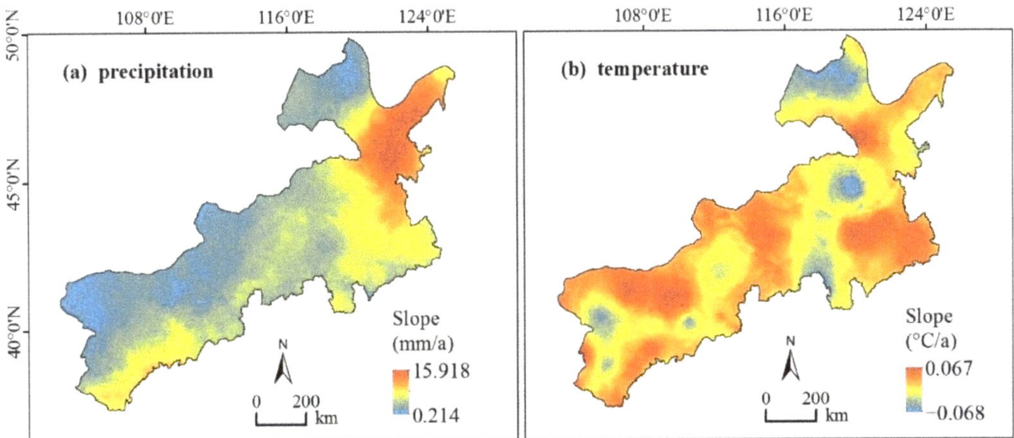

Figure 11. The slopes of changes in (**a**) precipitation and (**b**) temperature from 2000–2018.

5. Conclusions

In this research, we investigated the spatial and temporal variability in the mean growing season NDVI from 2000–2018 and quantified the individual and interactive influences of natural and human factors on NDVI change using the GeoDetector method in the Inner Mongolian grasslands. The results reveal that the NDVI increased at a rate of 0.003 a^{-1}. Both the natural and human factors had significant impacts on vegetation NDVI variations ($p < 0.05$), and the precipitation, livestock density, and wind speed had the greatest influences, while terrain slope and slope aspect had the lowest influences. The interactive impacts among factors often strengthened the impact of single factors.

Our study demonstrates that the GeoDetector method is an effective technique for disentangling the complicated driving factors of vegetation change. To effectively use the GeoDetector method, however, researchers need to carefully deal with the problem of spatial data discretization, which may introduce uncertainties and lead to misleading interpretations. The methodology used in this study can be applied to address the knowledge gap in the selection of the optimal discretization methods and the number of stratifications for further GeoDetector-based studies.

Supplementary Materials: The following supporting information can be downloaded at: https://www.mdpi.com/article/10.3390/rs14143320/s1, Figure S1: Diagram illustrating the detection of interaction; Figure S2: Inter-annual variations in the mean growing season NDVI in the Inner Mongolian grassland from 2000–2018; Figure S3: The q-statistics of impact factors for NDVI changes.

Author Contributions: Conceptualization, S.L. and X.L. (Xiaobing Li); methodology, S.L., D.D. and X.L. (Xin Lyu); data curation, S.L., H.D. and X.L. (Xin Lyu); formal analysis, S.L., D.D. and H.D.; resources, X.L.; writing—original draft preparation, S.L., D.D. and X.L. (Xiaobing Li); writing—review and editing, S.L. and X.L. (Xiaobing Li); visualization, S.L. and H.D.; supervision, X.L. (Xiaobing Li) and J.G.; project administration, X.L. (Xiaobing Li) and J.G.; funding acquisition, X.L (Xiaobing Li) and J.G. All authors have read and agreed to the published version of the manuscript.

Funding: This research was funded by the Key Science & Technology Special Program of Inner Mongolia (No. 2021ZD0011 and 2021ZD0015) and the Project Supported by State Key Laboratory of Earth Surface Processes and Resource Ecology (No. 2022-ZD-02).

Data Availability Statement: All data that support the findings of the study are available from the corresponding author upon reasonable request.

Acknowledgments: We would like to offer our sincere thanks to those who participated in the data processing and manuscript revisions.

Conflicts of Interest: The authors declare no conflict of interest.

References

1. Reynolds, J.F.; Stafford, D.M.; Lambin, E.F.; Turner, B.L., II; Mortimore, M.; Batterbury, S.P.J.; Downing, T.E.; Dowlatabadi, H.; Fernández, R.J.; Herrick, J.E.; et al. Global desertification: Building a science for dryland development. *Science* **2007**, *316*, 847–851. [CrossRef] [PubMed]
2. Berdugo, M.; Delgado-Baquerizo, M.; Soliveres, S.; Hernández-Clemente, R.; Zhao, Y.; Gaitán, J.J.; Gross, N.; Saiz, H.; Maire, V.; Lehmann, A.; et al. Global ecosystem thresholds driven by aridity. *Science* **2020**, *367*, 787–790. [CrossRef]
3. Maestre, F.T.; Eldridge, D.J.; Soliveres, S.; Kéfi, S.; Delgado-Baquerizo, M.; Bowker, M.A.; García-Palacios, P.; Gaitán, J.; Gallardo, A.; Lázaro, R.; et al. Structure and functioning of dryland ecosystems in a changing world. *Annu. Rev. Ecol. Evo. Syst.* **2016**, *47*, 215–237. [CrossRef]
4. UN. Transforming Our World: The 2030 Agenda for Sustainable Development. Available online: https://sustainabledevelopment.un.org/post2015/transformingourworld/publication (accessed on 15 March 2022).
5. Piao, S.; Fang, J. Seasonal changes in vegetation activity in response to climate changes in China between 1982 and 1999. *Acta Geogr. Sin.* **2003**, *58*, 119–125.
6. Sun, W.; Song, X.; Mu, X.; Gao, P.; Wang, F.; Zhao, G. Spatiotemporal vegetation cover variations associated with climate change and ecological restoration in the Loess Plateau. *Agr. For. Meteorol.* **2015**, *209–210*, 87–99. [CrossRef]
7. Fu, B.; Wang, S.; Liu, Y.; Liu, J.; Liang, W.; Miao, C. Hydrogeomorphic ecosystem responses to natural and anthropogenic changes in the Loess Plateau of China. *Annu. Rev. Earth Planet. Sci.* **2017**, *45*, 223–243. [CrossRef]
8. Easdale, M.H.; Bruzzone, O.; Mapfumo, P.; Tittonell, P. Phases or regimes? Revisiting NDVI trends as proxies for land degradation. *Land Degrad. Dev.* **2018**, *29*, 433–445. [CrossRef]
9. Leroux, L.; Bégué, A.; Seen, D.L.; Jolivot, A.; Kayitakire, F. Driving forces of recent vegetation changes in the Sahel: Lessons learned from regional and local level analyses. *Remote Sens. Environ.* **2017**, *191*, 38–54. [CrossRef]
10. Li, H.; Yang, X. Temperate dryland vegetation changes under a warming climate and strong human intervention—With a particular reference to the district Xilin Gol, Inner Mongolia, China. *Catena* **2014**, *119*, 9–20. [CrossRef]
11. Wan, H.; Bai, Y.; Hooper, D.U.; Schönbach, P.; Gierus, M.; Schiborra, A.; Taube, F. Selective grazing and seasonal precipitation play key roles in shaping plant community structure of semi-arid grasslands. *Landsc. Ecol.* **2015**, *30*, 1767–1782. [CrossRef]
12. Piao, S.; Wang, X.; Park, T.; Chen, C.; Lian, X.U.; He, Y.; Bjerke, J.W.; Chen, A.; Ciais, P.; Tømmervik, H.; et al. Characteristics, drivers and feedbacks of global greening. *Nat. Rev. Earth Environ.* **2020**, *1*, 14–27. [CrossRef]
13. Brandt, M.; Rasmussen, K.; Peñuelas, J.; Tian, F.; Schurgers, G.; Verger, A.; Mertz, O.; Palmer, J.R.; Fensholt, R. Human population growth offsets climate-driven increase in woody vegetation in sub-Saharan Africa. *Nat. Ecol. Evol.* **2017**, *1*, 0081. [CrossRef] [PubMed]
14. Liu, H.; Jiao, F.; Yin, J.; Li, T.; Gong, H.; Wang, Z.; Lin, Z. Nonlinear relationship of vegetation greening with nature and human factors and its forecast–a case study of Southwest China. *Ecol. Indic.* **2020**, *111*, 106009. [CrossRef]
15. Zhao, W.; Hu, Z.; Guo, Q.; Wu, G.; Chen, R.; Li, S. Contributions of climatic factors to interannual variability of the vegetation index in Northern China Grasslands. *J. Clim.* **2020**, *33*, 175–183. [CrossRef]
16. Motesharrei, S.; Rivas, J.; Kalnay, E.; Asrar, G.R.; Busalacchi, A.J.; Cahalan, R.F.; Cane, M.A.; Colwell, R.R.; Feng, K.; Franklin, R.S.; et al. Modeling sustainability: Population, inequality, consumption, and bidirectional coupling of the Earth and Human Systems. *Natl. Sci. Rev.* **2016**, *3*, 470–494. [CrossRef] [PubMed]
17. Chen, C.; Park, T.; Wang, X.; Piao, S.; Xu, B.; Chaturvedi, R.K.; Fuchs, R.; Brovkin, V.; Ciais, P.; Fensholt, R.; et al. China and India lead in greening of the world through land-use management. *Nat. Sustain.* **2019**, *2*, 122–129. [CrossRef]
18. Zhang, D.; Jia, Q.; Xu, X.; Yao, S.; Chen, H.; Hou, X. Contribution of ecological policies to vegetation restoration: A case study from Wuqi County in Shaanxi Province, China. *Land Use Policy* **2018**, *73*, 400–411. [CrossRef]
19. Yang, L.; Shen, F.; Zhang, L.; Cai, Y.; Yi, F.; Zhou, C. Quantifying influences of natural and anthropogenic factors on vegetation changes using structural equation modeling: A case study in Jiangsu Province, China. *J. Clean. Prod.* **2021**, *280*, 124330. [CrossRef]
20. Wang, Z.; Deng, X.; Song, W.; Li, Z.; Chen, J. What is the main cause of grassland degradation? A case study of grassland ecosystem service in the middle-south Inner Mongolia. *Catena* **2017**, *150*, 100–107. [CrossRef]
21. Verdoodt, A.; Mureithi, S.M.; Ye, L.; Van Ranst, E. Chronosequence analysis of two enclosure management strategies in degraded rangeland of semi-arid Kenya. *Agr. Ecosyst. Environ.* **2009**, *129*, 332–339. [CrossRef]

22. Evans, J.; Geerken, R. Discrimination between climate and human-induced dryland degradation. *J. Arid Environ.* **2004**, *57*, 535–554. [CrossRef]
23. Li, A.; Wu, J.; Huang, J. Distinguishing between human-induced and climate-driven vegetation changes: A critical application of RESTREND in inner Mongolia. *Landsc. Ecol.* **2012**, *27*, 969–982. [CrossRef]
24. Zhou, D.; Zhao, X.; Hu, H.; Shen, H.; Fang, J. Long-term vegetation changes in the four mega-sandy lands in Inner Mongolia, China. *Landsc. Ecol.* **2015**, *30*, 1613–1626. [CrossRef]
25. Wessels, K.J.; Van Den Bergh, F.; Scholes, R.J. Limits to detectability of land degradation by trend analysis of vegetation index data. *Remote Sens. Environ.* **2012**, *125*, 10–22. [CrossRef]
26. Cao, X.; Gu, Z.; Chen, J.; Liu, J.; Shi, P. Analysis of human_induced steppe degradation based on remote sensing in xilin gole, inner mongolia, China. *Chin. J. Plant Ecol.* **2006**, *30*, 268–277.
27. Xu, D.; Wang, Z. Identifying land restoration regions and their driving mechanisms in inner Mongolia, China from 1981 to 2010. *J. Arid Environ.* **2019**, *167*, 79–86. [CrossRef]
28. Piao, S.; Yin, G.; Tan, J.; Cheng, L.; Huang, M.; Li, Y.; Liu, R.; Mao, J.; Myneni, R.B.; Peng, S.; et al. Detection and attribution of vegetation greening trend in China over the last 30 years. *Glob. Chang. Biol.* **2015**, *21*, 1601–1609. [CrossRef]
29. Hunsicker, M.E.; Kappel, C.V.; Selkoe, K.A.; Halpern, B.S.; Scarborough, C.; Mease, L.; Amrhein, A. Characterizing driver–response relationships in marine pelagic ecosystems for improved ocean management. *Ecol. Appl.* **2016**, *26*, 651–663. [CrossRef] [PubMed]
30. Knapp, A.K.; Ciais, P.; Smith, M.D. Reconciling inconsistencies in precipitation–productivity relationships: Implications for climate change. *New Phytol.* **2017**, *214*, 41–47. [CrossRef]
31. Hickler, T.; Eklundh, L.; Seaquist, J.W.; Smith, B.; Ardö, J.; Olsson, L.; Sykes, M.T.; Sjöström, M. Precipitation controls Sahel greening trend. *Geophys. Res. Lett.* **2005**, *32*, L21415. [CrossRef]
32. Zhu, Z.; Piao, S.; Myneni, R.B.; Huang, M.; Zeng, Z.; Canadell, J.G.; Ciais, P.; Sitch, S.; Friedlingstein, P.; Arneth, A.; et al. Greening of the Earth and its drivers. *Nat. Clim. Chang.* **2016**, *6*, 791–795. [CrossRef]
33. Sitch, S.; Huntingford, C.; Gedney, N.; Levy, P.E.; Lomas, M.; Piao, S.L.; Betts, R.; Ciais, P.; Cox, P.; Friedlingstein, P.; et al. Evaluation of the terrestrial carbon cycle, future plant geography and climate-carbon cycle feedbacks using five Dynamic Global Vegetation Models (DGVMs). *Glob. Chang. Biol.* **2008**, *14*, 2015–2039. [CrossRef]
34. Wang, J.; Li, X.; Christakos, G.; Liao, Y.; Zhang, T.; Gu, X.; Zheng, X. Geographical detectors-based health risk assessment and its application in the neural tube defects study of the Heshun Region, China. *Int. J. Geogr. Inf. Sci.* **2010**, *24*, 107–127. [CrossRef]
35. Wang, J.; Zhang, T.; Fu, B. A measure of spatial stratified heterogeneity. *Ecol. Indic.* **2016**, *67*, 250–256. [CrossRef]
36. Luo, W.; Jasiewicz, J.; Stepinski, T.; Wang, J.; Xu, C.; Cang, X. Spatial association between dissection density and environmental factors over the entire conterminous United States. *Geophys. Res. Lett.* **2016**, *43*, 692–700. [CrossRef]
37. Shrestha, A.; Luo, W. Analysis of groundwater nitrate contamination in the Central Valley: Comparison of the geodetector method, principal component analysis and geographically weighted regression. *ISPRS Int. J. Geo. Inf.* **2017**, *6*, 297. [CrossRef]
38. Song, Y.; Wang, J.; Ge, Y.; Xu, C. An optimal parameters-based geographical detector model enhances geographic characteristics of explanatory variables for spatial heterogeneity analysis: Cases with different types of spatial data. *Gisci. Remote Sens.* **2020**, *57*, 593–610. [CrossRef]
39. Xu, L.; Du, H.; Zhang, X. Driving forces of carbon dioxide emissions in China's cities: An empirical analysis based on the geodetector method. *J. Clean. Prod.* **2021**, *287*, 125169. [CrossRef]
40. Kang, Y.; Guo, E.; Wang, Y.; Bao, Y.; Bao, Y.; Mandula, N. Monitoring vegetation change and its potential drivers in Inner Mongolia from 2000 to 2019. *Remote Sens.* **2021**, *13*, 3357. [CrossRef]
41. Batunacun, R.W.; Tobia, L.; Hu, Y.; Claas, N. Identifying drivers of land degradation in Xilingol, China, between 1975 and 2015. *Land Use Policy* **2019**, *83*, 543–559. [CrossRef]
42. Gao, S.; Shi, P.; Ha, S.; Pan, Y. Causes of rapid expansion of blown-sand disaster and long-term trend of desertification in northern China. *J. Nat. Disasters* **2000**, *9*, 31–37.
43. Liu, J.; Diamond, J. China's environment in a globalizing world. *Nature* **2005**, *435*, 1179–1186. [CrossRef] [PubMed]
44. Wu, J.; Zhang, Q.; Li, A.; Liang, C. Historical landscape dynamics of Inner Mongolia: Patterns, drivers, and impacts. *Landsc. Ecol.* **2015**, *30*, 1579–1598. [CrossRef]
45. Chen, H.; Shao, L.; Zhao, M.; Zhang, X.; Zhang, D. Grassland conservation programs, vegetation rehabilitation and spatial dependency in Inner Mongolia, China. *Land Use Policy* **2017**, *64*, 429–439. [CrossRef]
46. Jiang, C.; Nath, R.; Labzovskii, L.; Wang, D. Integrating ecosystem services into effectiveness assessment of ecological restoration program in northern China's arid areas: Insights from the Beijing-Tianjin Sandstorm Source Region. *Land Use Policy* **2018**, *75*, 201–214. [CrossRef]
47. Yin, H.; Pflugmacher, D.; Li, A.; Li, Z.; Hostert, P. Land use and land cover change in Inner Mongolia-understanding the effects of China's re-vegetation programs. *Remote Sens. Environ.* **2018**, *204*, 918–930. [CrossRef]
48. Bao, G.; Qin, Z.; Bao, Y.; Zhou, Y.; Li, W.; Sanjjav, A. NDVI-based long-term vegetation dynamics and its response to climatic change in the Mongolian Plateau. *Remote Sens.* **2014**, *6*, 8337–8358. [CrossRef]
49. Lu, Q.; Zhao, D.; Wu, S.; Dai, E.; Gao, J. Using the NDVI to analyze trends and stability of grassland vegetation cover in Inner Mongolia. *Theor. Appl. Climatol.* **2019**, *135*, 1629–1640. [CrossRef]
50. State Forestry Administration of China. *Annual Report of National Forestry (2000–2015)*; China Forestry Press: Beijing, China, 2015.

51. Zhu, W.; Pan, Y.; He, H.; Wang, L.; Mou, M.; Liu, J. A changing-weight filter method for reconstructing a high-quality NDVI time series to preserve the integrity of vegetation phenology. *IEEE Trans. Geosci. Remote* **2011**, *50*, 1085–1094. [CrossRef]

52. Liu, J.; Ning, J.; Kuang, W.; Xu, X.; Zhang, S.; Yan, C.; Li, R.; Wu, S.; Hu, Y.; Du, G.; et al. Spatiotemporal patterns and characteristics of land-use change in China during 2010–2015. *Acta Geogr. Sin.* **2018**, *73*, 789–802.

53. Ministry of Agriculture of the People's Republic of China. Agricultural Standards of the People's Republic of China: Calculation of Rangeland Carrying Capacity (NY/T 635-2015). Available online: http://www.jgj.moa.gov.cn/nybz/ (accessed on 19 March 2022).

54. Gocic, M.; Trajkovic, S. Analysis of changes in meteorological variables using Mann-Kendall and Sen's slope estimator statistical tests in Serbia. *Glob. Planet Chang.* **2013**, *100*, 172–182. [CrossRef]

55. Mann, H.B. Nonparametric test against trend. *Econometrica* **1945**, *13*, 245–259. [CrossRef]

56. Kendall, M.G. *Rank Correlation Methods*; Charles Griffin: London, UK, 1975.

57. Ju, H.; Zhang, Z.; Zuo, L.; Wang, J.; Zhang, S.; Wang, X.; Zhao, X. Driving forces and their interactions of built-up land expansion based on the geographical detector–a case study of Beijing, China. *Int. J. Geogr. Inf. Sci.* **2016**, *30*, 2188–2207. [CrossRef]

58. Wang, J.; Xu, C. Geodetector: Principle and prospective. *Acta Geogr. Sin.* **2017**, *72*, 116–134.

59. Zhu, L.; Meng, J.; Zhu, L. Applying Geodetector to disentangle the contributions of natural and anthropogenic factors to NDVI variations in the middle reaches of the Heihe River Basin. *Ecol. Indic.* **2020**, *117*, 106545. [CrossRef]

60. Han, J.; Huang, Y.; Zhang, H.; Wu, X. Characterization of elevation and land cover dependent trends of NDVI variations in the Hexi region, northwest China. *J. Environ. Manag.* **2019**, *232*, 1037–1048. [CrossRef]

61. Peng, W.; Kuang, T.; Tao, S. Quantifying influences of natural factors on vegetation NDVI changes based on geographical detector in Sichuan, western China. *J. Clean. Prod.* **2019**, *233*, 353–367. [CrossRef]

62. Cao, F.; Ge, Y.; Wang, J. Optimal discretization for geographical detectors-based risk assessment. *Gisci. Remote Sens.* **2013**, *50*, 78–92. [CrossRef]

63. Su, Y.; Li, T.; Cheng, S.; Wang, X. Spatial distribution exploration and driving factor identification for soil salinisation based on geodetector models in coastal area. *Ecol. Eng.* **2020**, *156*, 105961. [CrossRef]

64. Han, J.; Wang, J.; Chen, L.; Xiang, J.; Ling, Z.; Li, Q.; Wang, E. Driving factors of desertification in Qaidam Basin, China: An 18-year analysis using the geographic detector model. *Ecol. Indic.* **2021**, *124*, 107404. [CrossRef]

65. Ran, Q.; Hao, Y.; Xia, A.; Liu, W.; Hu, R.; Cui, X.; Xue, K.; Song, X.; Xu, C.; Ding, B.; et al. Quantitative assessment of the impact of physical and anthropogenic factors on vegetation spatial-temporal variation in Northern Tibet. *Remote Sens.* **2019**, *11*, 1183. [CrossRef]

66. Zhang, X.; Yue, Y.; Tong, X.; Wang, K.; Qi, X.; Deng, C.; Brandt, M. Eco-engineering controls vegetation trends in southwest China karst. *Sci. Total Environ.* **2021**, *770*, 145160. [CrossRef] [PubMed]

67. Wheeler, D.; Tiefelsdorf, M. Multicollinearity and correlation among local regression coefficients in geographically weighted regression. *J. Geogr. Syst.* **2005**, *7*, 161–187. [CrossRef]

68. Luo, L.; Mei, K.; Qu, L.; Zhang, C.; Chen, H.; Wang, S.; Di, D.; Huang, H.; Wang, Z.; Xia, F.; et al. Assessment of the Geographical Detector Method for investigating heavy metal source apportionment in an urban watershed of Eastern China. *Sci. Total Environ.* **2019**, *653*, 714–722. [CrossRef] [PubMed]

69. Wang, J.; Hu, Y. Environmental health risk detection with GeogDetector. *Environ. Modell. Softw.* **2012**, *33*, 114–115. [CrossRef]

70. Fan, Y.; Gan, L.; Hong, C.; Jessup, L.H.; Jin, X.; Pijanowski, B.C.; Sun, Y.; Lv, L. Spatial identification and determinants of trade-offs among multiple land use functions in Jiangsu Province, China. *Sci. Total Environ.* **2021**, *772*, 145022. [CrossRef]

71. Wang, Z.; Xu, D.; Peng, D.; Zhang, Y. Quantifying the influences of natural and human factors on the water footprint of afforestation in desert regions of northern China. *Sci. Total Environ.* **2021**, *780*, 146577. [CrossRef]

72. Qiao, P.; Yang, S.; Lei, M.; Chen, T.; Dong, N. Quantitative analysis of the factors influencing spatial distribution of soil heavy metals based on geographical detector. *Sci. Total Environ.* **2019**, *664*, 392–413. [CrossRef]

73. Ukkola, A.M.; Prentice, I.C.; Keenan, T.F.; Van Dijk, A.I.; Viney, N.R.; Myneni, R.B.; Bi, J. Reduced streamflow in water-stressed climates consistent with CO_2 effects on vegetation. *Nat. Clim. Chang.* **2016**, *6*, 75–78. [CrossRef]

74. Okin, G.S.; Murray, B.; Schlesinger, W.H. Degradation of sandy arid shrubland environments: Observations, process modelling, and management implications. *J. Arid Environ.* **2001**, *47*, 123–144. [CrossRef]

75. Zou, X.; Zhai, P. Relationship between vegetation coverage and spring dust storms over northern China. *J. Geophys. Res. Atmos.* **2004**, *109*, D03104. [CrossRef]

76. Akiyama, T.; Kawamura, K. Grassland degradation in China: Methods of monitoring, management and restoration. *Grassl. Sci.* **2007**, *53*, 1–17. [CrossRef]

77. Lin, Y.; Han, G.; Zhao, M.; Chang, S.X. Spatial vegetation patterns as early signs of desertification: A case study of a desert steppe in Inner Mongolia, China. *Landsc. Ecol.* **2010**, *25*, 1519–1527. [CrossRef]

78. Zhao, H.; Zhao, X.; Zhou, R.; Zhang, T.; Drake, S. Desertification processes due to heavy grazing in sandy rangeland, Inner Mongolia. *J. Arid Environ.* **2005**, *62*, 309–319. [CrossRef]

79. Inner Mongolia Bureau of Statistics. Inner Mongolia Statistical Yearbook, 2001–2019. Available online: https://data.cnki.net/Area/Home/Index/D05 (accessed on 10 April 2022).

80. Mu, S.; Zhou, S.; Chen, Y.; Li, J.; Ju, W.; Odeh, I. Assessing the impact of restoration-induced land conversion and management alternatives on net primary productivity in Inner Mongolian grassland, China. *Glob. Planet. Chang.* **2013**, *108*, 29–41. [CrossRef]

81. Li, J.; Liu, Z.; He, C.; Tu, W.; Sun, Z. Are the drylands in northern China sustainable? A perspective from ecological footprint dynamics from 1990 to 2010. *Sci. Total Environ.* **2016**, *553*, 223–231. [CrossRef]
82. Srivastava, A.; Saco, P.M.; Rodriguez, J.F.; Kumari, N.; Chun, K.P.; Yetemen, O. The role of landscape morphology on soil moisture variability in semi-arid ecosystems. *Hydrol. Processes* **2021**, *35*, e13990. [CrossRef]
83. Huete, A.; Didan, K.; Miura, T.; Rodriguez, E.P.; Gao, X.; Ferreira, L.G. Overview of the radiometric and biophysical performance of the MODIS vegetation indices. *Remote Sens. Environ.* **2002**, *83*, 195–213. [CrossRef]

remote sensing

Article

The Response of Vegetation to Regional Climate Change on the Tibetan Plateau Based on Remote Sensing Products and the Dynamic Global Vegetation Model

Mingshan Deng [1,2,3], Xianhong Meng [1,2,*], Yaqiong Lu [4], Zhaoguo Li [1,2], Lin Zhao [1,2], Hanlin Niu [1], Hao Chen [1,2], Lunyu Shang [1,2], Shaoying Wang [1,2] and Danrui Sheng [1,2,3]

[1] Key Laboratory of Land Surface Process and Climate Change in Cold and Arid Regions, Northwest Institute of Eco-Environment and Resources, Chinese Academy of Sciences, Lanzhou 730000, China; dengmingshan@lzb.ac.cn (M.D.); zgli@lzb.ac.cn (Z.L.); zhaolin_110@lzb.ac.cn (L.Z.); niuhanlin@lzb.ac.cn (H.N.); chenhao@lab.ac.cn (H.C.); sly@lzb.ac.cn (L.S.); wangshaoying@lzb.ac.cn (S.W.); shengdanrui@nieer.ac.cn (D.S.)
[2] Zoige Plateau Wetlands Ecosystem Research Station, Northwest Institute of Eco-Environment and Resources, Chinese Academy of Sciences, Lanzhou 730000, China
[3] University of Chinese Academy of Sciences, Beijing 100049, China
[4] Institute of Mountain Hazards and Environment, Chinese Academy of Sciences, Chengdu 610041, China; yaqiong@imde.ac.cn
[*] Correspondence: mxh@lzb.ac.cn

Citation: Deng, M.; Meng, X.; Lu, Y.; Li, Z.; Zhao, L.; Niu, H.; Chen, H.; Shang, L.; Wang, S.; Sheng, D. The Response of Vegetation to Regional Climate Change on the Tibetan Plateau Based on Remote Sensing Products and the Dynamic Global Vegetation Model. *Remote Sens.* **2022**, *14*, 3337. https://doi.org/10.3390/rs14143337

Academic Editors: Tinghai Ou, Wenxin Zhang, Youhua Ran and Xuejia Wang

Received: 30 May 2022
Accepted: 30 June 2022
Published: 11 July 2022

Publisher's Note: MDPI stays neutral with regard to jurisdictional claims in published maps and institutional affiliations.

Abstract: Changes in vegetation dynamics play a critical role in terrestrial ecosystems and environments. Remote sensing products and dynamic global vegetation models (DGVMs) are useful for studying vegetation dynamics. In this study, we revised the Community Land Surface Biogeochemical Dynamic Vegetation Model (referred to as the BGCDV_CTL experiment) and validated it for the Tibetan Plateau (TP) by comparing vegetation distribution and carbon flux simulations against observations. Then, seasonal–deciduous phenology parameterization was adopted according to the observed parameters (referred to as the BGCDV_NEW experiment). Compared to the observed parameters, monthly variations in gross primary productivity (GPP) showed that the BGCDV_NEW experiment had the best performance against the in situ observations on the TP. The climatology from the remote sensing and simulated GPPs showed similar patterns, with GPP increasing from northwest to southeast, although the BGCDV_NEW experiment overestimated GPP in the semi-arid and arid regions of the TP. The results show that temperature warming was the dominant factor resulting in the increase in GPP based on the remote sensing products, while precipitation enhancement was the reason for the GPP increase in the model simulation.

Keywords: dynamic vegetation; gross primary productivity; regional climate change; remote sensing products; Tibetan Plateau

1. Introduction

Terrestrial ecosystems have absorbed 25–30% of anthropogenic carbon dioxide emissions in the past five decades [1]. Vegetation is an essential part of terrestrial ecosystems and plays a crucial role in regulating regional climates and maintaining the surface energy balance [2,3]. Gross primary productivity (GPP) refers to the total amount of organic carbon fixed by plants in the ecosystem by absorbing solar energy and assimilating carbon dioxide through photosynthesis per unit time. GPP determines the total amount of initial energy and material entering terrestrial ecosystems. It reflects the vegetation productivity of terrestrial ecosystems, which profoundly influences the carbon cycle and global climate change.

The Tibetan Plateau (TP), with an average elevation of more than 4000 m, is often called the "third pole" [4]. Previous studies showed that 49% of streamflow in the Yellow

River, 15% of streamflow in the Lancang–Mekong River, and a considerable amount of streamflow in the Yangtze River originate from the TP [5], and the TP is often regarded as the "Asian water tower". With a large vertical height difference, rich landform types, diverse climatic environments, and complex ecosystem environments, the TP is a vital ecological barrier for China. Due to its unique geographical location, the TP is also a sensitive area of East Asia [6–8]. In particular, the TP shows a significant warming trend and increased precipitation [9,10]. The climate conditions on the TP are extremely variable, and its ecosystem pattern can easily be altered by external disturbances [11].

Land surface observation systems, mainly based on field observation and satellite remote sensing products, are critical for studying changes in land surface variables. With the development of global satellite remote sensing technology, vegetation index products retrieved by satellite remote sensing, such as the leaf area index (LAI), the normalized difference vegetation index (NDVI), and gross primary productivity (GPP), have provided effective support for studying global large-scale vegetation changes. However, due to the high elevation and harsh natural environment, there are few observation sites at high elevation on the TP, especially soil moisture and carbon flux observation sites [12,13]. Affected by high elevation, cloud cover, and auxiliary data, it is doubtful whether remote sensing products can provide a long-term stable vegetation index. Previous studies that included observations, remote sensing products, and simulations have been carried out for carbon flux and its responses to climate change on the TP and China [4,11,14–18]. Empirical models, remote sensing products, and numerical simulations are effective ways to estimate regional and global vegetation distributions. The next generation of land surface models includes the traditional hydrothermal transfer process, biogeochemical processes, and dynamic vegetation [19].

Dynamic global vegetation models (DGVMs) have been coupled with climate models to study the response of vegetation to climate change. In contrast to models of the fixed-satellite vegetation type, some vegetation characteristic parameters in DGVMs are calculated based on temperature, precipitation, and radiation; thus, the vegetation coverage simulated by DGVMs varies with climate change. DGVMs include: (1) simple biogeography rules to delineate the vegetation types based on climate; (2) carbon and nitrogen cycle modules; (3) vegetation dynamics modules. Current DGVMs include the Lund–Potsdam–Jena Dynamic Global Vegetation Model (LPJ DGVM), the Integrated Biosphere Simulator Model (IBIS), the BIOME4 model, the CLM-DGVM, and the Institute of Atmospheric Physics Dynamic Global Vegetation Model (IAP-DGVM). The LPJ model generally describes the temporal and spatial characteristics of regional vegetation, and it has effectively simulated the responses of vegetation patterns and functions to climate change [20]. IAP-DGVM simulations showed that the model could effectively reproduce the global distribution of shrubs, trees, grass, and bare soil, but it overestimated the fraction of temperate forests [21]. Ni et al. [4] used a modified BIOME DGVM to simulate the biome distribution on the TP; vegetation maps confirmed that the modified model did a better job of simulating biome patterns on the TP (grid cell agreement 52%) than the original BIOME4 model (35%). Song et al. [22] evaluated the CLM3.0-DGVM model, showing that the CLM3.0-DGVM predicted a relatively high population density with small individual trees in boreal forests. DGVMs perform differently in simulating vegetation distribution, and the models still have certain defects that cause deviations in simulations.

Vegetation phenology considers the influence of climate and other environmental factors and shows the time pattern of vegetation seasonal growth and decadency. Murray-Tortarolo et al. [23] validated leaf area index (LAI) estimates from multiple land surface models (LSMs) and found that all the LSMs overestimated the mean LAI and the length of the vegetation growing season, primarily due to late dormancy as a result of late summer phenology. A more complete vegetation phenology parameterization in DGVMs is significant for simulating the vegetation distribution and carbon fluxes accurately.

With their development and improvement, vegetation models have been added to Earth system models to study future changes in the global climate and terrestrial ecosystems. Dynamic vegetation models play a major role in simulating carbon, nitrogen, and water cycle processes in the ecosystem. Improving their accuracy is one of the focuses of vegetation model developers. Due to the high altitude and unique natural environment, DGVMs still encounter great uncertainty in simulating the vegetation distribution on the TP. In this study, we validated the performance of vegetation distribution and ecosystem productivity simulations on the TP by using the CLM5.0-DGVM. The CLM5.0-DGVM runs the CLM biogeochemistry model in combination with the LPJ-derived DGVM introduced in CLM3.0. The accuracy of soil moisture and temperature simulations in the model is significant for improving the ability to simulate ecosystem productivity. Following the work of Deng et al. [24], in which the soil moisture and temperature simulations in CLM5.0 were improved for the TP, the revised soil water and heat transfer schemes were used in all the simulations in this paper. Firstly, we modified the original code in the establishment and survival module, replaced the default vegetation parameters with observations (referred to as the BGCDV_CTL experiment), and evaluated the performance of the BGCDV_CTL experiment. In addition, we modified the vegetation phenology parameterization in the BGCDV_CTL experiment to conduct a comprehensive sensitivity analysis for the modified dynamic vegetation model. Secondly, we evaluated the vegetation productivity and vegetation distribution results from both remote sensing and simulations. Lastly, based on the improved model and the remote sensing GPP products, we investigated vegetation, climate change, and the response of the vegetation to regional climate change on the TP. Following this introduction Section 1, the data and model are briefly described in Section 2. An evaluation of the CLM5.0-DGVM and the GPP remote sensing products is conducted in Section 3. Section 4 analyzes the response of the vegetation to regional climate change on the TP. The conclusions are included in Section 5.

2. Materials and Methods

2.1. Study Area

In this study, we mainly focused on an area 25–40°N and 75–105°E (Figure 1) that covers the whole TP. The TP covers an area of approximately 2.5×10^6 km^2; with more than half of its area over 4000 m above sea level, it is the highest and largest highland in the world. There are five main plant types on the TP, from southeast to northwest: forests, shrubs, crops, alpine meadows, and sparse alpine grasslands. The alpine meadows and sparse alpine grasslands are the dominant land surface types on the TP.

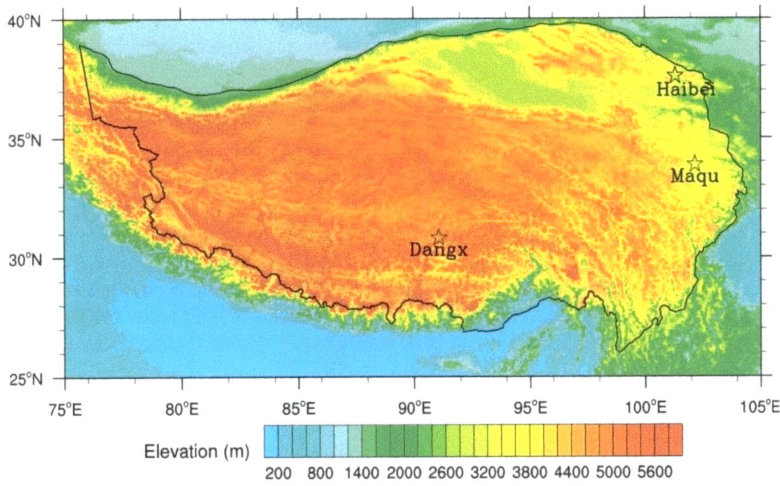

Figure 1. Distribution of vegetation and locations of the flux observational sites over the TP (the pentagrams represent the sites with flux observations).

2.2. Data

2.2.1. Observation Data

The Maqu site is one of the sites set up by the Zoige Plateau Wetlands Ecosystem Research Station, which is located in the eastern region of the TP [14]. The vegetation type at the Maqu site is typically alpine meadow, with an average height of about 0.2 m. The climate at the Maqu site is a sub-humid climate, with an average annual temperature of 1.9 °C, and the multi-year average precipitation is 593 mm [25]. The soil type at Maqu is silt loam, which is rich in organic matter. The observation data from Dangx and Haibei that are used in this study were derived from the China flux network (http://www.cnern.org.cn/) (accessed on 16 December 2020) [26,27]. The vegetation type at the Haibei and Dangx sites is alpine meadow, and the altitudes are 3148 m and 4250 m above sea level, respectively. The Haibei site has a sub-humid climate, the average annual temperature is −1.2 °C, and the average annual precipitation is 535.2 mm; precipitation in the growing season (May to September) accounts for 82% of the annual precipitation. The vegetation coverage at the end of the growing season can reach more than 98%, and the soil organic matter content is 5.85% in the top 40 cm. The climate at the Dangx site is a semi-arid climate, with an average annual temperature of 1.3 °C; the average annual precipitation is 450 mm, of which 85% is distributed from June to August. The vegetation coverage is 50–80%, and the soil organic matter content is 0.9–2.79%. The Maqu, Dangx, and Haibei sites provide long-term conventional meteorological data and flux data. The carbon flux data include gross primary productivity (GPP), net ecosystem exchange productivity (NEE), and ecosystem respiration (Re) (detailed information of the sites see Table 1).

Table 1. Detailed information on the observation sites on the TP used in this study.

Site	Lat (°N)	Lon (°E)	Elevation (m)	Vegetation Type	Years
Maqu	33.9	102.17	3471	Alpine meadow	2010–2016
Dangx	30.85	91.08	4250	Alpine meadow	2007–2010
Haibei	37.62	101.3	3148	Alpine meadow	2003–2007

2.2.2. Remote Sensing Product

The GLASS (Global Land Surface Satellite) GPP product was provided by the Beijing Normal University Center for Global Change Data Processing and Analysis (http://www.bnu-datacenter.com/) (accessed on 5 April 2021). Its time resolution is 8 days, and the spatial resolution is 0.05°. This product has been quality-controlled and had its accuracy verified, and it has the advantages of high accuracy, long time series, and spatial integrity [28,29].

The FLUXCOM GPP product (www.fluxcom.org) (accessed on 12 April 2021) provides carbon fluxes at a high spatial resolution of 0.05°, with a temporal resolution of 8 days, and it is available for the Moderate Resolution Image Spectroradiometer (MODIS, 2000-present). FLUXCOM combines carbon and energy fluxes and meteorological measurements from 224 global FLUXNET sites, using machine learning techniques to scale up these fluxes to a global extent [30,31]. Tramontana et al. [32] evaluated FLUXCOM GPP products against observations and found that the FLUXCOM dataset adequately estimated global carbon fluxes. Long-term time-series GLASS GPP products were used to evaluate the climatology and trend of the GPP simulated using CLM5.0-BGCDV.

The improvements in GPP products, including near-infrared reflectance (NIRv) and solar-induced chlorophyll fluorescence (SIF), provide a method to estimate global GPP [33]. Limited by its short duration, satellite SIF can hardly be used to monitor long-term GPP [34]. In this study, the global long-term (1982–2018) GPP dataset [35] generated by the proxy of GPP used satellite-based near-infrared reflectance to study the response of vegetation to regional climate change. Previous studies showed that the long-term datasets derived from NIRv better capture the seasonal and inter-annual variations in terrestrial GPP at the global scale [35].

In this study, a GPP dataset derived from the Moderate Resolution Imaging Spectro-radiometer (MODIS) instruments on the National Aeronautics and Space Administration (NASA) Terra and Aqua satellites was used (https://search.earthdata.nasa.gov/) (accessed on 15 March 2022). The MODIS GPP product provides global daily estimates of GPP at a 0.05° resolution for the period 2000–2003 [36]. MCD43C4v006 Nadir Bidirectional Reflectance Distribution Function (BRDF)-Adjusted Reflectance (NBAR) products were used as inputs to neural networks that were used to upscale the GPP estimated from the FLUXNET 2015 eddy covariance tower sites.

The fractions of land cover types used in this study were sourced from the Global Vegetation Continuous Fields product (MOD44B, hereafter called MEaSURES), based on observations from the MODIS [37]. This product comprises three different land cover types: (1) bare soil, (2) trees, and (3) non-trees. It can be downloaded from NASA's Earth data portal (https://search.earthdata.nasa.gov/) (accessed on 10 May 2020). Specific descriptions of the remote sensing products used in this study are presented in Table 2.

Table 2. Detailed information on the remote sensing products used in this study.

Product	Time Period	Temporal Resolution	Spatial Resolution
NIRv	1982–2016	8 d	0.05° × 0.05°
GLASS	1982–2016	8 d	0.05° × 0.05°
FLUXCOM	2001–2016	8 d	0.05° × 0.05°
FLUXCOM1	1982–2013	daily	0.5° × 0.5°
MODIS	2000–2016	daily	0.05° × 0.05°
MEaSURES	1982–2013	yearly	0.05° × 0.05°

2.2.3. Meteorological Data

The observational meteorological datasets used in this study include daily precipitation and temperature with a spatial resolution of 0.5° from 1979 to 2016. The grid precipitation and temperature datasets originated from daily and monthly precipitation and temperature data at 2472 meteorological sites. The data were processed by quality control and then interpolated from station data to grid data, with only stations with data available from 1961 used. A digital elevation model (Global 30 Arc-Second Elevation, GTOPO30) was introduced in order to weaken the effects of elevation on the interpolation precision of temperature and precipitation [38].

2.3. Description of CLM

The land surface model used in this study is provided by the National Center for Atmospheric Research (NCAR) and is the land component of the Community Earth System Model (CESM). The CLM5.0-DGVM is one of the models that simulate the structure and distribution of natural vegetation dynamically, primarily using mechanistic parameterizations of large-scale vegetation processes [39,40]. The dynamic vegetation model of the Lund–Potsdam–Jena model (LPJ) was coupled with the NCAR's Land Surface Model prior to the first release of the CLM [41]. In addition, the CLM-DGVM has a set of routines that allow vegetation structure and cover to be simulated instead of prescribed from data.

2.3.1. Establishment and Survival

Plant functional type (PFT) survival in a grid cell requires the 20-year running mean of the minimum monthly temperature, T_c, to exceed the PFT-dependent 20-year running mean of the coldest minimum monthly temperature, $T_{c,min}$ (Table 3). Existing PFTs that can survive in the current climate continue to exist without change. PFTs do not present in a grid cell unless they can establish. Establishment criteria are stricter than those for survival, additionally requiring that T_c be less than the PFT-dependent 20-year running mean of the warmest minimum monthly temperature, $T_{c,max}$ (Table 3), that growing degree days ($GDD_{5°C}$) be greater than the PFT-dependent GDD_{min} (Table 3), and that $GDD_{23°C}$ be equal to 0. Establishment also requires the 365-day running mean of precipitation to be greater than 100 mm/year.

Table 3. Rules that delineate PFT biogeography according to climate.

PFT	Survival		Establishment
	$T_{c,min}(°C)$	$T_{c,max}(°C)$	GDD_{min}
Tropical broadleaf evergreen tree (BET)	15.5	No limit	0
Tropical broadleaf deciduous tree (BDT)	15.5	No limit	0
Temperate needleleaf evergreen tree (NET)	−2.0	22.0	900
Temperate broadleaf evergreen tree (BET)	3.0	18.8	1200
Temperate broadleaf deciduous tree (BDT)	−17.0	15.5	1200
Boreal needleleaf evergreen tree (NET)	−32.5	−2.0	600
Boreal deciduous	No limit	−2.0	350
C4	15.5	No limit	0
C3	−17.0	15.5	0
C3 arctic	No limit	−17.0	0

2.3.2. Seasonal–Deciduous Phenology

The onset trigger for the seasonal–deciduous phenology algorithm is based on an accumulated growing-degree-day (GDD) approach [42]. The GDD summation is initiated ($GDD_{sum} = 0$) when the phenological state is dormant and the model timestep crosses the winter solstice. Once these conditions are met, GDD_{sum} is updated at each timestep as

$$\begin{cases} GDD_{sum}^{n-1} + (T_{s,3} - TKFRZ)f_{day} & T_{s,3} > TKFRZ \\ GDD_{sum}^{n-1} & T_{s,3} < TKFRZ \end{cases} \tag{1}$$

where $T_{s,3}(K)$ is the temperature of the third soil layer, and $f_{day} = \Delta t/86400$ (unit : day). The onset period is initiated if $GDD_{sum} > GDD_{sum\_crit}$ (unit : day), where

$$GDD_{sum\_crit} = \exp(4.8 + 0.13(T_{2m,ann\_avg} - TKFRZ)) \tag{2}$$

where $T_{2m,ann\_avg}(K)$ is the annual average of the 2 m air temperature, and TKFRZ is the freezing point of water (273.15 K).

The offset period is triggered when the day length is <39,300 s.

Previous studies found that current DGVMs overestimated the length of the active vegetation-growing season, mostly due to a late dormancy resulting from late summer phenology [23]. The beginning of the growing season in observation is judged when the 5 cm soil temperature is greater than 5 °C. In this study, we modified GDD_{sum} as follows:

$$\begin{cases} GDD_{sum}^{n-1} + (T_{s,2} - TKFRZ)f_{day} & T_{s,2} > TKFRZ \\ GDD_{sum}^{n-1} & T_{s,2} \leq TKFRZ \end{cases} \tag{3}$$

Additionally, the offset period is triggered when the day length of <39,300 s changes to a day length of <42,000 s.

2.4. Experimental Design

In this study, the CLM model we used in simulating ecosystem productivity and vegetation distribution was CLM5.0-BGCDV. To achieve a steady state for the CLM-BGCDV model, we first ran it from a cold state using the "accelerated decomposition spin-up (AD spin-up)" mode for 400 years. Afterward, we saved the last restart file from the AD spin-up simulation and took it as a "finidat" file to use in the normal mode simulations.

To investigate the performance of CLM5.0-DGVM in simulating ecosystem productivity and vegetation distribution over the TP, we conducted the following set of regional simulations on the TP using CLM5.0-DGVM. (1) Following Deng et al. [24], we replaced the soil property data with the Beijing Normal University soil property data, the Balland

and Arp, and the virtual temperature schemes were used to replace the original code in the CLM5.0-DGVM. In addition, we modified the establishment and survival code so that PFTs could establish on the TP, and we replaced the vegetation parameters with observations. (2) The seasonal–deciduous phenology parameterization was used to replace the original code (see Table 4 for details). (3) We compared the difference in soil water and heat transfer modeling using CLM5.0-SP and CLM5.0-BGCDV.

Table 4. Designs of the regional experiments by CLM.

Experiment	Soil Property Data	Parameterization	Model
BGCDV_CTL	BNU soil property data	Balland and Arp; virtual temperature; establishment and survival	CLM5.0-BGCDV
BGCDV_NEW	BNU soil property data	Balland and Arp; virtual temperature; establishment and survival; seasonal–deciduous phenology	CLM5.0-BGCDV

2.5. Analytical Method

In this study, four statistical features were calculated in order to quantify the differences between the simulated and observed parameters, which were calculated at each station: bias (Bias), root mean square error (RMSE), correlation coefficient (Corr), and ratio of standard deviation (RSD). PBIAS was used to assess the performance of simulations regarding the tendency of the simulated carbon fluxes to be overestimated or underestimated [43]. It is considered unacceptable if PBIAS is greater than 20% [44,45]. When the Corr is high and the RMSE is low, the simulation is considered robust and more desirable [46].

$$\text{PBIAS} = \frac{1}{N}\sum_{i=1}^{N}(M_i - O_i) \Big/ \sum_{i=1}^{N}(O_i) \tag{4}$$

$$\text{RMSE} = \left(\frac{1}{N-1}\sum_{i=1}^{N}(M_i - O_i)^2\right)^{\frac{1}{2}} \tag{5}$$

$$\text{Corr} = \frac{\frac{1}{N}\sum_{i=1}^{N}(M_i - \overline{M})(O_i - \overline{O})}{\sqrt{\frac{1}{N}\sum_{i=1}^{N}(M_i - \overline{M})^2}\sqrt{\frac{1}{N}\sum_{i=1}^{N}(O_i - \overline{O})^2}} \tag{6}$$

$$\text{RSD} = \frac{\sqrt{\frac{1}{N-1}\sum_{i=1}^{N}(M_i - \overline{M})^2}}{\sqrt{\frac{1}{N-1}\sum_{i=1}^{N}(O_i - \overline{O})^2}} \tag{7}$$

3. Evaluation of Remote Sensing Products and GPP Simulations

Figure 2 shows the time series of the observed and the simulated GPP, NEE, and Re at the Dangx, Haibei, and Maqu sites. After revising the establishment and survival parameterizations, the simulated vegetation coverage was about 50–80% from 2007 to 2010 at the Dangx site, 70–90% from 2003 to 2010 at the Haibei site, and 90–100% from 2010 to 2016 at the Maqu site, which corresponds to the actual observations at each site. Compared to the seasonal variation in the daily GPP, NEE, and Re at Dangx, the observed GPP, NEE, and Re showed different characteristics. Both the absolute GPP and NEE increased steadily from the beginning of the growing season until July, then increased rapidly from July and reached a peak value at the end of August and the beginning of September. The simulated GPP generally coincided with the observations, and the simulated GPP of the BGCDV_CTL experiment started earlier and ended later than that of the observed GPP. At the Haibei site, GPP, NEE, and Re showed similar variations during the growing season, increasing beginning in May and reaching a peak value in August (Figure 2). Before modifying the establishment and survival parameterizations, the PFTs in dynamic vegetation simulations could not establish at the Haibei site. After modifying the establishment and survival scheme (BGCDV_CTL experiment), the simulated Re generally coincided with the observations, while the dynamic vegetation model underestimated GPP and NEE from July to September. In addition, the simulated start time of the growing season was advanced, and the end time of the growing season was delayed. The increasing trend of the simulated GPP and NEE was faster than that of the observed GPP and NEE. Seasonal variations in GPP, NEE, and Re at the Maqu site showed a similar pattern to those at the Haibei site. The CLM5.0-DGVM model overestimated Re during the growing season and underestimated NEE. In addition, the model underestimated GPP, NEE, and Re in a vigorous-growth period. In the BGCDV_CTL experiment, variations in the ecosystem productivity simulations were consistent with those in the observations, while the end time of the growing season was delayed.

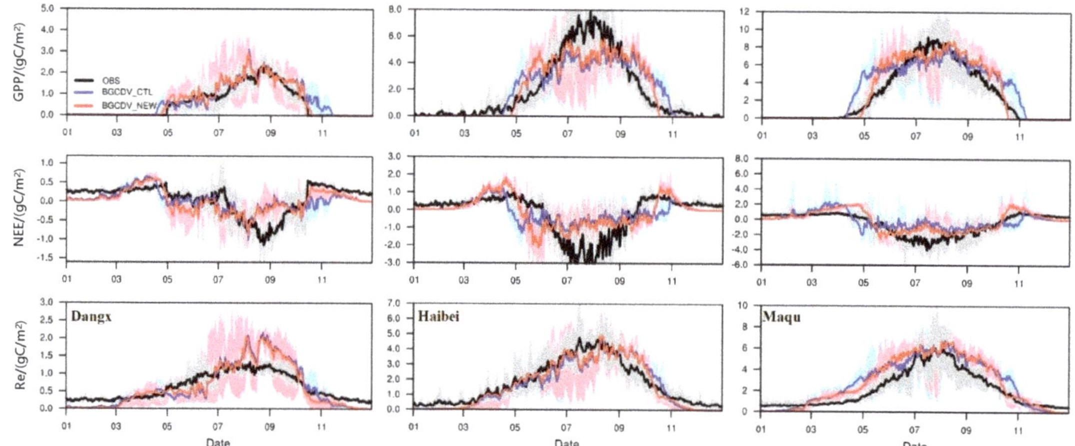

Figure 2. Model simulations from CLM5.0-DGVM versus observations of daily GPP (unit: gC·m^{-2}·d^{-1}), NEE (unit: gC·m^{-2}·d^{-1}), and Re (unit: gC·m^{-2}·d^{-1}) at Dangx from 2007 to 2010, Hiabei from 2003 to 2007, and Maqu from 2010 to 2019; the black, blue, and red lines are the means from observation, the BGCDV_CTL experiment, and the BGCDV_NEW experiment, respectively.

After replacing the original seasonal–deciduous phenology scheme (the BGCDV_NEW experiment), the model effectively shortened the growing season length at the Dangx site. The average Bias and RMSE decreased from 0.116 gC·m^{-2}·d^{-1} and 0.321 gC·m^{-2}·d^{-1} to 0.092 gC·m^{-2}·d^{-1} and 0.261 gC·m^{-2}·d^{-1}, respectively, and Corr increased from 0.922 to 0.995 (Table 5). At the Dangx site, the simulated NEE tended to underestimate the observed NEE from August to October, mainly caused by overestimating the simulated Re. Compared to the BGCDV_CTL experiment, the simulated NEE in the BGCDV_NEW experiment was smaller. The RMSE decreased from 0.316 gC·m^{-2}·d^{-1} to 0.299 gC·m^{-2}·d^{-1}, and Corr increased from 0.572 gC·m^{-2}·d^{-1} to 0.640 gC·m^{-2}·d^{-1}. Re began to increase gradually when the growing season started and reached a peak value in August. With the leaves beginning dormancy, Re decreased beginning in September. During the growing season, the simulated Re overestimated the observed Re, and there was a dramatic decline in mid-August in the simulated Re. The average Bias values for the BGCDV_CTL experiment and the BGCDV_NEW experiment were 0.003 gC·m^{-2}·d^{-1} and −0.017 gC·m^{-2}·d^{-1}, respectively. After modifying the seasonal–deciduous phenology parameterization, the simulated Re tended to reduce at the Dangx site. The RMSE reduced from 0.323 $gC·m^{-2}·d^{-1}$ to 0.300 $gC·m^{-2}·d^{-1}$, and Corr increased from 0.899 to 0.915. Compared to the BGCDV_CTL experiment, the average Bias and RMSE of the simulated GPP in the BGCDV_NEW experiment reduced from -0.136 $gC·m^{-2}·d^{-1}$ and 1.227 $gC·m^{-2}·d^{-1}$ to -0.109 $gC·m^{-2}·d^{-1}$ and 1.127 $gC·m^{-2}·d^{-1}$, respectively, and Corr increased from 0.863 to 0.882. The dramatic increase in the simulated GPP from May to June was mainly caused by the increase in the simulated NEE. The simulated NEE from the BGCDV_NEW experiment had a better performance than that from the BGCDV_CTL experiment. At the Haibei site, the simulated Re underestimated the observed Re during the non-growing seasons and overestimated Re in autumn (September to November). The average Bias of the simulated Re by the BGCDV_NEW experiment decreased by 21%, and seasonal variations in the BGCDV_NEW experiment coincided better with observations. The simulated NEE from the BGCDV_NEW experiment tended to overestimate NEE during the non-growing season and reduced the biases of NEE during the growing season at the Maqu site. The average Bias and RMSE of the simulated NEE in the BGCDV_NEW experiment reduced from 0.366 $gC·m^{-2}·d^{-1}$ and 1.114 $gC·m^{-2}·d^{-1}$ to 0.355 $gC·m^{-2}·d^{-1}$ and 0.877 $gC·m^{-2}·d^{-1}$, respectively, and Corr increased from 0.634 to 0.814. The BGCDV_CTL experiment tended to overestimate Re, and the average Bias between the BGCDV_CTL experiment simulations and observations was 0.606 $gC·m^{-2}·d^{-1}$. In addition, obvious systematic errors existed at the Dangx and Haibei sites from July to September, which may have been caused by the imperfections of the water use efficiency parameterization scheme. The simulated carbon flux of CLM5.0-DGVM is sensitive to water content and less sensitive to temperature. According to PBIAS, the simulated GPP at the Dangx, Haibei, and Maqu sites are considered acceptable. The Re simulations are satisfactory at the Dangx and Haibei sites, and the simulated NEE at the Dangx, Haibei, and Maqu sites are considered unacceptable.

Table 5. Statistical results of the simulated GPP, NEE, and Re.

Sites		PBIAS (%)		RMSE ($gC·m^{-2}·d^{-1}$)		Corr		RSD	
		CTL	NEW	CTL	NEW	CTL	NEW	CTL	NEW
Dangx	GPP	20.9	16.6	0.321	0.261	0.922	0.952	1.100	1.121
	NEE	-71.1	-87.8	0.316	0.299	0.572	0.640	0.676	0.745
	Re	0.5	2.7	0.323	0.300	0.899	0.915	1.676	1.636
Haibei	GPP	-6.9	-5.5	1.227	1.127	0.863	0.882	0.780	0.892
	NEE	-45.8	-40.7	0.917	0.882	0.571	0.615	0.480	0.716
	Re	-10.9	-8.6	0.580	0.516	0.927	0.945	1.054	1.085
Maqu	GPP	18.2	11.3	1.401	1.028	0.903	0.956	0.946	1.087
	NEE	-82.0	-79.7	1.114	0.877	0.634	0.814	0.691	0.924
	Re	28.0	26.2	1.008	0.963	0.936	0.954	1.238	1.288

The Maqu and Haibei sites are located in the eastern and northern regions of the TP, respectively, and both sites are obvious carbon sinks, while the Dangx site is a weaker carbon source that is located in the central region of the TP and at a high elevation. At the Maqu and Haibei sites, with rich soil organic matter, the photosynthetic carbon absorption during the growing season is higher than that of the Dangx site. Compared to the Maqu and Haibei sites, the observed GPP and Re are smaller at the Dangx site.

Figure 3 shows the monthly variations in the simulated GPP from CLM5.0-DGVM and the NIRv, GLASS, FLUXCOM, and MODIS products versus observations at the Dangx, Haibei, and Maqu sites. The GPP provided by NIRv underestimated the growing season length at the Dangx site, mostly due to a delayed start of the growing season. At the Haibei and Maqu sites, the NIRv GPP product overestimated GPP in the growing season. FLUXCOM assimilated observations at the Dangx site into the product, but deviations still existed in the descriptions of the GPP characteristics at the Dangx site. The FLUXCOM product overestimated the length of the vegetation growing season, mostly due to an early start of the growing season. Compared to the FLUXCOM product, the start of the growing season of the GLASS product was closer to the observation, while the product seriously overestimated the observed GPP, especially in 2007, 2009, and 2010. The GPP provided by MODIS overestimated the GPP at Dangx and Haibei at the start of the growing season and the growing season at the Haibei site. At the Maqu site, the MODIS GPP generally coincided with the observed GPP. The simulated GPP from CLM5.0-DGVM displayed better performance and simulated the variation characteristics of GPP. As seen in Table 6, at the Dangx site, the average Bias values of the BGCDV_NEW experiment and the NIRv, GLASS, FLUXCOM, and MODIS products were 0.091 $gC·m^{-2}·d^{-1}$, -0.174 $gC·m^{-2}·d^{-1}$, 0.840 $gC·m^{-2}·d^{-1}$, 0.242 $gC·m^{-2}·d^{-1}$, and 0.370 $gC·m^{-2}·d^{-1}$, respectively. At the Haibei site, the GLASS product overestimated the observed GPP, while the seasonal variations generally coincided with observations. The FLUXCOM product underestimated GPP during the growing season and overestimated the observed GPP during the non-growing season. The bias between simulations from CLM5.0-DGVM and observations was smaller than that between the GLASS and FLUXCOM products. However, there was a valley value in the simulated GPP during the growing season, which did not exist in the observations. Compared to the NIRv, GLASS, FLUXCOM, and MODIS products, the RMSE of the simulated GPP from the BGCDV_NEW experiment decreased by 74%, 36%, 30%, and 39%, respectively. The GLASS product at Maqu generally coincided with observations, while the product overestimated GPP in 2011, 2014, and 2015. The years 2011, 2014, and 2015 were years with less precipitation, which led to lower GPP values. The MODIS GPP product had the best performance in simulating GPP at the Maqu site. Compared to the NIRv, GLASS, FLUXCOM, and MODIS products, the average Bias of the simulated GPP from the BGCDV_NEW experiment decreased by 2.02 $gC·m^{-2}·d^{-1}$, 1.265 $gC·m^{-2}·d^{-1}$, 0.143 $gC·m^{-2}·d^{-1}$, and -0.159 $gC·m^{-2}·d^{-1}$, respectively. Overall, the PBIAS values of the GPP simulations from the BGCDV_NEW experiment at Dangx, Haibei, and Maqu are within 20%, which is considered to be acceptable. In addition, the FLUXCOM and MODIS data are satisfactory at Maqu.

Table 6. Statistical results of the simulated GPP, GLASS GPP, and FLUXCOM GPP against observations at a monthly scale.

	PBIAS (%)			RMSE ($gC·m^{-2}·d^{-1}$)			Corr		
Experiment	Dangx	Haibei	Maqu	Dangx	Haibei	Maqu	Dangx	Haibei	Maqu
NEW	16.4	−11.0	11.3	0.322	1.107	1.084	0.940	0.890	0.947
GLASS	151.4	45.9	23.1	1.587	1.723	1.239	0.740	0.948	0.968
FLUXCOM	43.6	44.1	16.8	0.520	1.582	1.472	0.779	0.881	0.923
NIRv	−31.4	118.7	89.6	0.659	4.184	3.554	0.637	0.931	0.959
MODIS	66.7	45.7	5.2	0.569	1.727	0.905	0.812	0.947	0.957

Figure 4 shows the spatial distribution of the bare soil proportion from the BGCDV_NEW experiment simulations and the MEaSURES vegetation continuous field in 1982, 1992, 2002, and 2012. Compared to MEaSURES, the simulated percentage of vegetation from the BGCDV_NEW experiment was greater in the semi-arid region of the TP in 1980. Compared to 1982, the simulated proportion of bare soil from the BGCDV_NEW experiment tended to decrease in the semi-arid and arid regions of the TP. As shown in Figure 4, the simulated proportion of bare soil from MEaSURES tended to increase in the southeast region of the TP from 1982 to 2012. Overall, the simulated percentage of vegetation from the BGCDV_NEW experiment was greater than that from MEaSURES.

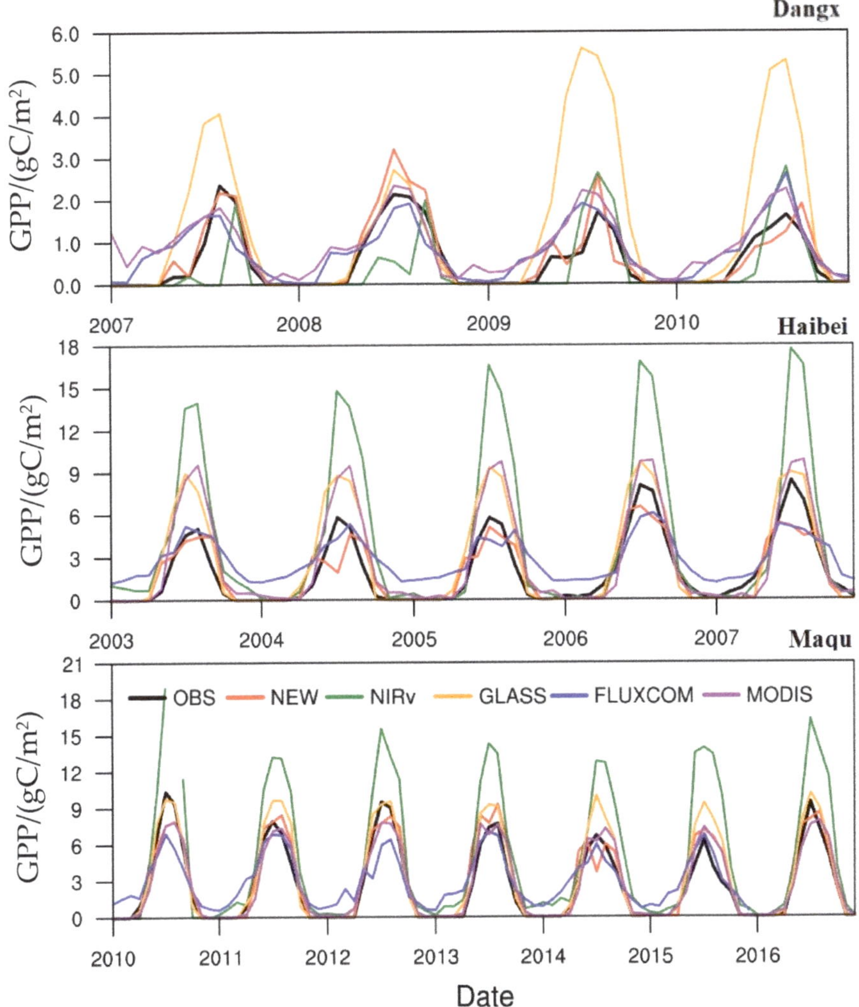

Figure 3. Model simulations from CLM5.0-DGVM and remote sensing GPP products versus observations of monthly GPP (unit: gC·m^{-2}·d^{-1}), NEE (unit: gC·m^{-2}·d^{-1}), and Re (unit: gC·m^{-2}·d^{-1}) at the Dangx, Haibei, and Maqu sites.

Figure 5 shows the climatology of the GLASS, NIRv, FLUXCOM, FLUXCOM1, and MODIS GPPs and the simulated GPP from the BGCDV_NEW experiment. As shown in Figure 5, a large area of the arid region of the TP has missing values for the NIRv, FLUXCOM, FLUXCOM, and MODIS GPP products. All remote sensing GPP products and the GPP simulated using CLM5.0-DGVM show a similar pattern, with GPP increasing from northwest to southeast and with the maximum at the southeast edge of the TP. Compared to the MODIS GPP product, the NIRv GPP product overestimated GPP in the east of the TP, and the FLUXCOM1 GPP product underestimated GPP in sub-humid regions. The simulated GPP from the BGCDV_NEW experiment was underestimated in the sub-humid regions and overestimated in the semi-arid and arid regions compared to the remote sensing GPP products.

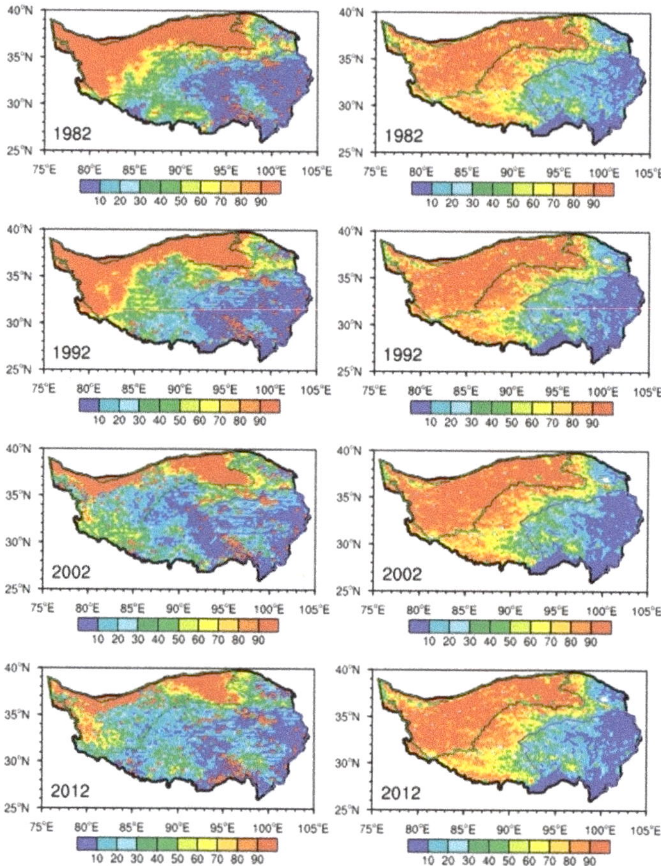

Figure 4. Proportion of bare soil (unit: %) on the TP. Left: CLM-BGCDV; right: MEaSURES.

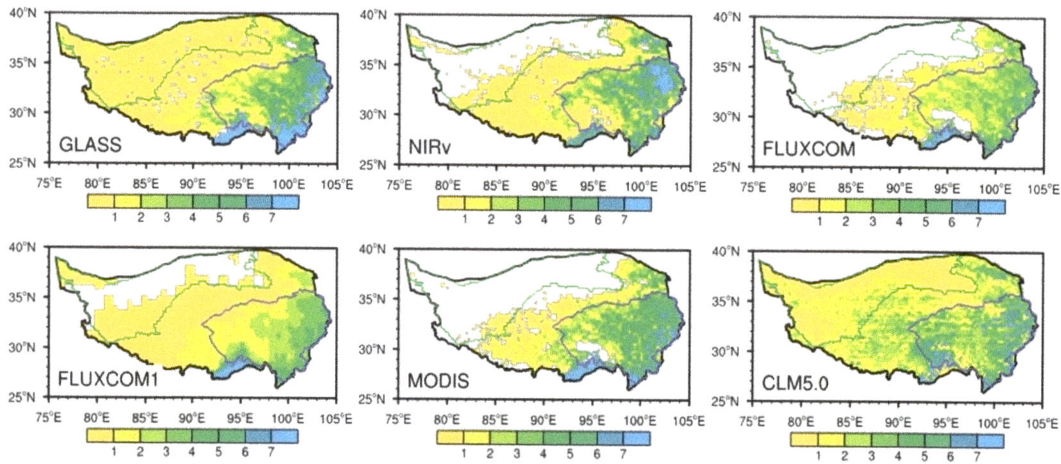

Figure 5. Climatology and trend of GPP (gC·m^{-2}·d^{-1}) from May to October.

4. Response of Vegetation to Regional Climate Change on the TP

4.1. Climatology and Trend of Temperature, Precipitation, and GPP

Figure 6 shows the climatology and trend of precipitation and temperature from May to October for the period from 1980 to 2016. As is shown in Figure 6, precipitation increased from the northwest to the southeast of the TP, with the maximum in the southeast of the TP. Temperature decreased from the low latitude/altitude region to the high latitude/altitude region, with the maximum in the south of the TP and the Tarim Basin. Both precipitation and temperature showed increasing trends in most regions of the TP. However, precipitation increased in the central region of the TP and decreased on the eastern edge of the TP, while temperature increased in the whole TP.

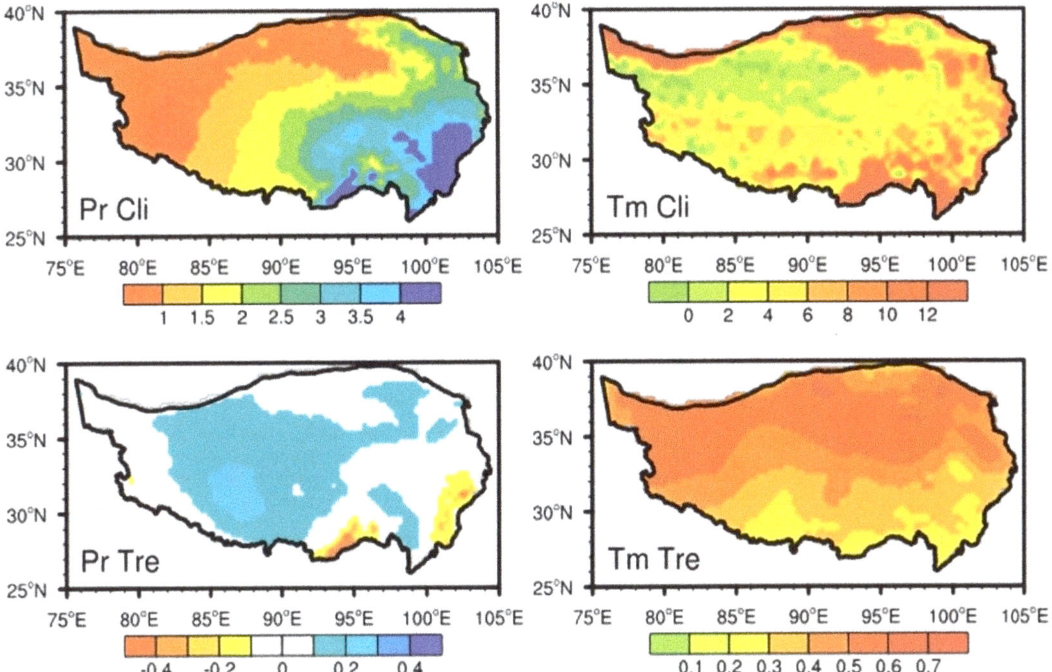

Figure 6. Climatology and trend (unit: mm/day, °C) of precipitation (unit: mm/day) and temperature (unit: °C), May to October from 1982 to 2016 on the TP.

Figure 7 shows the trend of the GLASS, NIRv, FLUXCOM, FLUXCOM1, and MODIS GPPs and the simulated GPP from the BGCDV_NEW experiment. From 1982 to 2016, the GPP products provided by GLASS and NIRv showed an increasing trend in most regions of the TP, with the maximum in the sub-humid regions. In addition, the increasing trend of the GLASS product was larger than that of the NIRv product. From 1982 to 2013, the GPP provided by the FLUXCOM1 product displayed no significant change trend on the TP. From 2001 to 2016, the FLUXCOM GPP product showed a decreasing trend in the sub-humid and semi-arid regions, while the MODIS GPP product showed an increasing trend in the northeast of the TP. The simulated GPP in the BGCDV_NEW experiment showed an increasing trend in the semi-arid and arid regions of the TP and decreased in the sub-humid regions. In the semi-arid and sub-humid regions of the TP, both the GLASS and NIRv GPP products, as well as temperature, showed consistent trends. However, the GPP simulated using CLM5.0-DGVM, as well as temperature, showed inconsistent trends, with temperature increasing in the whole TP while the simulated GPP decreased in the sub-humid regions of the TP. The simulated GPP in the BGCDV_NEW experiment showed a similar trend to precipitation, with the maximum in the central region of the TP.

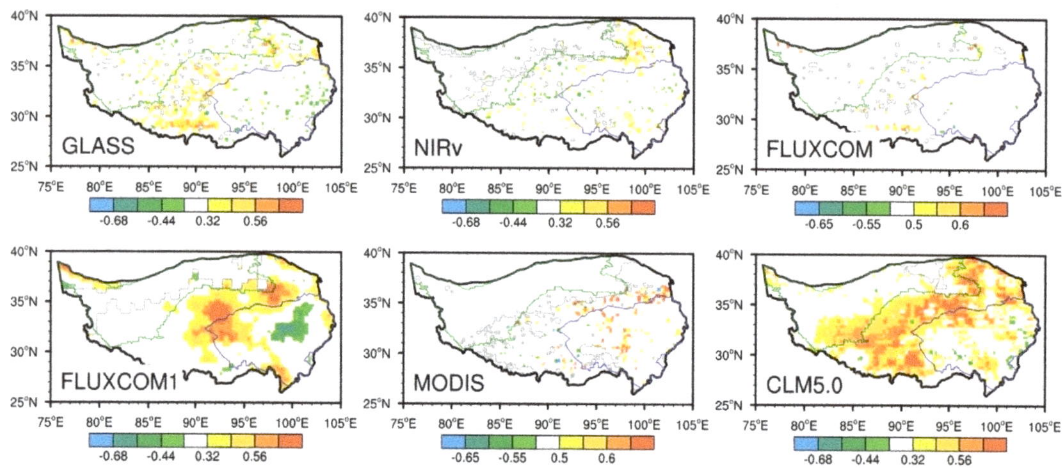

Figure 7. Trend of GPP from May to October on the TP.

4.2. Response of GPP to Climate Change

Figure 8 shows partial correlations between different GPP products and precipitation. The results indicated that the GLASS, NIRv, FLUXCOM, FLUXCOM1, and MODIS GPP products did not significantly correlate with precipitation changes in most regions of the TP. Compared to the sub-humid regions of the TP, the remote sensing GPP products showed a relatively high correlation in the semi-arid regions of the TP, especially the FLUXCOM1 GPP product. The GPP simulated using CLM5.0-DGVM significantly correlated with precipitation changes in most regions of the TP. Among all the remote sensing products, the GLASS and FLUXCOM1 GPP products were most sensitive to precipitation in the semi-arid regions.

Figure 8. Partial correlations between GPP and precipitation.

Figure 9 shows partial correlations between different GPP products and temperature. Limited by the period covered by the FLUXCOM and MODIS GPP products, there was no significant correlation between GPP and temperature for these products. As shown in Figure 9, the GLASS GPP product significantly correlated with temperature changes in most regions of the TP. Partial correlations between the GLASS GPP product and temperature show that the increase in GPP was mainly caused by climate warming on the TP. The NIRv and FLUXCOM1 GPP data show positive correlations with temperature in the sub-humid and semi-arid regions of the TP, while the CLM5.0-DGVM GPP simulation shows a positive correlation in the arid regions of the TP. The overall results indicate that the increase in the remote sensing GPP was mainly caused by temperature, while the increase in the GPP simulated in CLM5.0-DGVM was mainly caused by precipitation in most regions of the TP.

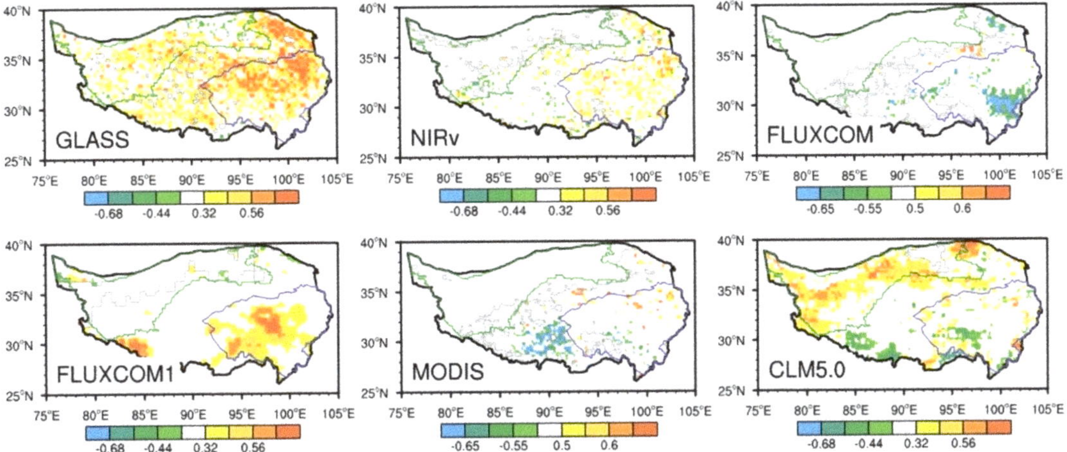

Figure 9. Partial correlations between GPP and temperature.

Figure 10 shows the annual change in the GPP anomalies based on the GLASS, NIRv, and FLUXCOM1 products, as well as the BGCDV_NEW experiment, with precipitation and temperature anomalies averaged over the TP. The correlations between the GLASS, NIRv, FLUXCOM, and BGCDV_NEW GPP anomalies and the temperature anomaly were 0.83, 0.80, 0.53, and 0.57, respectively, showing significant positive correlations between the GPP calculated by these products and temperature. The variances in precipitation and temperature were weaker for the GLASS and NIRv GPP products than in the GPP simulation. The correlations between the remote sensing product GPP anomalies and the temperature anomalies were greater than the correlations between the remote sensing GPP products and precipitation. Compared to precipitation, the remote sensing GPP products were more sensitive to temperature. The correlations between the GPP simulation from the BGCDV_NEW experiment and precipitation and temperature were 0.52 and 0.57, respectively. As shown in Figure 10, the correlation between the BGCDV_NEW experiment GPP simulation and precipitation was smaller than that of the GLASS GPP. In addition, the variances in the remote sensing GPPs were weaker than in the simulated GPP from the BGCDV_NEW experiment. The correlation between the BGCDV_NEW experiment GPP and temperature was smaller than those of the GLASS and NIRv GPPs, which indicates that the GLASS and NIRv GPPs were more sensitive to temperature than the simulated GPP from the BGCDV_NEW experiment. A negative precipitation anomaly was recovered from 1998, while the simulated GPP from the BGCDV_NEW experiment dramatically increased from 1996, which mainly caused the significantly positive precipitation anomaly in 1996.

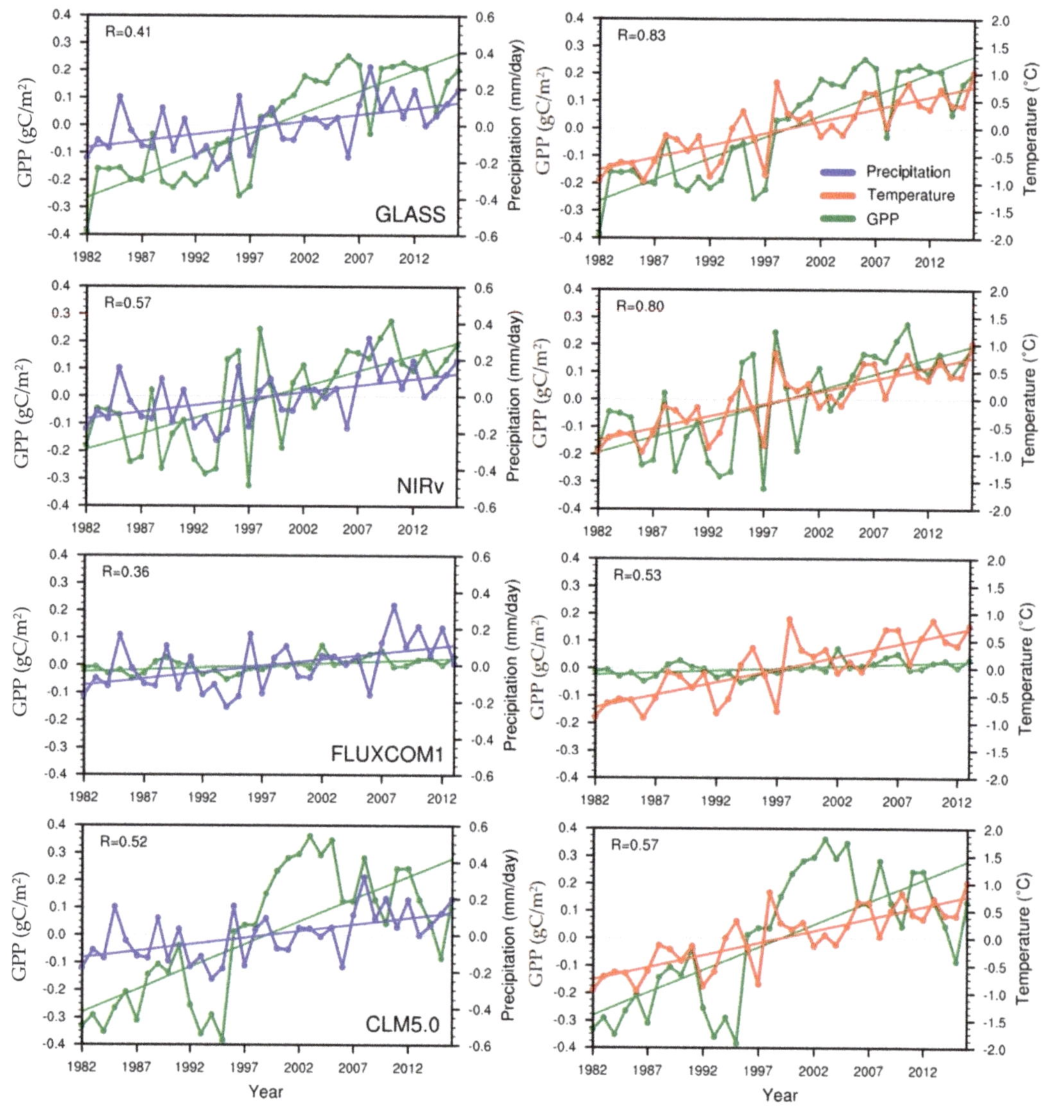

Figure 10. Annual changes in the GLASS GPP anomaly, simulated GPP anomaly in CLM5.0-BGCDV, temperature anomaly, and precipitation anomaly.

5. Conclusions

According to the CLM5.0 technical note, the DGVM was not tested in the development of the CLM5.0 and is no longer scientifically supported. The Functionally Assembled Terrestrial Ecosystem Simulator (FATES) is an actively developed DGVM for the CLM5.0. FATES was derived from the CLM Ecosystem Demography Model and is a cohort model of vegetation competition and co-existence that is mainly used to study trees of various PFTs, while the study region in this paper is mainly covered by grass. In this paper, after modifying the establishment and survival parameterizations (BGCDV_CTL parameterization), PFTs were able to establish over the whole TP. We evaluated simulations for ecosystem productivity by using CLM5.0-DGVM on the TP. The results showed that the simulated ecosystem productivity generally coincided with the observations, while overestimating the length of the growing season. Then, we modified the seasonal–deciduous phenology parameterization (BGCDV_NEW experiment) and compared it to the BGCDV_CTL experiment; the BGCDV_NEW experiment simulations reduced the biases in the simulated length of the growing season. Monthly variations in GPP showed that the BGCDV_CTL experiment displayed the best performance at the three sites considered. Compared to the GLASS, NIRv, FLUXCOM, and MODIS GPP products, the average bias of the simulated GPP in the BGCDV_NEW experiments for these three sites was reduced by 0.502 gC·m^{-2}·d^{-1}, 1.409 gC·m^{-2}·d^{-1}, 0.424 gC·m^{-2}·d^{-1}, and 0.376 gC·m^{-2}·d^{-1}, respectively. Compared to MEaSURES, the percentage of vegetation from the BGCDV_NEW experiment was greater in the semi-arid regions of the TP. The simulated proportion of bare soil from the BGCDV_NEW experiment tended to decrease in the semi-arid and arid regions of the TP. The climatology of the remote sensing GPP products and the simulated GPP showed similar patterns, with GPP increasing from northwest to southeast, while the simulations overestimated GPP in the semi-arid and arid regions of the TP.

From 1982 to 2018, the climate on the TP showed an overall warming and wetting trend, and GPP showed an increasing trend in most regions of the TP. Partial correlations between the remote sensing products and temperature indicated that the increase in GPP on the TP was mainly affected by climate warming. Partial correlations between the simulated GPPs and precipitation indicated that precipitation increased in the central region of the TP, leading to the increase in the simulated GPPs in the central region of the TP. The annual changes in the remote sensing GPP anomalies and the BGCDV_NEW experiment GPP anomaly indicated that the variance in the remote sensing GPPs was weaker than that in the simulated GPP from the BGCDV_NEW experiment. The correlation between the BGCDV_NEW experiment GPP and temperature was smaller than that of the GLASS and NIRv GPPs, which indicates that the GLASS and NIRv GPPs were more sensitive to temperature than the simulated GPP from the BGCDV_NEW experiment.

The simulated ecosystem productivity from the revised CLM5.0-DGVM model generally coincided with the observations, although the evaluation of ecosystem productivity was only conducted at three sites on the TP. There is a lack of validation of the carbon flux caused by using the revised CLM5.0-DGVM model in the northwest of the TP. In addition, the pattern of the simulated GPP from the revised CLM5.0-DGVM model was similar to that of the remote sensing products, while the BGCDV_NEW experiment overestimated GPP in the semi-arid and arid regions of the TP and underestimated GPP on the southeast edge of the TP. Great uncertainty still exists regarding the performance of remote sensing in the high-altitude regions of the TP. We need more carbon flux data to further validate the performance of carbon flux simulations by using DGVMs and by studying the responses of carbon flux to climate change. Otherwise, the BGCDV_NEW experiment was more sensitive to water content than the GLASS or NIRv GPP products. The parameterization of water use efficiency is imperfect in the current CLM5.0-DGVM, and the development of vegetation parameterization is still the main challenge to be faced in further work.

Author Contributions: Data curation, M.D., H.N. and D.S.; formal analysis, M.D., X.M., Y.L. and Z.L.; project administration, X.M.; software, M.D.; writing—original draft, M.D.; writing—review and editing, X.M., Y.L., L.Z., H.C., L.S. and S.W. All authors have read and agreed to the published version of the manuscript.

Funding: This research was supported by the National Natural Science Foundation of China (41930759), the Second Scientific Expedition to Qinghai-Tibet Plateau (2019QZKK0102), and the Science and Technology Research Plan of Gansu Province (20JR10RA070).

Data Availability Statement: Not applicable.

Acknowledgments: We would like to thank the observation data provider from the Key Laboratory of Land Process and Climate Change in Cold and Arid Regions, Northwest Institute of Eco-Environment and Resource, Chinese Academy of Sciences; observation data at Maqu were obtained from the Zoige Plateau Wetland Ecosystem Research Station of the Northwest Institute of Eco-Environmental Resources, Chinese Academy of Sciences (http://tpwrr.nieer.cas.cn/) (accessed on 1 March 2020); observation data at the Dangx and Haibei sites were provided by http://www.cnern.org.cn/; the China Meteorological Forcing Dataset was provided by the Institute of Tibetan Plateau Research, Chinese Academy of Sciences (http://www.tpedatabase.cn/portal/index.jsp) (accessed on 15 March 2019); temperature and precipitation data were provided by the China Meteorological Administration (http://data.cma.cn/) (accessed on 15 March 2019). We would also like to thank the CESM provider; the CESM 2.1 is freely available at http://www.cesm.ucar.edu/models/ (accessed on 15 December 2019). We would also like to acknowledge the data support from the National Earth System Science Data Center, National Science and Technology Infrastructure of China (http://www.geodata.cn) (accessed on 5 April 2021).

Conflicts of Interest: The authors declare no conflict of interest.

References

1. Reichstein, M.; Bahn, M.; Ciais, P.; Frank, D.; Mahecha, M.D.; Seneviratne, S.I.; Zscheischler, J.; Beer, C.; Buchmann, N.; Frank, D.C.; et al. Climate extremes and the carbon cycle. *Nature* **2013**, *500*, 287–295. [CrossRef] [PubMed]
2. Zhang, Y.Q.; You, Q.L.; Chen, C.C.; Li, X. Flash droughts in a typical humid and subtropical basin: A case study in the Gan River Basin, China. *J. Hydrol.* **2017**, *551*, 162–176. [CrossRef]
3. Sun, Y.L.; Shan, M.; Pei, X.R.; Zhang, X.K.; Yang, Y.L. Assessment of the impacts of climate change and human activities on vegetation cover change in the Haihe River Basin, China. *Phys. Chem. Earth Parts A/B/C* **2020**, *115*, 102834. [CrossRef]
4. Ni, J.; Herzschuh, U. Simulating biome distribution on the tibetan plateau using a modified global vegetation model. *Arct. Antarct. Alp. Res.* **2011**, *43*, 429–441. [CrossRef]
5. Zhang, Y.Y.; Zhang, S.F.; Zhai, X.Y.; Xia, J. Runoff variation in the three rivers source region and its response to climate change. *Acta Geogr. Sin.* **2012**, *67*, 71–82. [CrossRef]
6. Kang, S.C.; Li, J.J.; Yao, T.D.; Yan, Y.P. A study of the climate variation in the tibetan plateau during the last 50 years. *J. Glaciolgy Geocryol.* **1998**, *20*, 381–387.
7. Zhong, L.; Ma, Y.M.; Salama, M.S.; Su, Z.B. Assessment of vegetation dynamics and their response to variations in precipitation and temperature in the Tibetan Plateau. *Clim. Chang.* **2010**, *103*, 519–535. [CrossRef]
8. Yao, T.; Thompson, L.; Yang, W.; Yu, W.; Gao, Y.; Guo, X.; Yang, X.; Duan, K.; Zhao, H.; Xu, B.; et al. Different glacier status with atmospheric circulation in Tibetan Plateau and surroundings. *Nat. Clim. Chang.* **2012**, *2*, 663–667. [CrossRef]
9. Niu, T.; Chen, L.X.; Zhou, Z.J. The characteristic of climate change over the Tibetan Plateau in the last 40 years and the detection of climatic jumps. *Adv. Atmos. Sci.* **2004**, *21*, 193–203. [CrossRef]
10. Meng, X.; Li, R.; Luan, L.; Lyu, S.; Zhang, T.; Ao, Y.; Han, B.; Zhao, L.; Ma, Y. Detecting hydrological consistency between soil moisture and precipitation and changes of soil moisture in summer over the Tibetan Plateau. *Clim. Dyn.* **2017**, *51*, 4157–4168. [CrossRef]
11. Gao, Q.; Guo, Y.; Xu, H.; Ganjurjav, H.; Li, Y.; Wan, Y.; Qin, X.; Ma, X.; Liu, S. Climate change and its impacts on vegetation distribution and net primary productivity of the alpine ecosystem in the Qinghai-Tibetan Plateau. *Sci. Total Environ.* **2016**, *554*, 34–41. [CrossRef] [PubMed]
12. Bi, H.Y.; Ma, J.W.; Zheng, W.J.; Zeng, J.Y. Comparison of soil moisture in gldas model simulations and in-situ observations over the Tibetan Plateau. *J. Geophys. Res. Atmos.* **2016**, *121*, 2658–2678. [CrossRef]
13. Chen, Y.Y.; Yang, K.; Qin, J.; Zhao, L.; Tang, W.J.; Han, M.L. Evaluation of amsr-e retrievals and gldas simulations against observations of a soil moisture network on the central Tibetan Plateau. *J. Geophys. Res. Atmos.* **2013**, *118*, 4466–4475. [CrossRef]
14. Shang, L.Y.; Zhang, Y.; Lyu, S.H.; Wang, S.Y. Seasonal and inter-annual variations in carbon dioxide exchange over an alpine grassland in the eastern Qinghai-Tibetan Plateau. *PLoS ONE* **2016**, *11*, e0166837. [CrossRef] [PubMed]
15. Zhang, T.; Zhang, Y.; Xu, M.; Xi, Y.; Zhu, J.; Zhang, X.; Wang, Y.; Li, Y.; Shi, P.; Yu, G.; et al. Ecosystem response more than climate variability drivers the inter-annual variability of carbon fluxes in three Chinese grasslands. *Agric. For. Meteorol.* **2016**, *225*, 48–56. [CrossRef]
16. Song, M.H.; Zhou, C.P.; Ouyang, H. Simulated distribution of vegetation types in response to climate change on the Tibetan Plateau. *J. Veg. Sci.* **2005**, *16*, 341–350. [CrossRef]
17. Li, Y.; Piao, S.; Li, L.Z.; Chen, A.; Wang, X.; Ciais, P.; Zhou, L. Divergent hydrological response to large-scale afforestation and vegetation greening in China. *Sci. Adv.* **2018**, *4*, eaar4182. [CrossRef]
18. Piao, S.; Ciais, P.; Huang, Y.; Shen, Z.; Peng, S.; Li, J.; Fang, J. The impacts of climate change on water resources and agriculture in China. *Nature* **2010**, *467*, 43–51. [CrossRef]

19. Niu, G.Y.; Yang, Z.L.; Mitchell, K.E.; Chen, F.; Ek, M.B.; Barlage, M.; Xia, Y. The community noah land surface model with multiparameterization options (noah-mp): 1. model description and evaluation with local: Cale measurements. *J. Geophys. Res. Atmos.* **2011**, *116*, D12109. [CrossRef]

20. Liu, R.G.; Li, N.; Su, H.X.; Sang, W.G. Simulation and analysis on future carbon balance of three deciduous forests in Beijing mountain area, warm temperate zone of China. *Chin. J. Plant Ecol.* **2009**, *33*, 516–534.

21. Zeng, X.D.; Fang, L.; Song, X. Development of the IAP dynamic global vegetation model. *Adv. Atmos. Sci.* **2014**, *31*, 505–514. [CrossRef]

22. Song, X.; Zeng, X.D.; Zhu, J.W. Evaluating the Tree Population Density and Its Impacts in CLM-DGVM. *Adv. Atmos. Sci.* **2013**, *30*, 116–124. [CrossRef]

23. Murray-Tortarolo, G.; Anav, A.; Friedlingstein, P.; Sitch, S.; Piao, S.; Zhu, Z.; Zeng, N. Evaluation of Land Surface Models in Reproducing Satellite-Derived LAI over the High-Latitude Northern Hemisphere. Part I: Uncoupled DGVMs. *Remote Sens.* **2013**, *5*, 4819–4838. [CrossRef]

24. Deng, M.; Meng, X.; Lu, Y.; Li, Z.; Zhao, L.; Hu, Z.; Li, Q. Impact and Sensitivity Analysis of Soil Water and Heat Transfer Parameterizations in Community Land Surface Model on the Tibetan Plateau. *J. Adv. Modeling Earth Syst.* **2021**, *13*, e2021MS002670. [CrossRef]

25. Luo, S.Q.; Fang, X.W.; Lyu, S.H.; Zhang, Y.; Chen, B.L. Improving CLM4.5 simulations of land-atmosphere exchange during freeze-thaw processes on the Tibetan Plateau. *J. Meteorol. Res.* **2017**, *31*, 916–930. [CrossRef]

26. Yu, G.R.; Wen, X.F.; Sun, X.M.; Tanner, B.D.; Lee, X.; Chen, J.Y. Overview of ChinaFLUX and evaluation of its eddy covariance measurement. *Agric. For. Meteorol.* **2006**, *137*, 125–137. [CrossRef]

27. Wei, D.; Qi, Y.; Ma, Y.; Wang, X.; Ma, W.; Gao, T.; Wang, X. Plant uptake of CO_2 outpaces losses from permafrost and plant respiration on the Tibetan Plateau. *Proc. Natl. Acad. Sci. USA* **2021**, *118*, e2015283118. [CrossRef]

28. Yuan, W.; Liu, S.; Yu, G.; Bonnefond, J.M.; Chen, J.; Davis, K.; Verma, S.B. Global estimates of evapotranspiration and gross primary production based on MODIS and global meteorology data. *Remote Sens. Environ.* **2010**, *114*, 1416–1431. [CrossRef]

29. Yuan, W.; Cai, W.; Xia, J.; Chen, J.; Liu, S.; Dong, W.; Wohlfahrt, G. Global comparison of light use efficiency models for simulating terrestrial vegetation gross primary production based on the LaThuile database. *Agric. For. Meteorol.* **2014**, *192*, 108–120. [CrossRef]

30. Bastos, A.; Fu, Z.; Ciais, P.; Friedlingstein, P.; Sitch, S.; Pongratz, J.; Zaehle, S. Impacts of extreme summers on European ecosystems: A comparative analysis of 2003, 2010 and 2018. *Philos. Trans. R. Soc. B Biol. Sci.* **2020**, *375*, 20190507. [CrossRef]

31. Jung, M.; Koirala, S.; Weber, U.; Ichii, K.; Gans, F.; Camps-Valls, G.; Reichstein, M. The FLUXCOM Ensemble of global land-atmosphere energy fluxes. *Sci. Data* **2019**, *6*, 74. [CrossRef] [PubMed]

32. Tramontana, G.; Jung, M.; Schwalm, C.R.; Ichii, K.; Camps-Valls, G.; Ráduly, B.; Papale, D. Predicting carbon dioxide and energy fluxes across global fluxnet sites with regression algorithms. *Biogeosciences* **2016**, *13*, 4291–4313. [CrossRef]

33. Badgley, G.; Field, C.B.; Berry, J.A. Canopy near-infrared reflectance and terrestrial photosynthesis. *Sci. Adv.* **2017**, *3*, e1602244. [CrossRef] [PubMed]

34. Zhang, Y.; Joiner, J.; Gentine, P.; Zhou, S. Reduced solar-induced chlorophyll fluorescence from GOME-2 during Amazon drought caused by dataset artifacts. *Glob. Chang. Biol.* **2018**, *24*, 2229–2230. [CrossRef]

35. Wang, S.; Zhang, Y.; Ju, W.; Qiu, B.; Zhang, Z. Tracking the seasonal and inter-annual variations of global gross primary production during last four decades using satellite near-infrared reflectance data-sciencedirect. *Sci. Total Environ.* **2021**, *755*, 142569. [CrossRef]

36. Joiner, J.; Yoshida, Y. *Global MODIS and FLUXNET-Derived Daily Gross Primary Production, V2*; ORNL DAAC: Oak Ridge, TN, USA, 2021.

37. Martens, B.; Miralles, D.G.; Lievens, H.; Van Der Schalie, R.; De Jeu, R.A.; Fernández-Prieto, D.; Verhoest, N.E. Gleam v3: Satellite-based land evaporation and root-zone soil moisture. *Geosci. Model Dev.* **2017**, *10*, 1903–1925. [CrossRef]

38. Zhao, Y.F.; Zhu, J.; Xu, Y. Establishment and assessment of the grid precipitation datasets in China for Recent 50 years. *J. Meteorol. Sci.* **2014**, *34*, 414–420.

39. Foley, J.A.; Levis, S.; Prentice, I.C.; Pollard, D.; Thompson, S.L. Coupling dynamic models of climate and vegetation. *Glob. Chang. Biol.* **1998**, *4*, 561–579. [CrossRef]

40. Sitch, S.; Smith, B.; Prentice, I.C.; Arneth, A.; Bondeau, A.; Cramer, W.; Venevsky, S. Evaluation of ecosystem dynamics, plant geography andterrestrial carbon cycling in the LPJ dynamic global vegetation model. *Glob. Chang. Biol.* **2003**, *9*, 161–185. [CrossRef]

41. Bonan, G.B.; Levis, S.; Sitch, S.; Vertenstein, M.; Oleson, K.W. A dynamic global vegetation model for use with climate models: Concepts and description of simulated vegetation dynamics. *Glob. Chang. Biol.* **2003**, *9*, 1543–1566. [CrossRef]

42. White, M.A.; Thornton, P.E.; Running, S.W. A continental phenology model for monitoring vegetationresponses to interannual climatic variability. *Glob. Biogeochem. Cycle* **1997**, *11*, 217–234. [CrossRef]

43. Gllewski, P.; Nawalany, M. Inter-Comparison of Rain-Gauge, Radar, and Satellite (IMERG GPM) Precipitation Estimates Performance for Rainfall-Runoff Modeling in a Mountainous Catchment in Poland. *Water* **2018**, *10*, 1665. [CrossRef]

44. Fang, G.; Yuan, Y.; Gao, Y.; Huang, X.; Guo, Y. Assessing the Effects of Urbanization on Flood Events with Urban Agglomeration Polders Type of Flood Control Pattern Using the HEC-HMS Model in the Qinhuai River Basin, China. *Water* **2018**, *10*, 1003. [CrossRef]

45. Gilewski, P. Impact of the Grid Resolution and Deterministic Interpolation of Precipitation on Rainfall-Runoff Modeling in a Sparsely Gauged Mountainous Catchment. *Water* **2021**, *13*, 230. [CrossRef]
46. Ahmad, A.; Zhang, Y.; Nichols, S. Review and evaluation of remote sensing methods for soil-moisture estimation. *J. Photonics Energy* **2011**, *2*, 028001.

MDPI

St. Alban-Anlage 66

4052 Basel

Switzerland

Tel. +41 61 683 77 34

Fax +41 61 302 89 18

www.mdpi.com

Remote Sensing Editorial Office

E-mail: remotesensing@mdpi.com

www.mdpi.com/journal/remotesensing